Modern Optimization Methods for Science, Engineering and Technology

Modern Optimization Methods for Science, Engineering and Technology

Edited by
G R Sinha
Myanmar Institute of Information Technology Mandalay, Myanmar

IOP Publishing, Bristol, UK

ISBN 978-0-7503-2404-5 (ebook)
ISBN 978-0-7503-2402-1 (print)
ISBN 978-0-7503-2403-8 (mobi)

DOI 10.1088/978-0-7503-2404-5

Version: 20191101

IOP ebooks

British Library Cataloguing-in-Publication Data: A catalogue record for this book is available from the British Library.

Published by IOP Publishing, wholly owned by The Institute of Physics, London

IOP Publishing, Temple Circus, Temple Way, Bristol, BS1 6HG, UK

US Office: IOP Publishing, Inc., 190 North Independence Mall West, Suite 601, Philadelphia, PA 19106, USA

Dedicated to my late grandparents, my teachers and Revered Swami Vivekananda.

Contents

Preface

Optimization is generally defined as the process by which an optimum solution of a problem is achieved. Optimization methods are designed to provide the best possible solution or values in engineering problems or system design. These methods are intended to improve the performance of a system in terms of several performance evaluation factors, such as cost, time, computational complexity, raw materials, etc. Genetic algorithms are one of the most popular areas of study and are based on an optimization theory which works by utilizing the concept of evolution and natural selection. Achieving better solutions and improving the performance of existing system designs is an ongoing and continuous process on which scientists, engineers, mathematicians, philosophers and researchers have been working for many years. Optimization techniques are widely used in a wide range of applications, such as robotics, artificial intelligence (AI) based applications, the chemical, electrical and manufacturing industries, and many others.

This book focuses on the following: an introduction and background; linear programming; multivariable methods for risk assessment; an overview of nonlinear methods; implementation of the traveling salesman problem using modified ant colony optimization; the application of particle swarm optimization; multi-criterion and topology optimization; learning classifiers; case studies on six sigma real-time steel industry applications; performance measures and evaluation; multi-objective optimization problems; machine learning approaches; genetic algorithms and their application; QoS optimization in cellular networks; and the future directions of optimization methods and applications.

The purpose of this book is to present the fundamentals, background and theoretical concepts of optimization principles in a comprehensive manner, along with potential applications and implementation strategies. This book covers case studies, real-time applications, development objectives and research directions, in addition to the basic fundamentals. The book will be very useful for a wide spectrum of readers, such as research scholars, academics and industry professionals, in particular for those who are working on solving optimization issues, challenges and problems.

Acknowledgements

I begin by expressing my sincere thanks to my wife Shubhra, my daughter Samprati and my wonderful parents for their great support and encouragement throughout the completion of this important book. This book is the outcome of focused and sincere efforts that could be given to the book only due to the great support of my family.

I am grateful to my teachers who have left no stone unturned in empowering and enlightening me, in particular Shri Bhagwati Prasad Verma who is like a godfather to me. I extend my heartfelt thanks to the Ramakrishna Mission order and Revered Swami Satyaroopananda of Ramakrishna Mission, Raipur, India.

I extend my sincere thanks to all the contributors for writing on the relevant theoretical background and real-time applications of optimization methods and entrusting upon me the role of editor.

I also wish to thank all my friends, well-wishers and all those who keep me motivated to do more and more, better and better (as is the objective of any optimization method).

My reverence with folded hands to Swami Vivekananda who has been the source of inspiration for all my work and achievements.

Last, but most importantly, I express my humble thanks to Dr John Navas, Senior Commissioning Manager of IOP Publishing for his great support, necessary help, appreciation and quick responses. It has been wonderful experience working with John. My Thanks to Daniel Heatley of IOP for the necessary support and many others in the IOP team who helped me directly or indirectly. I also wish to thank IOP Publishing for giving me this opportunity to contribute on a relevant topic with a reputed publisher.

Editor biography

G R Sinha

Dr G R Sinha is Adjunct Professor at the Institute of Information Technology (IIIT) Bangalore and is currently deputed as a Professor at Myanmar Institute of Information Technology (MIIT), Mandalay, Myanmar. He obtained his BE (electronics engineering) and MTech (computer technology) with a Gold Medal from the National Institute of Technology, Raipur, India. He received his PhD in electronics and telecommunications engineering from Chhattisgarh Swami Vivekanand Technical University (CSVTU), Bhilai, India.

He has published 227 research papers in various international and national journals and conferences. He is an active reviewer and editorial member of more than 12 reputed international journals such IEEE's *Transactions on Image Processing*, Elsevier's *Computer Methods and Programs in Biomedicine*, etc. He has been Dean of Faculty and Executive Council Member of CSVTU India and is currently a member of the Senate of MIIT. Dr Sinha has been appointed as an ACM Distinguished Speaker in the field of DSP for the years 2017–20. He has also been appointed as an Expert Member for the Vocational Training Program by Tata Institute of Social Sciences (TISS) for two years (2017–19). He has been the Chhattisgarh Representative of the IEEE MP Sub-Section Executive Council for the last three years. He has served as a Distinguished Speaker in Digital Image Processing for the Computer Society of India (2015). He also served as Distinguished IEEE Lecturer on the IEEE India council for the Bombay section. He has been a Senior Member of IEEE for many years.

He is the recipient of many awards, such as the TCS Award 2014 for Outstanding Contributions in the Campus Commune of TCS, R B Patil ISTE National Award 2013 for Promising Teacher by ISTE New Delhi, Emerging Chhattisgarh Award 2013, Engineer of the Year Award 2011, Young Engineer Award 2008, Young Scientist Award 2005, IEI Expert Engineer Award 2007, ISCA Young Scientist Award 2006, and the nomination and awarding of the Deshbandhu Merit Scholarship for five years. He has authored six books, including *Biometrics* published by Wiley India, a subsidiary of John Wiley, and *Medical Image Processing*, published by Prentice Hall of India. He is a consultant for various skill development initiatives of NSDC, Government of India. He is a regular referee of project grants under the DST-EMR scheme and several other schemes of the Government of India. He has delivered many keynote/invited talks and chaired many technical sessions at international conferences in Singapore, Myanmar, Bangalore, Mumbai, Trivandrum, Hyderabad, Mysore, Allahabad, Nagercoil, Nagpur, Kolaghat, Yangon, Meikhtila and many other places. His special session on 'Deep Learning in Biometrics' was included in the IEEE International Conference on Image Processing in 2017. He is a Fellow of IETE New Delhi and a member of international professional societies such as IEEE, ACM and many other national professional bodies such as ISTE, CSI, ISCA and IEI. He is a member of various committees of the university and has been Vice President of the Computer Society of India for the Bhilai chapter for two consecutive years. He has supervised eight PhD scholars and 15 MTech scholars. His research interests include image processing and computer vision, optimization methods, employability skills, outcome based education (OBE), etc.

List of contributors

Sirajuddin Ahmed
Jamia Millia Islamia
New Delhi
India

Rajesh Chamorshikar
Bhilai Steel Plant
Bhilai
Chhattisgarh
India

Siddharth Choubey
SSTC-SSGI
CSVTU
Bhilai
Chhattisgarh
India

Abha Choubey
SSTC-SSGI
CSVTU
Bhilai
Chhattisgarh
India

Sien Deng
Department of Mathematical Sciences
Northern Illinois University
DeKalb, IL
USA

Santosh R Desai
Electronics and Instrumentation Engineering
BMS College of Engineering
Basavangudi
Bangalore
India

Somesh Kumar Dewangan
SSTC-SSGI
CSVTU
Bhilai
Chhattisgarh
India

Vladimir Gorbunov
Moscow Institute of Electronic Technology
Moscow
Russia

Shankru Guggari
Department of Computer Applications
BMS College of Engineering
Bengaluru
Karnataka
India

Sailesh Kumar Gupta
Darjeeling Government College
Darjeeling
West Bengal
India

Glenn Harris
Department of Mathematical Sciences
Northern Illinois University
DeKalb, IL
USA

Zar Chi Su Su Hlaing
University of Computer Studies (Magway)
Magway
Myanmar

Nadeem Ahmad Khan
Jamia Millia Islamia
New Delhi
India

Vandana Khare
CMR College of Engineering and Technology
Hyderabad
Telangana
India

Myo Khaing
University of Computer Studies (Magway)
Magway
Myanmar

Ajay Kulkarni
Medi-Caps University
Indore
Madhya Pradesh
India

Rahul Kumar
Department of Information Technology
National Institute of Technology Raipur
Raipur
Chhattisgarh
India

Bonya Mukherjee
Bhilai Steel Plant
Bhilai
Chhattisgarh
India

Kapil Kumar Nagwanshi
MPSTME Shirpur Campus
SVKM's NMIMS University
Mumbai
Maharashtra
India

Pushkala Narasimhan
PG Department of Commerce
NMKRV College for Women
Bangalore
Karnataka
India

Jyotiprakash Patra
SSTC-SSGI
CSVTU
Bhilai
Chhattisgarh
India

Jyothi Pillai
Department of Information Technology
Bhilai Institute of Technology
Durg
Chhattisgarh
India

Rajendra Prasad
Indian Institute of Technology Roorkee
Roorkee
Uttarakhand
India

Sachin Puntambekar
Medi-Caps University
Indore
Madhya Pradesh
India

Subrahmanian Ramani
Bhilai Steel Plant
Bhilai
Chhattisgarh
India

K C Raveendranathan
Rajadhani Institute of Engineering and Technology
Thiruvananthapuram
Kerala
India

Arpana Rawal
Bhilai Institute of Technology
Durg
Chhattisgarh
India

Mridu Sahu
Department of Information Technology
National Institute of Technology Raipur
Raipur
Chhattisgarh
India

Mamta Singh
Sai College
Bhilai
Chhattisgarh
India

G R Sinha
Myanmar Institute of Information Technology
Mandalay
Myanmar

Kostiantyn Tkachuk
Igor Sikorsky Kyiv Polytechnic Institute
Kiev
Ukraine

Oksana Tverda
Igor Sikorsky Kyiv Polytechnic Institute
Kiev
Ukraine

Sergij Vambol
State Ecological Academy of Postgraduate Education and Management
Berdyansk State Pedagogical University
Berdyansk
Ukraine

Viola Vambol
Berdyansk State Pedagogical University
Berdyansk
Ukraine

K A Venkatesh
Myanmar Institute of Information Technology
Mandalay
Myanmar

Chapter 1

Introduction and background to optimization theory

G R Sinha

This chapter presents an overview and brief background of optimization methods which are very popular in almost all applications of science, engineering, technology and mathematics. The historical background of optimization is studied, making a distinction between optimization and robustness. A few important definitions such as design variables, constraints and optimization objectives are highlighted followed by optimization problems and methods. Structural, control, signal processing and other important optimization methods are introduced, and finally the importance of the mathematical background which is necessary for implementing optimization methods for various tasks is briefly presented.

1.1 Historical development

Optimization is a much talked about topic that has long been challenging. Much literature suggests that the concept of optimization was used in 100 BC for calculating the appropriate distance between two points. There are several such literature sources, articles and stories about optimization theory and practice. In the developments and advancements in any area of science, engineering, technology, mathematics, philosophy and many others, an attempt is always made to achieve better results, products or outcomes with better performance. The improvement of performance over existing methods, solutions, products, devices or any advances of science, technology, engineering and mathematics (STEM), is always studied by numerous scholars and research groups. Research continues in all areas of STEM for further improvement of the performance of existing work or discovering some better method. In fact, robustness and optimization are two important terms which can be considered as analogous because both are attempts to continuously improve.

doi:10.1088/978-0-7503-2404-5ch1

Wikipedia suggests that the optimization problem was studied in the seventeenth and eighteenth centuries, including with regard to Kepler's law, Bernoulli's theorem, Leibniz's calculus of variations, etc. In calculus, probability theory, mechanics, chemical equilibrium and many other fields, optimization theorems and principles evolved in the nineteenth and twentieth centuries. In the twentieth century, the traveling salesman problem, which is one of the most popular among optimization problems, was studied and worked upon. Now, optimization theory is not only limited to STEM and allied branches but the concept is widely used in all fields, such as financial management, economics, life sciences, genetics, population studies and several others. We can use the evolution of human beings as an example, with human ancestors being more ape-like. Evolution then continued and human beings can be seen in the present context as the best possible form of the species of its kind that natural selection could produce. The human species is now educated and is trying become even faster through computing speed, although the thinking capacity of the brain is unlimited. Thought processes and their outcomes are being improved continuously. This is a simple case of optimization. A person can become a better intellectual or individual using some methods, thought processes or techniques—employing some type of optimization concept to perform better.

Let us understand optimization problems with the help of an example. A system is defined by the following equations:

$$M = ax_1 + bx_2 \tag{1.1}$$

$$P = Mx_1 + c, \tag{1.2}$$

where x_1 and x_2 are two input variables, a and b are constants, M is an intermediate output, c is one more constant and P is the final output or response of the system.

Now, with a given set of conditions we are interested in obtaining the optimum value of P, which can be achieved in two different ways:

- Optimizing system performance by optimizing the internal parameters and constants which are characteristics of the system.
- Optimizing by adjusting or optimizing all individual variables or selected variables present in the system.

Optimization is expected to produce the best possible final value otherwise some more optimization methods need to be applied to further improve the results—there arises the need for a robust approach that can suit all requirements and conditions. Another situation for using optimization in the above example may be due to some error introduced in the final value. If the error in the final value is more than the tolerance limit (the allowed limit) then the error can be minimized by using a suitable optimization method.

1.1.1 Robustness and optimization

Optimization and robustness are strongly related in terms of a common objective, which is to achieve some improvement. Optimization always aims to improve something, some parameter, so that a method, system or device performs better. The term robustness relates to attempting a process more and more times since the solution can never be optimum, rather it can be optimal. So, researchers will be always engaging in experiments, simulations, modeling and study so that a better result or performance is obtained. Robustness has long been a challenging research problem in all fields of science and technology irrespective of application. This could be in the field of automobiles, manufacturing, computing, or information and communication technology (ICT) enabled services, all of which need continuous improvement. If we are to discuss any particular example of robustness then let us discuss image processing and computer vision based applications [1, 2]. This is an important application because computer vision is used in almost all modern advancements. The prominent authors of image processing admit that there is no general theory in image processing, which means that whatever is done in a novel manner becomes new in the area of image processing. Obtaining a robust approach for any components of image processing such as image de-noising, segmentation or pattern matching is challenging, and it is very difficult to claim that the obtained solution is robust one. The robustness depends on optimizing one or a few or all the parameters involved in the research. The extensive research on image processing suggests that robustness is a permanent research problem, a few research contributions can be seen in [1–6] where optimization of performance was attempted but not as a robust approach.

Combining the two important terms, robust optimization is an emerging area of study and research where robustness is investigated among a number of optimization problems and solutions in real-time industries and practice. The optimization process employs a number of methods for improving the performance of a particular task and different optimization methods produce different performance metrics and therefore robustness needs to be investigated among the optimization methods. The best possible optimization method that can result in optimum performance of a system will be referred to as a robust optimization method. If the method is robust then it should also be used for a variety of tasks irrespective of different performance evaluation parameters.

1.2 Definition and elements of optimization

Oxford Dictionary defines optimization as a process or a method that can make something perfect and effective[1]. Something here may include a design, system or decision which becomes better and better using the methodology called optimization. Cambridge Dictionary defines optimization as process that makes something as good as possible, or as effective as possible[2]. We can refer to many other standard definitions and find one thing common, which is the improvement in some process, method, design or decision to make something more effective.

Although optimization is defined as a process or methodology, it is actually made up of several components, such as decision variables, constraints and objectives.

[1] The definition is taken from: https://www.lexico.com/en/definition/optimization and slightly modified.

[2] The definition is taken from https://dictionary.cambridge.org/dictionary/english/optimization and slightly modified.

Variables are the most important and guiding factors deciding the result being produced by the optimization and constraints are certain conditions under which optimization achieves its desired goal (its objective).

1.2.1 Design variables and parameters

Variables are design parameters used in any system which are actually responsible for obtaining a desired output or a product. Generally, the design variables are set of one-dimensional or multi-dimensional vectors, such as

$$X = (x_1, x_2, x_3, \dots , x_n), \tag{1.3}$$

where x_1, x_2, x_3, ..., are n number of variables and constitute together a single vector X. Here, the vector X is one-dimensional but it can be two-dimensional, three-dimensional or more.

These variables are controlled, manipulated and varied in the process of optimization. In some cases, only one variable is varied or controlled and performance improvement results and optimization works, but in other cases a greater number of variables may also change in order to achieve the desired objective in the system design. In the applications where the theorem of equilibrium or inequalities is used to control and obtain equilibrium, many variables are used. If optimization is desired, then there is a wide range of possibilities to optimize the behavior of a mechanical system [7]. We sometimes use the terms parameters and variables interchangeably, but there is a slight difference between the two. Variables are vectors or values which actually vary in the system, whereas parameters may be some standard operating parameters, constants or the attributes that interrelate the design variables. So, both of them are attributes, and their small difference lies in the role of parameters in obtaining some relationship between the variables.

1.2.2 Objectives

An objective means a goal to be achieved. Every STEM performance improvement will have a certain goal to achieve, maybe an error to minimize at some specific level. In all such cases, optimization is used with the important purpose to achieve the goal that is set by the system response or method. Thus, the optimization method is assigned a minimum goal that is required to be achieved. As an example, a few images are presented in figure 1.1 which are the results of breast cancer segmentation and detection. The results were obtained using an optimized method that combined gray level clustering enhancement and adaptive threshold segmentation (GLCE-ATS). If we look at four different segmentation results which all highlight breast cancer masses, they appear the same and it is very difficult to determine the cases in which more cancerous elements (malignant elements) are present. The optimization method has the objective to achieve the maximum number of malignant masses present in the breast mammograms [8].

The objective of optimization is also to minimize the error or value of a suitable function $e(x)$ or $f(x)$, y = minimize $\{f(x)\}$ or minimize $\{e(x)\}$.

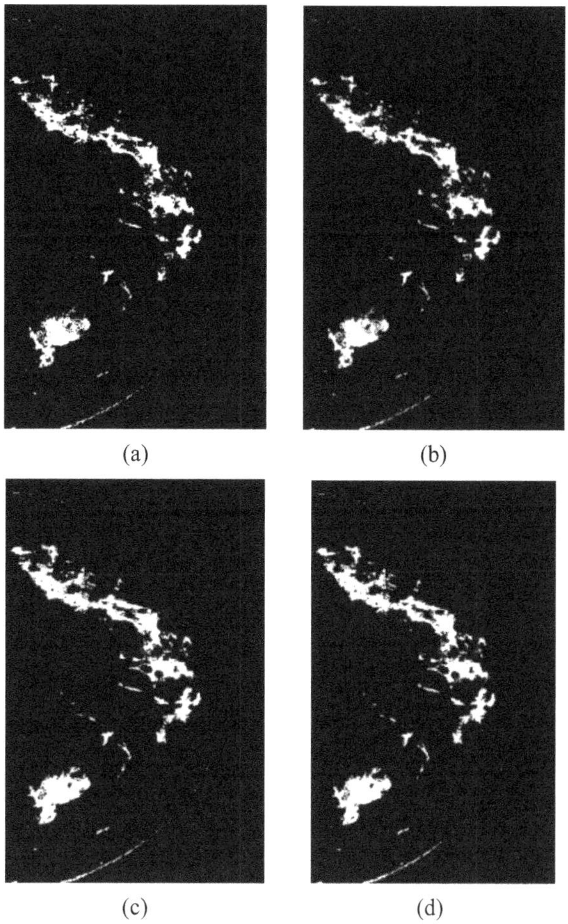

Figure 1.1. Results of breast cancer detection, with different results for different optimization parameters used in the GLCE-ATS method.

1.2.3 Constraints and bounds

Constraints are some limitations and conditions under which the optimization technique or method works. The method successfully improves the system performance provided a certain set of constraints are satisfied. The limitations include some bounds as well, which may be upper bounds, lower bounds or intermediate.

The optimization operates on a set of data or samples subjected to some limitations that may depend on certain characteristics or features of the system being optimized, or some external factors affecting the system design or performance. The bounds are generally based on design variables but may also depend on parameters. For example,

$$Y = \text{minimize } \{e(x)\} \text{ under } x_2 < x_0 \text{ and } x_1 > x_p. \quad (1.4)$$

Here the goal is to minimize the value of the error signal. The minimum or optimum value has to be obtained as Y and the optimization has to be achieved under the conditions given in equation (1.4). These conditions are bounds which are apparently lower and upper bounds in the case of x_2 and x_1, respectively. The error signal or value needs to be minimized between x_1 and x_2. In different types of optimization problems and applications, various types of bounds and constraints are used. The optimization methods can be generalized but the constraints are specific to the applications or problems being addressed. Therefore, the constraints are subjective, whereas the optimization methods are objective for the applications. In [7], the solution to achieving equilibrium requires a number of constraints in the implementation of optimization in the work.

1.3 Optimization problems and methods

In applications related to STEM or allied fields, there are a variety of problems that need to be addressed in terms of improvement of system design and performance, or minimization of error, reduction of labor and cost, etc. These problems are solved using suitable optimization methods [9]. In the determination of optimization problems, several standard aspects are used, such as:

1. Optimal input number of arguments.
2. Maximum and minimum values of functions.
3. Characteristics of the system.
4. Constraints and bounds.

The number of arguments or attributes plays an important role in choosing the optimization problem. In a few cases, the optimization performs in such a manner that the computational complexity decreases with the increase in the number of input arguments, whereas in some other cases the complexity is also increased. In fact, the goal should also be to reduce the computational complexity in addition to improved performance or system design. The number of arguments is to be selected wisely so that optimization produces the best possible result while expending the least possible time and fewest operations. The bounds are limits to certain conditions, which are also important points to be kept in mind while designing an optimization method and implementing it. So, the problems are governed by the upper and lower limits of constraints and bounds, and also by the minimum and maximum possible values of the functions being optimized [7, 9]. The targets for optimization are also decided on the basis of specific characteristics of system response, design or behavior.

The number of methods for optimization is large and these can be employed for solving a variety of problems.

1.3.1 Workflow of optimization methods

Any optimization problem and its possible solution has some typical workflow diagram, one such workflow for the execution of any general-purpose optimization method is shown in figure 1.2. This figure highlights the major steps that are

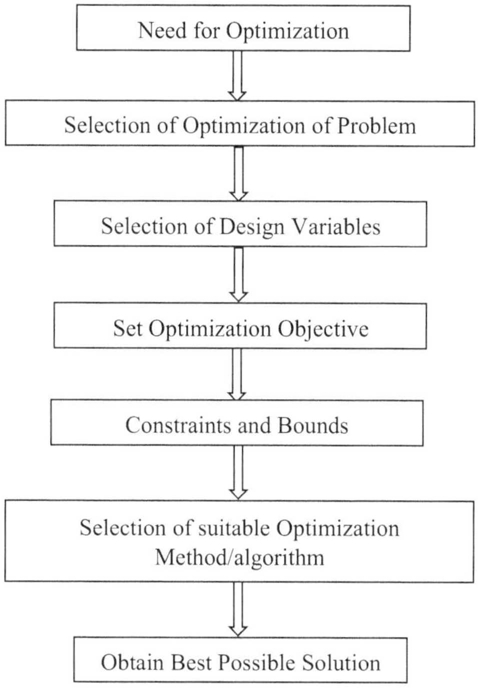

Figure 1.2. Major steps of the workflow of a typical optimization method.

executed in the implementation of any general-purpose optimization method. The process begins with the need for optimization, which is the very first step in any application. The need for optimization helps in the identification of possible optimization problems. There may be multiple problems in an application that could be optimized using a single method or multi-method based optimization. Once the selection of problems is finally made, the design variables are selected which further determine the main objective of employing a suitable optimization method. The optimization has to be carried out under certain constraints and boundary values, and therefore before we actually choose an appropriate method, we set various design constraints and limitations.

Then, the optimization method is decided based on the design variables, requirements, optimization problem, constraints and bounds. The results of the suitable optimization method are assessed as to whether they are the best possible results which were expected or not. If not, then the workflow of the implementation may have several reverse operations (the reverse direction arrows in the diagram are not shown as they depend on what is actually needed). If the design variables need some modification or manipulation, or similarly the constraints require some changes along with set-in limit values, these aspects are corrected so as to achieve the best possible response or outcome produced by the optimization method. Thus, it can be seen in the flow diagram that the optimization problem is chosen in the beginning as per the need and the optimization method is selected after certain steps, such as defining design variables, constraints and bounds.

1.3.2 Classification of optimization methods

As we have discussed, the optimization method is decided based on several factors, and constraints may be linear as well as non-linear. Accordingly, the methods are broadly categories based on:

- Linear constraint terms and bounds.
- Non-linear constraint terms and bounds.

The methods of optimization will fall under different categories, but the basis of choosing a suitable method will be the above two factors, that is, the constraints and bounds being either linear or non-linear. The optimization methods are further classified as:

- Single variable or multi-variable methods.
- Constrained or non-constrained methods.
- Linear or non-linear optimization methods.
- Single objective or multi-objective optimization methods.
- Specialized or general-purpose methods.
- Traditional or non-traditional methods.

The classification of optimization methods is also based on the characteristics or properties of the objective function of optimization [9]. For example, if control variables are continuous and real numbers, then the optimization method is called continuous optimization. Similarly, for integers it is known as discrete optimization and for a combination of continuous real numbers and integers the optimization is known as mixed integer. Based on different types of design variables, quadratic, linear and non-linear, the optimization method is referred to accordingly as quadratic, linear or non-linear. If the optimization problem is subject to some limitations or constraints then the optimization method is called a constrained optimization method and if there are no such constraints, the method is called an unconstrained optimization method.

Conventional or traditional methods of optimization include mainly linear programming, non-linear programming, separable programming and integer programming optimization methods. The methods which are based on evolutionary concepts are referred to as evolutionary methods of optimization, which include a few main methods such as the heuristic search, genetic algorithm, evolutionary programming, particle swarm optimization (PSO), simulated annealing, bacterial foraging and differential equation based methods. Examples of unconstrained optimization methods are the gradient method, Newton's method, quasi-Newton method, conjugate directive method, etc. Another emerging area of optimization is global optimization methods, that include some probabilistic and deterministic methods. These methods are based on probability theory and similar mathematical fundamentals. Global optimization methods are widely used in numerical analysis and feature based analysis in various signal processing and control applications.

In the fast changing dynamic environments of various emerging applications of signal processing, control, transport, etc, optimization is employed where the methods, models and problems are dynamic in nature. Telecommunications, artificial intelligence based advancements, financial analysis and several others fields are rapidly changing with a wide range of dynamics wherein the problems and models of optimization used also change rapidly. In such applications, the methods which are traditionally used also need modification and dynamic updates [10]. There are a number of optimization methods which are recommended for such applications, a few of which are: ant colony optimization, evolutionary optimization, genetic algorithms, neural networks and swarm intelligence based methods, etc. In the context of dynamic optimization, the quantitative measures of performance to assess how the optimization method is working become essential. Such measures are required in all optimization methods, but in a dynamically changing environment the assessment or evaluation is particularly important [1–3, 9, 10]. Some of the important measures which are widely used, irrespective of type of optimization problem and application, are:

- Cumulative fitness and mean fitness.
- Off-line error.
- Robustness.
- Diversity.
- Standard deviation.
- Time.
- Computational complexity.
- Average error.
- Average best function value.
- Current best evolution.

1.4 Design and structural optimization methods

This is an era of technological and industrial revolutions that keep on changing dynamically. In order to meet the ever-changing needs of technology enabled industries, optimization is required to achieve better design and structural performance. For example, numerous studies suggest that airfoil design plays an important role in the aviation sector and a number of significant research contributions can be found in the literature in which airfoils, blades and others components of flight are designed in such an optimal manner that the performance of airplanes improves considerably. The structure of any system mainly depends on the shape, size and other similar attributes, and optimization works on optimizing one or more of these attributes.

1.4.1 Structural optimization

Structural optimization deals directly with the variables related to the structure of the system or a product, such as size, shape, area, perimeter, volume, etc. If the

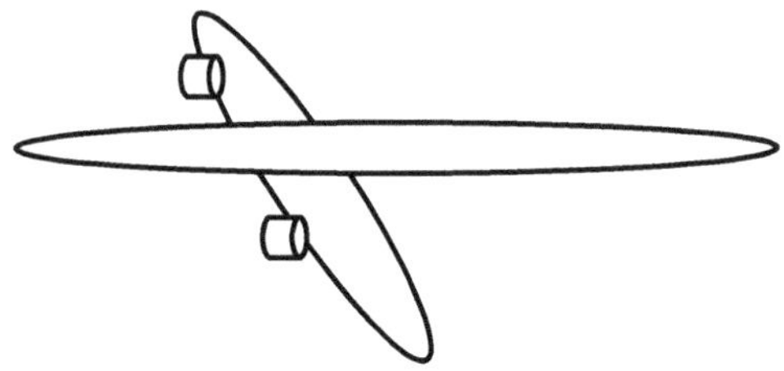

Figure 1.3. A typical structural design of the outer surface of an airplane.

comfort level has to be increased in the field of ergonomics, then the optimization utilizes the structural properties of the system or device whose ergonomics require improvement. Figure 1.3 shows a simple and typical structural diagram of the outer surface of an airplane where the airfoil or blade design and their optimization are important factors in the performance of the flight. This is just an symbolic diagram, as there are thousands of small, medium and high level structures present in an airplane.

Structural optimization is divided into various types based on the structural attributes. A few major structural optimization methods or techniques are as follows:

1. *Shape optimization*: This is based on the different shapes of systems or products.
2. *Area and volume optimization*: Area and volume decide the performance of a huge number of products in the transport and control based applications, and therefore the performance could be improved by optimizing either the area or volume or both or some other similar variables.
3. *Size optimization*: This may include length, width and other similar dimensional features in the optimization of the system's structural performance.
4. *Topological optimization*: Since the topology represents the overall interconnection of various components in a system and how those components are structured together determines the system's performance, in particular in the automobile and similar sectors, and the role of optimization is to improve the topology of the system architecture [11].

   The topological optimization of structures such as seismic loads, structural optimization for steel plants and reliability optimization are examples of structural optimization methods. Topological optimization is implemented following a certain workflow of operations; one such flow diagram can be seen in figure 1.4. The need for optimization arises and structural properties are studied. The optimization problem is identified and a suitable method is implemented that utilizes the structural attributes in the process of optimization. The optimization results in some post-processing with the application of an optimization method which is sensitive to the structural features of the system.

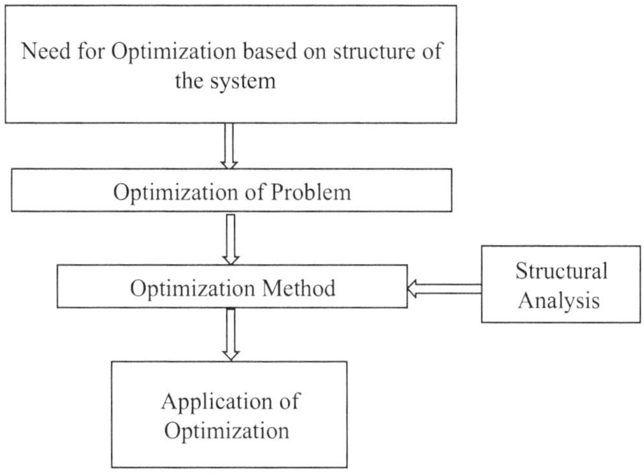

Figure 1.4. Topological optimization design flow.

1.4.2 Design optimization

Design optimization is not entirely different from structural optimization because the design of a system and its structural properties are interrelated. The improvement in system performance, reduction in cost and other benefits of optimization can be achieved using structural optimization methods. The design optimization techniques is mainly associated with computer-aided design (CAD) for various applications. Software tools are embedded within the CAD systems and any design modification or manipulation can easily be done so as to obtain the desired system characteristics.

1.5 Optimization for signal processing and control applications

Signal processing has become an essential component of almost all modern applications and implementations of STEM and allied fields, and in all such applications signal processing and certain signal operations are decisive factors for the performance of application based systems. We can mention communication systems, robotics, self-driving cars, remote sensing, navigational aids, global positioning systems, aerodynamics, mechanics, geographic information systems, weather forecasting, financial analysis, mobile communications, satellite communications, etc, in which signal processing and some suitable mathematical operations are very important factors in the assessment of the performance of the systems. If the performance improvement of the system is intended for such applications, then applying optimization methods in some or all signal processing variables can help in improving performance.

For example, in the area of biometrics and CAD systems, there are a significant number of research publications, but still no robust approach is available and therefore optimization can help in optimizing some parameters so that the system behaves in a robust way.

1.5.1 Signal processing optimization

Signal processing involves digital signal processing (DSP) or DSP processors where numerous signal processing operations are performed. We need to understand the scope of optimization in signal processing applications by analyzing DSP, since DSP is the core part of signal processing and therefore let us examine the main components of DSP. DSP includes some major steps:

1. Data acquisition and sampling.
2. Quantization.
3. Encoding.
4. Signal processing operations.

We have to understand where the optimization could be applied so that signal processing becomes better and contributes to the overall improvement of system performance in which signal processing is present. If there is scope to apply a suitable optimization method in sample processing, sampling rate, or a data acquisition step then this has to be identified. Quantization plays an important role in DSP and therefore optimization can also be implemented with a quantization process. Encoding is the step that converts the data into a digital format which is used in all DSP, and thus optimization can be applied in the process of encoding as well. In all such optimization processes, the following types of optimization could be broadly required:

- *Window size optimization*: In applications where DSP is required, several filtering operations will also be required, which further employ certain types of windowing operations. In DSP applications, several types of window functions are used, but even after choosing a suitable window function, its operation could be improved by using some optimization methods. Window size along with execution time, memory requirements and threshold value can also be optimized so that the windowing effect ultimately contributes to the system in optimal ways.
- *System throughput optimization*: Signal processing is implemented using computers and a number of operations are associated with signal processing. DSP implementation affects memory requirements, central processing unit (CPU) time, memory utilization, parallelism, CPU scheduling, etc. The goal of optimization at the throughput level is to obtain more and better output from computing systems.

Best candidate selection is a factor that plays an important role in operation research, where the core of the performance lies in signal processing. How to choose an appropriate sorting method for best candidate selection in operation research tasks matters a lot in achieving reduced complexity and shorter computation times. Sometimes it is very difficult to achieve both reduced complexity and shorter computation times, and one needs to sacrifice something to achieve better values elsewhere. This balance, which is essential, can be obtained using an optimization method [12].

1.5.2 Communication and control optimization

We have seen that signal processing flow is communicated to various stages of the system design and that the control of signal flow also affects the system's performance. Therefore, we need to also understand the optimization concept for communication and control related operations. There are many optimization methods in the literature that are suitable for the applications, but these methods are all subject to one concept in optimization theory which is a common factor in all communication related applications, namely the convex theory of optimization. One example of such convex optimization for communication and signal processing can be seen in [13], which highlights the concepts of the theory and some examples of optimization methods utilizing convex theory as central theme of optimization.

The main aim of using convex optimization in signal processing and communication is to deal with convex constraints that are obtained from convex mathematical functions. Conic optimization and interior point method optimization are examples of convex theory based optimization. Convex itself means to achieve a suitable objective function that can help in obtaining a local optimal function in a problem. This optimization concept is used in obtaining both global and local optima. Several engineering applications employ convex optimization methods. Linear sub-space, second order cone and Affine sub-space are a few major examples of the convex functions used for optimization tasks [13]. In fact, unconstrained methods can also be converted into some finite constrained methods by using the convex concept. Analog band pulse amplitude modulation and filter design are the cases where convex optimization is useful in communication related applications.

1.6 Design vectors, matrices, vector spaces, geometry and transforms

In applications related to search engine optimization [14], or any STEM application, the implementation of optimization requires a strong mathematical background because optimization uses variables, operations, constraints, minima and maxima functions, etc. Therefore, system design optimization in any applications of optimization needs sufficient consideration of the vectors, matrices and vector spaces that support the mathematical implementation of the optimization method. This section provides a brief overview of these concepts, because all the methods discussed and implemented in the other chapters use realistic and actual mathematical functions and variables.

The advances in technology, evolution of the internet, increased use of computing devices and large amounts of data are some of the factors that require optimization of various types of system designs and operations. In the flow diagram of implementation of optimization, we have discussed that the process depends on design variables, constraints and some final conditions to check if the optimization is working or not. All of these variables and their operations make use of linear algebra, matrices and vector spaces [15] for different computations involved in the system, which are presented in this section.

1.6.1 Linear algebra, matrices and design vectors

Linear algebra is the basis for linear optimization methods and the algebra utilizes linear equations. A typical example of set of linear equations is

$$X_1 + 2Y_1 + 1.5\,Z_1 = P_1 \tag{1.5}$$

$$X_2 + 2Y_2 + 3Z_2 = P_2 \tag{1.6}$$

$$X_3 + 1.5Y_3 + 5Z_3 = P_3, \tag{1.7}$$

where X_1, X_2, X_3, Y_1, Y_2, Y_3, Z_1, Z_2, Z_3 are variables and P_1, P_2 and P_3 are constants. The equations are linear equations and the values of P_1, P_2 and P_3 may be any real values. The representation as shown by the three equations can also be represented as a vector:

- $X = [X_1 \times 2 \times 3]$; $Y = [Y_1\ Y_2\ Y_3]$; $Z = [Z_1\ Z_2\ Z_3]$; and $P = [P_1\ P_2\ P_3]$.
- The coefficients used in the three equations can also be expressed as vectors: $A_1 = [1\ 1\ 1]$; $B = [2\ 2\ 1.5]$; and $C = [1.5\ 3\ 5]$.

We can also write the above equations as

$$\begin{bmatrix} 1 & 2 & 1.5 \\ 1 & 2 & 3 \\ 1 & 1.5 & 5 \end{bmatrix}\begin{bmatrix} X \\ Y \\ Z \end{bmatrix} = \begin{bmatrix} P_1 \\ P_2 \\ P_3 \end{bmatrix}. \tag{1.8}$$

There are many other ways in which vector representation can replace the set of linear equations. This representation of equations is more appropriately called a matrix representation. When matrix representation comes into the analysis of any optimization theory then all of its properties, such as determinant, rank, characteristic equation, eigenvalues, eigenvectors, etc, become very important to understand. Since the matrix and its properties at this level fall under the elementary knowledge of matrices we will not discuss them further here. Thus, linear optimization can work on:

- Linear equations.
- Vector representations of linear equations.
- Matrix representations.

Matrix representations provide an efficient way to express the operations and also better help in the analysis of systems compared to the other approaches as far as the implementation of linear optimization is concerned [16]. Linear representation based optimization includes:

- The steepest descent method.
- The conjugate gradient method [17].
- The normal equation method.

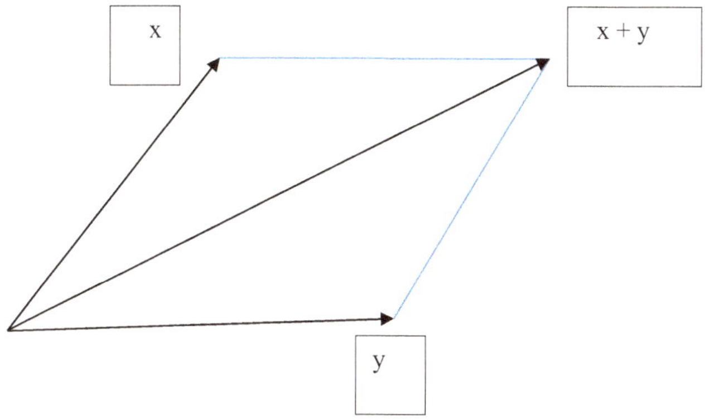

Figure 1.5. Vector addition.

1.6.2 Vector spaces

Vector space represents a set function whose elements are called vectors and vector space also includes another field which is a scalar quantity (generally a real number). Two important operations are characterized by vector space operation, plus (+) and multiplication (·), which are vector addition and scalar multiplication, respectively. If x and y are vectors of vector space X then $x + y$ will also belong to the vector space X. The following are the important principles of vector addition:

- Cumulative law: $x + y = y + x$, for all x, y in X.
- Associative law: $x + (y + z) = (x + y) + z$, where x, y and z are vectors in vector space X.
- Additive identity and additive inverse: If a vector is added with additive identity then it remains the same vector, $x + 0 = x$, and if the vector is added with additive inverse then it results in an additive identity: $x + x' = 0$.

Similarly, scalar multiplication satisfies the following:

- Distributive law: $p \cdot (x + y) = p \cdot x + p \cdot y$ and $(p + q) \cdot x = p \cdot x + q \cdot x$ where p ad q are real numbers.
- Associative law: $p \cdot (q \cdot x) = (p \cdot q) \cdot x$.
- Unitary law: $1 \cdot x = x$ for all x in X.

Any vector space can also have its vector sub-spaces. The concepts of vector space and sub-spaces are utilized in the optimization methods for linear optimization in several applications [18–24]. Vector space includes vector addition which is a common mathematical tool in vector algebra. This can be seen in figure 1.5.

1.6.3 Geometry, transforms, binary and fuzzy logic

Mathematical geometry and transforms are used in a number of optimization problems and solutions. Different types of operations such as rotation, scaling, shift, resizing and reformatting all come under the geometry and transforms that are used

Figure 1.6. Rotation by a certain angle.

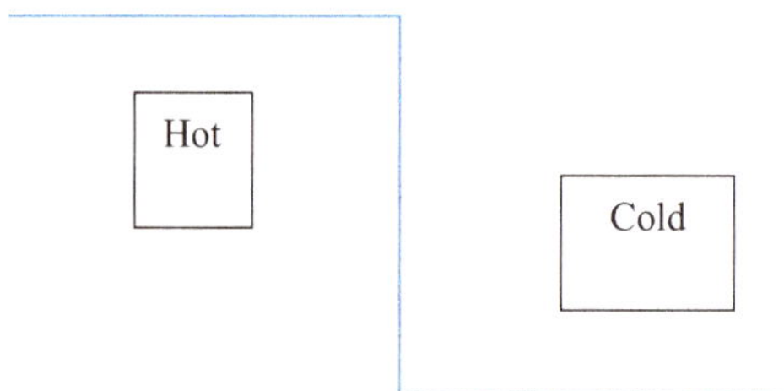

Figure 1.7. Two possible values of temperature as hot and cold.

some steps of optimization methods. In image processing, for example, rotation is an important translation operation which can be performed to any extent; one such example is shown in figure 1.6.

There are several examples of operations which can be included in this section, but the idea of showing one example is to just highlight what transform and geometry operations mean.

Now in the age of DSP, binary operations are essential components in all stages of DSP. When it comes to decision making and path planning control then a new logic is preferred in all STEM applications, which is called fuzzy logic. The concept of fuzzy logic is now utilized in almost all applications, including optimization methods. Neural network and other machine leaning methods are used for various feature extraction and training methods for samples, but a hybrid approach is recommended for better performance of neuro-fuzzy logic to reduce the uncertainty. Figure 1.7 shows the temperature of an object as hot and cold.

The problem arises when we want to know the degree of being hot or cold, which means how hot or cold the object is. Here, the binary concept produces some amount of uncertainty which is addressed by using fuzzy logic. This logic is characterized by using some suitable membership function, for example in the current case of temperature, if the membership function is defined properly then the degree of hotness or coldness can be determined (see figure 1.8).

Hot

0.1 0.2 0.3 0.4 0.5 0.6 0.7 0.8 0.9 1.0

Membership value (function), m

Figure 1.8. Membership function in defining a fuzzy variable.

If we describe the temperature with a suitable value of membership function, say the maximum temperature is 1100 °C and the minimum is 100 °C, then if the membership function (value) is 0.3 it means that the temperature of the object is 100 (minimum) + 0.3 × (1100 – 100) = 400 °C. So, the object is hot but the certainty is there as to what extent the object is hot. The concept of fuzzy logic is very popular among scientists and researchers for various science and engineering applications.

References

[1] Sinha G R, Raju K S, Patra R, Aye D W and Khin D T 2018 Research studies on human cognitive ability *Int. J. Intell. Defen. Supp. Syst.* **5** 298–304

[2] Sinha H, Meshram M R and Sinha G R 2018 BER performance analysis of MIMO-OFDM over wireless channel *Int. J. Pure Appl. Math.* **118** 195–206

[3] Patel B C and Sinha G R 2011 Comparative performance evaluation of segmentation methods in breast cancer images *Int. J. Mach. Intell.* **3** 130–3

[4] Sinha G R 2015 *Fuzzy Based Medical Image Processing Advances in Medical Technologies and Clinical Practice (AMTCP) Book Series* (Hershey, PA: IGI Global)

[5] Patel B C and Sinha G R 2010 Early detection of breast cancer using self-similar fractal method *Int. J. Comput. Appl.* **10** 39–43

[6] Sinha K and Sinha G R 2013 Comparative analysis of optimized K-means and C-means clustering methods for segmentation of brain MRI images for tumor extraction *Proc. of Int. Conf. on Emerging Research in Computing, Information, Communication and Applications* (Amsterdam: Elsevier), pp 619–25

[7] Prepoka A 1980 On the development of optimization theory *Am. Math. Month.* **87** 527–42

[8] Patel B C and Sinha G R 2018 Mass segmentation and feature extraction of mammographic breast images in computer-aided diagnosis *PhD Thesis* Chhattisgarh Swami Vivekanand Technical University Bhilai

[9] Chong E K P and Zak S H 2001 *An Introduction to Optimization* (New York: Wiley)

[10] Cruz C, González J R and Pelta D A 2011 Optimization in dynamic environments: a survey on problems, methods and measures *Soft Comput.* 15 1427–48

[11] Labanada S R 2015 Mathematical programming methods for large-scale topology optimization problems *PhD Thesis* Technical University of Denmark

[12] Ahmad H A 2012 The best candidates method for solving optimization problems *J. Comput. Sci.* **8** 711–5

[13] Luo Z Q and Yu W 2006 An introduction to convex optimization for communications and signal processing *IEEE J. Select. Area Commun.* **24** 1426–38

[14] Bezhovski Z 2015 The historical development of search engine optimization *Inform. Know. Manage.* **5** 91–6

[15] Luenberger D G 1969 *Optimization by Vector Space Methods* (New York: Wiley)

[16] Absil P A, Mahony R and Sepulchre R 2008 *Optimization Algorithms on Matrix Manifolds* (Princeton, NJ: Princeton University Press)

[17] Abrudan T, Eriksson J and Koivunen V 2009 Conjugate gradient algorithm for optimization under unitary matrix constraint *Signal Process.* **89** 1704–14

[18] Ehrgott M and Gandibleux X 2003 *Multiple Criteria Optimization: State of the Art Annotated Bibliographic Surveys* (New York: Kluwer/Academic)

[19] Boumal N 2014 Optimization and estimation on manifolds *PhD* Princeton University https://web.math.princeton.edu/~nboumal/papers/boumal_optimization_and_estimation_on_manifolds_phd_thesis.pdf

[20] Hartley R I and Kahl F 2009 Global optimization through rotation space search *Int. J. Comput. Vis.* **82** 64–79

[21] Gallier J and Quaintance J 2018 Fundamentals of linear algebra and optimization *Technical Report* University of Pennsylvania

[22] Nocedal J and Wright S J 2006 *Numerical Optimization* 2nd ed (Berlin: Springer)

[23] Dubourg V 2011 Adaptive surrogate models for reliability analysis and reliability-based design optimization *PhD Thesis* Université Blaise Pascal https://www.phimeca.com/IMG/pdf/these_dubourg_2011.pdf

[24] Simon D 2013 *Evolutionary Optimization Algorithms: Biologically-Inspired and Population-Based Approaches to Computer Intelligence* (Hoboken, NJ: Wiley)

Chapter 2

Linear programming

K A Venkatesh

Optimization is the key for success in any field. In general, every resource, such as time, availability of skilled work force, etc, is finite; allocating scarce resources to the requested needs in an optimal way is one of the important tasks in every operation. Linear programming is one of the mathematical programming models which has roots in ancient mathematics, namely algebra, in particular the solving of systems of simultaneous linear equations, linear algebra and specifically matrix operations. The potential applications of the linear programming problem (LPP) are enormous: supply chain management, organ donor and recipient matching, communications network design, aviation industries, financial engineering, network optimization, smart grids, decision making and so on. LPP is a special case of optimization problems, which look for maximum or minimum values for a single objective function or multiple objective functions with a set of constraints; the constraints may be equations or inequalities. In LPP, the variables in the objective function and constraints are linearly related.

2.1 Introduction

The study of modeling with linear equations began its journey with the birth of algebra. As a natural extension, many real-world problems can be modeled as a system of simultaneous linear equations. Let us begin with the example of a high school level mathematics problem.

Example 2.1. Three pens and two pencils cost \$190 and two pens and three pencils cost \$180. Find the cost of each pen and pencil.

To model as a system, we use the letters x, y to denote the unknown quantities and these are known as variables. The above problem can be written as

doi:10.1088/978-0-7503-2404-5ch2

$$3x + 2y = 190$$
$$2x + 3y = 180.$$

The above system in matrix notion, can be written as $AX = b$, where A is the coefficient matrix of order $2(n)$, and X and b are column vectors of order 2×1 $(n \times 1)$. To solve the system, there are many methods from the theory of matrices. In general, the methods are classified into two categories, namely direct methods and iterative methods. Methods such as Cramer's, Gauss elimination, Jordon and matrix decomposition are examples of direct methods and Gauss–Jacobi, Gauss–Seidel and SOR methods are examples of iterative methods. All the mentioned methods work well as long as the coefficient matrix is a square one. Suppose that the coefficient matrix is a rectangular one, that is the system has a greater number of variables than the number of equations, we cannot then solve in the usual sense. Suppose the system has k variables in m equations $(k > m)$, set $k - m$ variables as zero then solve for the remaining variables. Such solutions are called basic solutions. The number of basic solutions is $\binom{k}{m}$.

A single entity which consists of a linear expression, called the objective function, with finitely many linear inequalities, called linear constraints and non-negative restrictions, is called the linear programming problem (LPP). The general form of the LPP is as follows.

Objective function:

$$\text{Optimize } Z = \sum_{i=1}^{n} c_i x_i. \tag{2.1}$$

Subject to the constraints

$$AX\{\leqslant, =, \geqslant\}b \tag{2.2}$$

$$X \geqslant 0, \tag{2.3}$$

where A is an $m \times n$ matrix, and X and b are column vectors of order $m \times 1$.

The basic solutions which satisfy the given set of constraints are called *basic feasible solutions*. The basic feasible solution which optimizes the objective function is called the *optimal basic feasible solution*.

All LPPs fall into one of three categories: feasible and bounded, unbounded, and infeasible. If the given LPP is feasible and bounded then it has at least one optimal solution.

Prior to Dantzig's simplex method to solve the LPP, there was an enumerative algorithm to find a feasible solution to the given set of constraints, but this method was not a viable model because it inherently generates additional inequalities. In 1963, Dantzig developed the simplex method and led the development of a new branch of study, called LPP.

The simplex method principle is that rather than working with inequalities, you transform the inequalities into equalities by introducing slack variables (if the sign of the inequality is $\leqslant$) or surplus (if the sign of the inequality is $\geqslant$) variables. This augmented form of LPP is called the standard form. If the type of objective function is maximization, then convert all constraints which are of $\geqslant$ type.

2.2 Applicability of LPP

From the time of the development of the simplex method and with advancements in technology, a lot of applications of LPP have evolved, in particular, a major portion of management science for decision making and analytics is based on LPP. In this section we look at some classical examples such as the product mix and diet problems, and the applicability of LPP in transportation problems, transshipment problems and portfolio optimization.

2.2.1 The product mix problem

The product mix problem is one of the common problems in any book on LPPs. The important characteristic of a product mix problem is the available finite amount of resources; one has to supply these resources to competing products. Suppose there are n products and m resources, the requirement and availability can be given in an $m \times n$ table as m rows and n columns of the corresponding coefficients. The values in a row of the table are the coefficients of a constraint in the LPP. In addition, the table shows the profit per unit and associated with each available resource. The aim is to identify the number of units to be produced to maximize the profit using the available resources.

Example 2.2. A manufacturer produces three products P_1, P_2 and P_3 in one of its plants. Each product requires a certain amount of time on each of two machines, given in table 2.1 (h unit^{-1}).

The goal of the problem is to identify the optimal number of units of P_1, P_2, P_3 to produce so that the earned profit is maximized. Let x, y, z be the number of units of P_1, P_2, P_3 to be produced, respectively.

From the given scenario, the objective is to maximize the profit. Therefore the objective function is given by

$$\text{Maximize } Z = 3x + 3y + 3z$$

Table 2.1. Product mix.

Product	Machines		Profit per unit
	M_1	M_2	
P_1	7	8	\$3
P_2	8	12	\$4
P_3	5	7	\$2
Available m/c time (h week^{-1})	40	35	

Subject to the constraints (machines)

$$7x + 8y + 5z \leqslant 40 \text{ (M1)}$$
$$8x + 12y + 7z \leqslant 35 \text{ (M2)}$$
$$x, y, z \geqslant 0.$$

2.2.2 Diet problem

This is another common and important problem in LPP. The goal of the diet problem is to find the optimal combination of food products available such that the daily requirement of the nutritional component is met with minimal cost. The diet problem is given as: suppose there are k foods available say $F_1, F_2, \ldots, F_k$ in the market that has l nutritional components say $N_1, N_2, \ldots, N_l$. Let dj be the daily requirement of N_j and C_i be the cost price per unit of the food F_i. Let m_{ij} be the amount of nutrition N_j in food F_i.

Let x_i be the unit of food F_i to be purchased to meet the objective. The objective function is given by

$$\text{Minimize} \sum_{i=1}^{k} c_i x_i.$$

Subject to the constraints

$$\sum_{i=1}^{k} m_{ij} x_i \geqslant d_j \quad \text{for } j = 1, 2, \ldots, l$$
$$x_i \geqslant 0, \forall\, i.$$

2.2.3 Transportation problem

A chain of stores in any city needs to establish the brand and the goodwill of the customers. The daily problem in such a scenario is to identify the routes to transport the commodities from the various warehouses to their retail stores. Suppose there are m warehouses say $W_1, W_2, \ldots, W_m$ and n retail stores say $R_1, R_2, \ldots, R_n$ and the unit cost of transporting the commodity from the warehouse W_i to the retail store R_j is denoted by C_{ij}. The commodities must be supplied as per the demand at the retail store and the availability at the warehouse. If the sum of the availability/supply is equal to the sum of the demand then it is called a balanced transportation problem. In the case that the sum of the demand is not equal to the sum of supply, then the transportation problem is called an unbalanced transportation problem. The unbalanced transportation problem can be solved after converting it to a transportation problem by adding a dummy row(s)/column(s) with cost 0 in each of its cells. A balanced transportation problem and its entries are shown in table 2.2. The transportation model is also represented as a complete bi-partite graph. The transportation problem can be formulated as an LPP.

Table 2.2. Transportation cost table.

Warehouse/retail store	R_1	R_2	R_3	...	...	R_n	Supply
W_1	C_{11}	C_{12}	C_{13}	...	...	C_{1n}	b_1
W_2	C_{21}	C_{22}	C_{23}	...	...	C_{2n}	b_2
W_3	C_{31}	C_{32}	C_{33}	...	...	C_{3n}	b_3
...	...	...	...	...	...		
...	...	...	...	...	...		
W_m	C_{m1}	C_{m2}	C_{m3}	...	...	C_{mn}	b_m
Demand	a_1	a_2	a_3			a_n	

The objective function of the transportation problem is

$$\text{Minimize } Z = \sum_{i=1}^{m}\sum_{j=1}^{n} C_{ij}x_{ij}.$$

Subject to the constraints

$$\sum_{j=1}^{n} C_{ij}x_{ij} \leqslant b_i \text{ for } i = 1, 2, \dots, m \text{ (constraints due to supply)}$$

$$\sum_{i=1}^{m} C_{ij}x_{ij} \geqslant a_j \text{ for } j = 1, 2, \dots, n \text{ (constraints due to demand)}$$

$$x_{ij} \geqslant 0, \; 1 \leqslant i \leqslant m, \; 1 \leqslant j \leqslant n.$$

Note that here the warehouses and retail stores are known as pure supply nodes and pure consumption nodes. The transportation problem can be represented by a complete bi-partite graph. In certain scenarios, some nodes can be supply as well as consumption nodes, such models are called transshipment models and the nodes which are supply and demand nodes are called *transshipment nodes*.

Example 2.3. A firm has three warehouses W_1, W_2 and W_3 and four retail outlets R_1, R_2, R_3 and R_4. The unit transportation cost of shifting from W_i to R_j and the maximum possible supply and the demand requirements are given in the table 2.3. The company wishes to minimize the transportation cost.

Solution: total demand is 5700 = total supply and hence the problem is balanced.
The corresponding LPP model is

$$\text{Mininmize } Z = 6x_{11} + 4x_{12} + 8x_{13} + 4x_{14} + 7x_{21} + 21x_{22} + 4x_{23} + 6x_{24} + 10x_{31} + 4x_{32} + 9x_{32} + 7x_{34}.$$

Subject to constraints:

Table 2.3. Transportation cost table.

	R_1	R_2	R_3	R_4	Supply
W_1	6	4	8	4	2000
W_2	7	21	4	6	1200
W_3	10	4	9	7	2500
Demand	1500	2000	1000	1200	

Table 2.4. Result for the transportation cost problem.

	R_1	R_2	R_3	R_4			Supply
W_1	1300	0	0	700	2000	Leq	2000
W_2	200	0	1000	0	1200	Leq	1200
W_3	0	2000	0	500	2500	Leq	2500
	1500	2000	1000	1200			
	Geq	Geq	Geq	Geq			
Demand	1500	2000	1000	1200			27 500

Supply constraints

$$6x_{11} + 4x_{12} + 8x_{13} + 4x_{14} \leqslant 2000$$
$$7x_{21} + 21x_{22} + 4x_{23} + 6x_{24} \leqslant 1200$$
$$10x_{31} + 4x_{32} + 9x_{32} + 7x_{34} \leqslant 2500.$$

Demand constraints

$$6x_{11} + 7x_{21} + 10x_{31} \geqslant 1500$$
$$4x_{12} + 21x_{22} + 4x_{32} \geqslant 2000$$
$$8x_{13} + 4x_{23} + 9x_{33} + 4x_{14} \geqslant 1000$$
$$4x_{14} + 6x_{24} + 7x_{34} \geqslant 1200$$
$$x_{ij} \geqslant 0;\ 1 \leqslant i \leqslant 3\ \&\ 1 \leqslant j \leqslant 4.$$

Using the Excel solver we obtain the optimal values of x_{ij} and the minimum transportation cost and the same is presented in table 2.4.

The minimal transportation cost is 27 500.

The shipment details are:

- W_1 to R_1 is 1300 units and W_1 to R_4 is 700 units.
- W_2 to R_1 is 200 units and W_2 to R_3 is 1000 units.
- W_3 to R_2 is 2000 units and W_3 to R_4 is 500 units.

The variations of the transportation problem are:

1. Unbalanced transportation.
2. Maximization of the objective function.
3. Unacceptable routes.

Example 2.4. (*Transshipment problem*) A firm has two plants P_1 and P_2 which produce switch gears; two carry and forward (C&F) agents (C_1, C_2 and C_3) and dealers (D_1, D_2 and D_3). P_1 and P_2 produce 1500 switch gears and 2500 switch gears, respectively, and the demands at the dealers are 1900, 1200 and 900, respectively. All goods can reach the dealers through any of the two C&F agents. The unit transportation costs between nodes are shown in the network diagram (figure 2.1).

The objective of the organization is to minimize the transportation cost.

Here P_1 and P_2 are pure supply nodes, C_1, C_2, D_3 are transshipment nodes and D_1, D_2 are pure demand nodes. Using the buffer, we will convert this problem into a transportation problem.

Now we have nodes P_1, P_2, C_1, C_2, D_3 as supply nodes and C_1, C_2, D_1, D_2, D_3 are demand nodes. The supply at the transshipment nodes equals the buffer, and demand at the transshipment nodes equals demand plus buffer, where the buffer can be total supply or total demand. The objective function is the same as in the transportation problem. The transshipment cost table is presented in table 2.5.

The assignment problem is a special case of the transportation problem. There are m jobs and n bidders who are available to complete the jobs. Assume that $n \geqslant m$ or vice versa. Every job is assigned to only one bidder and every bidder will do only one job. Hence this can be modeled as square matrices and hence in both cases we add dummy rows/columns as per the need and assign unity as the supply and demand for both nodes. Let C_{ij} be the cost quoted by the ith bidder for doing the jth job. While formulating this scenario, as a linear programming model, change the inequality in both constraints as equality and solve as usual.

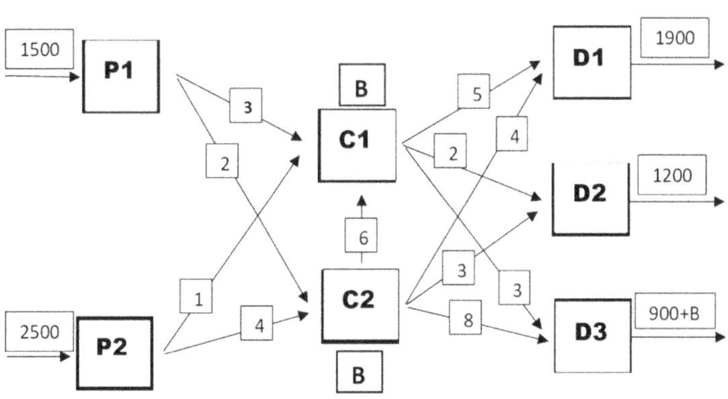

Figure 2.1. Transshipment network model.

Table 2.5. Transshipment cost table.

Supply/demand nodes	C_1	C_2	D_1	D_2	D_3	Supply
P_1	3	2	*	*	*	1500
P_2	1	4				2500
C_1	0	*	5	2	3	B
C_2	6	0	4	3	8	B
D_3	*	*	*	4	0	B
Demand	B	B	1900	1200	$900 + B$	B = Buffer

Replace (*) with a sufficiently large value and solve as in the transportation problem.

Table 2.6. Assignment cost.

Contractors	Jobs		
	J_1	J_2	J_3
Contractor C_1	123	78	67
Contractor C_2	158	65	75
Contractor C_3	200	50	35
Contractor C_4	100	30	40
Contractor C_5	225	80	68

Example 2.5. Assignment problem.
Mandalay City Corporation called for tenders to do three jobs from five authorized contractors. All eight contractors are eligible to perform all the jobs, but every contractor will be assigned only one job and every job will be assigned to only one contractor, as per the city corporation norms. Let C_{ij} be the quote by the ith contractor to perform the jth job. All jobs must be allocated with a minimal cost. The quote submitted by the contractors is given in table 2.6.

To convert the assignment problem to a transportation problem, put unity as the demand and supply, then solve as in the transportation problem (table 2.7).

Clearly this problem is a unbalanced problem, so add two dummy columns as J_4 and J_5 with 0 as values in each cell of J_4 and J_5 columns, to make the problem a balanced one.

The optimal cost is 188 and J_1 is assigned to contractor C_1, J_2 is assigned to contractor C_4 and J_3 is assigned to contractor C_3.

The solution obtained using the Excel solver is given in table 2.8.

2.2.4 Portfolio optimization

The mundane job of the fund manager is to identify the various instruments to invest in, in the right proportions as per the needs of the investor/firm. The objective may be minimizing the risk or maximizing the return on investment.

Table 2.7. Assignment problem as a transportation problem.

Contractors	Jobs					Supply
	J_1	J_2	J_3	J_4	J_5	
Contractor C_1	123	78	67	0	0	1
Contractor C_2	158	65	75	0	0	1
Contractor C_3	200	50	35	0	0	1
Contractor C_4	100	30	40	0	0	1
Contractor C_5	225	80	68	0	0	1
Demand	1	1	1	1	1	

Table 2.8. Answer table for example 2.5.

Contractors	Jobs							Supply
	J1	J2	J3	J4	J5			
Contractor C1	1	0	0	0	0	1	Eq	1
Contractor C2	0	0	0	1	0	1	Eq	1
Contractor C3	0	0	1	0	0	1	Eq	1
Contractor C4	0	1	0	0	0	1	Eq	1
Contractor C5	0	0	0	0	1	1	Eq	1
Demand	1	1	1	1	1			188
	Eq	Eq	Eq	Eq	Eq			
	1	1	1	1	1			

An investment management firm assigned one of its fund managers to select a portfolio to invest \$1 000 000 as per the firm's policy. The firm's policy states that all new investments must be made in the information technology sector, insurance sector and banking sector in the coming quarters. The diligence report of the firm identified the investment opportunities in certain companies and the expected annual rate of return as given in table 2.9.

Moreover, the restriction based on the diligence report is as follows:

- The investment in any industry sector should not exceed 50% of the funds available.
- Investment made in the banking industry must be at least 20% of the total investment in the IT sector.
- Investment in Info Systems will yield a high return but the risk is high and hence the investment in Info Systems must be 55% of the total investment in the insurance sector. Find the proportion of the amount invested in each sector which would yield the maximum return.

Table 2.9. Predicted annual return.

#	Firm	Rate of return (%)
1	Info Systems	6
2	BTS	11
3	Advanced Systems	9
4	Pearless Insurance	5.50
5	Loyal Insurance	6
6	Commercial Bank	4.30
7	Agro Bank	4.10

Let us formulate the given scenario as LPP. Let I, B, A, P, L, CB and AB be the decision variables to represent the proportions of the amount to be invested in the respective firms.

The objective function is

$$\text{Maximize } Z = 0.11I + 0.06B + 0.09A + 0.055P + 0.06L + 0.043CB + 0.041AB.$$

Constraints:

$$\begin{aligned} I + B + A + P + L + CB + AB &= 1000000 \\ I + B + A &\leqslant 500000 \\ CB + AB &\leqslant 500000 \\ P + L &\leqslant 500000 \\ CB + AB &\leqslant 0.2(I + B + A) \\ I &\leqslant 0.55(CB + AB) \\ I, B, A, P, L, CB, AB &\geqslant 0. \end{aligned}$$

2.3 The simplex method

This method was the first to obtain the solution of the system of linear inequalities in a large scale and efficient method. We deploy the Excel add-in Solver to solve the given LPP in this chapter. The fundamental idea of this method is to start with a basic feasible solution and obtain the optimal basic feasible solution by moving from one boundary point to other adjacent boundary points. In a nut-shell, the simplex method can be described as 'moving from feasibility to optimality'.

The various steps in the simplex method (in the case of the maximization problem) are:

1. Convert into the standard form (converting all constraints into $\leqslant$ type).
2. Introduce the slack variables (to change inequalities into equalities).
3. Set up the initial table (with the objective function as Z – rhs = 0).
4. Check for the optimality (in the z-row all values are $\geqslant 0$).

5. Identify the entering/pivot variable (the variable that has the most negative value in the z-row, the denominator should be positive).
6. Identify the leaving variable (among the ratios between the solution column values and the entering variable column values, choose the minimal one).
7. Create the new table as in the Gauss–Jordon method for the next iteration
8. If not, repeat step 4–6 until optimality is reached.

Let us consider an example of an LPP.

Example 2.6. Max $Z = 3x + 2y + z$.
Subject to

$$\begin{aligned} 2x + 4y - 3z &\leqslant 25 \\ -x + 2y + 2z &\leqslant 35 \\ x, y, z &\geqslant 0. \end{aligned}$$

Using the Excel Solver add-in, we solve this problem. Tables 2.10 and 2.11 show the entry, the solution obtained and the answer for the problem.

From the tables, the optimal values for $x = 155$, $y = 0$ and $z = 95$ and the maximum value is 560. In this problem, both constraints are of a binding nature, that is the obtained solution of the decision variables satisfies the constraints and hence the solution obtained is the optimal basic feasible solution. The solution report generated by the Excel Solver is given in table 2.12.

Table 2.10. The LPP for example 2.6.

	x	y	z			
Coefficients of the objective function	3	2	1			
Optimal values of *D* variables						**RHS**
Constraint 1	2	4	-3	0	Leq	25
Constraint 2	−1	2	2	0	Leq	35

Table 2.11. Solution.

	x	y	z			
Coefficients of the objective function	3	2	1			
Optimal values of *D* variables	155	0	95		560	RHS
Constraint 1	2	4	-3	25	Leq	25
Constraint 2	−1	2	2	35	Leq	35

Table 2.12. Answer report for example 2.6.

		Objective Cell (Max)			
Cell	Name	Original Value	Final Value		
\$H\$3	Optimal values of *D* variables	0	560		
		Variable Cells			
Cell	Name	Original Value	Final Value	Integer	
\$D\$3	Optimal values of *D* variables *x*	0	155	Contin	
\$E\$3	Optimal values of *D* variables *y*	0	0	Contin	
\$F\$3	Optimal values of *D* variables *z*	0	95	Contin	
		Problem-Constraints-Set			
Cell Add	Name of the Constraint	Cell Value—RHS	Formula-Type	Status/Result	Slack
\$G\$4	Constraint 1	25	\$G\$4 <= \$I\$4	Binding	0
\$G\$5	Constraint 2	35	\$G\$5 <= \$I\$5	Binding	0

2.4 Artificial variable techniques

There are two methods that fall into this category, namely the big-M method/ Charne's penalty method and the two-phase method. This section deals only with the two-phase method. As the name suggests, there are two phases to solve the given LPP, the initial basic feasible solution is found in phase 1; if such an initial basic feasible solution exists, the optimal basic feasible solution is obtained in phase 2.

Steps involved in the two-phase simplex method:

1. Convert the given problem into standard form by adding slack/surplus variables to the set of constraints.
2. For the constraints of $=\geqslant$ and $=$ type, add an artificial variable, in addition to the surplus variables (artificial variables play the role of slack variables).

   Phase 1: Ignore the original objective function.
3. Set the new objective function so as to minimize the sum of artificial variables and express the basic variables in terms of non-basic variables subject to the given constraints and solve as in the simplex method.
4. Consider the final iteration of phase 1 and start with phase 2.

5. In phase 2 start with the original objective function subject to the constraints in the final iteration of phase 1, without the artificial variables column and solve as usual to obtain the optimal basic feasible solution.

Example 2.7. Obtain the optimal basic feasible solution for the LPP using the two-phase method.

$$\text{Minimize } Z = 4x + y,$$

subject to

$$\begin{aligned} 3x + y &= 3 \\ 4x + 3y &\geqslant 6 \\ x + 2y &\leqslant 4 \\ x, y &\geqslant 0. \end{aligned}$$

Phase 1

Minimize $w = A_1 + A_2$,
subject to the constraints

$$\begin{aligned} 3x + y + A_1 &= 3 \\ 4x + 3y - S_1 + A_2 &= 6 \\ x + 2y + S_2 &= 4 \\ x, y, S_1, S_2, A_1, A_2 &\geqslant 0. \end{aligned}$$

The initial table is shown below.

We have to modify the objective function that is expressed by the basic variables A_1, A_2 and S_2 in terms of non-basic variables x, y, S_1.

The new objective function is $w = 7x + 4y - S_1 - 9$. The initial table is presented in table 2.13.

Solve as usual to obtain the optimal solution of phase 1 and the final iteration table is presented in table 2.14.

Phase 2

Minimize $Z = 4x + y$,
subject to the constraints

Table 2.13. Initial table.

Basic variables	x	y	S_1	S_2	A_1	A_2	Solutions
W	7	4	0	0	0	0	9
A_1	3	1	0	0	1	0	3
A_2	4	3	−1	0	0	1	4
S_2	1	2	0	1	0	0	4

Table 2.14. Final optimal table—phase 1.

B variables	x	y	S_1	S_2	A_1	A_2	Solution
Z	0	0	1/5	0	1	1	18/5
x	1	0	1/5	0	3/5	1/5	3/5
y	0	1	−3/5	0	4/5	3/5	6/5
S_2	0	0	1	1	1	1	1

Table 2.15. Initial table—phase 2.

B variables	x	y	S_1	S_2	Solution
Z	0	0	1/5	0	18/5
X	1	0	1/5	0	3/5
Y	0	1	−3/5	0	6/5
S_2	0	0	1	1	1

$$x + 0y + \frac{1}{5}S_1 + 0S_2 = \frac{3}{5}$$
$$0x + y - \frac{3}{5}S_1 + 0S_2 = \frac{3}{5}$$
$$0x + 0y + S_1 + S_2 = 1$$
$$x, y, S_1, S_2 \geqslant 0.$$

In the final iteration of phase 1, the basic variables identified are x, y and S_2.

Again express the basic variables in terms of non-basic variables to obtain the new objective function of phase 2. The initial table of phase 2 is shown in table 2.15.

On solving this, we obtain $x = 2/5$, $y = 9/5$ and the minimum value of Z is 17/5.

2.5 Duality

Each LPP is associated with another LPP called the dual linear problem, whereas the original problem is called the primal problem. In this section, we will see the conversion of primal to dual and dual to primal. If one of the primal or dual problems has an optimal solution the other also has an optimal solution. One can obtain the optimal values of a primal problem by solving the dual problem and vice versa.

Steps involved in writing the dual of the given LPP:

1. Convert all constraints into $\leqslant$ type, if the objective function is maximization; otherwise convert all constraints into $\geqslant$ type.
2. Convert all the constraints into the equalities type by adding slack/surplus variables.

3. If the objective function of the primal is maximization then the objective function of the dual is minimization and vice versa.
4. The number of dual variables is equal to the number of primal constraints.
5. The number of dual constraints is equal to the number of primal variables.
6. The coefficients of the objective function of the dual are the right-hand side of the primal constraints.
7. Reverse the inequality of the primal to obtain the dual problem constraints with dual variables.
8. The coefficient of the dual constraints are the primal variable coefficients (column).

Note that dual of a dual is a primal.

Example 2.8. Write the dual of the given LPP.

$$\text{Max } Z = 2x + 3y + u,$$

subject to

$$\begin{aligned} x + 2y - u &\leqslant 15 \\ 2x + 3y + 15u &\leqslant 50 \\ x, y, u &\geqslant 0. \end{aligned}$$

The dual of the given problem is

$$\text{Minimize } W = 15r + 50s,$$

subject to

$$\begin{aligned} r + 2s &\geqslant 2 \\ 2r + 3s &\geqslant 3 \\ -r + 15s &\geqslant 1 \\ r, s &\geqslant 0. \end{aligned}$$

Here r and s are the dual variables.

2.6 Sensitivity analysis

The examination of the effect of marginal changes in the problem on the optimal solution is known as sensitivity analysis. That is, how small changes in the coefficients of the decision variables and/or right-hand side values of the constraints will affect the optimal solution. Also, one could use sensitivity analysis to find the range of optimality and a reduced cost for the coefficients of the objective function by allowing a small change in the objective function coefficients. Similarly a small change in the right-hand side of the constraint helps us to study how this change affects the optimal solution. This is called a shadow price.

Example 2.9. Solve the following LPP and answer the following questions,

$$\text{Maximize } Z = 8x + 9y + 6z,$$

subject to

$$5x + 4y + 2z \leqslant 120$$
$$x + 2y + z \leqslant 55$$
$$x + 2y + 3z \leqslant 30$$
$$x, y, z \geqslant 0.$$

1. Find the range of optimality of the coefficients of the objective function.
2. Find the shadow prices of all the constraints.
3. If the rhs of the first constraint is changed to 120, what will happen to the optimal solution?

The obtained optimal solution from solving using Excel is given in table 2.16. The optimal values of $x = 21.67$, $y = 4.17$ and $z = 0$. The maximum value of $Z = 210.83$. The first and third constraints are binding, whereas the second one is not binding and it indicates that there are 25 units of unused resources available.

The sensitivity report generated by the Excel Solver is given in table 2.17.

Table 2.16. Answer report for example 2.9.

		Solver Engine—Excel Solver Options			
		Objective Cell (Max)			
Cell Add	Name	Original Value	Opt/Final Value		
G3		0	210.83		
		Variable Cells			
Cell Add	Name	Original Value	Final Value	Integer	
C3	x	0	21.67	Contin	
D3	y	0	4.17	Contin	
E3	Z	0	0	Contin	
		Constraints—Set			
Cell	Name	Cell Value	Formula	Status	Slack
F4		125	F4 <= H4	Binding	0
F5		30	F5 <= H5	Not binding	25
F6		30	F6 <= H6	Binding	0

Table 2.17. Report for the sensitivity analysis.

Problem-Variable Cells						
Cell	Name	Final Value	Reduced Cost	Objective Coefficient	Max. Allowable Increase	Max. Allowable Decrease
C3	x	21.67	0	8	2.13	3.5
D3	y	4.17	0	9	7.00	1.31
E3	z	0	−2.83	6	2.83	1E+30
Constraints Set						
Cell	Name	Final Value	Shadow Price	Constraint R.H. Side	Max. Allowable Increase	Max. Allowable Decrease
F4		125	1.17	125	25	65
F5		30	0.00	55	1E+30	25
F6		30	2.17	30	25	5

The optimality range for C_1 is $4.5 \leqslant C_1 \leqslant 10.125$. The optimality range for C_2 is $7.69 \leqslant C_2 \leqslant 16$. The optimality range for C_3 is $C_1 \leqslant 8.83$ and no lower bound. Within these ranges the optimal solution remains unchanged.

The shadow price for each constraint is given in the table and the range for optimality is also presented. Any change in the rhs of the first constraint is within the range, but the optimal value will be decreased by $5 \times 1.17 = 5.85$. This sensitivity report helps the decision maker in a better way.

2.7 Network models

Any communication/transport network can be viewed as a graph $G = (V, E)$, where V is the set of nodes/vertices and E is the set of links/edges.

2.7.1 Shortest path problem

Let G be the given network. A link/edge from the i-node to the j-node is denoted by $(i, j) \in E$ and C_{ij} is the link cost/maximum capacity the link can hold/distance/duration between i and j. Let x_{ij} be the amount of flow in the link (i, j). We wish to find the shortest distance from an s-node to a t-node. The LPP formulation of this problem is given as:

- The objective function is given by mininmize $Z = \sum_{(i,j)\in E} C_{ij} x_{ij}$, where x_{ij} can assume either 1 if (i, j) is on the shortest path, otherwise 0.
- The constraints are defined for each node and they must satisfy the conservation of flow. That is total in-flow = total out-flow.

Example 2.10. Find the shortest path between nodes 1–4 in the network, the link cost/distance between the nodes are given as: Cost(1,2) = 3; Cost(1,3) = 4; Cost (2,3) = 2; Cost(2,4) = 6; Cost(3,4) = 4 (figure 2.2).

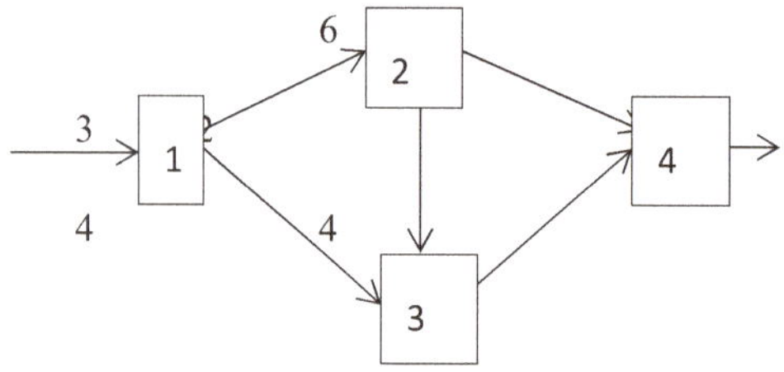

Figure 2.2. Network diagram.

Table 2.18. Optimal solution table.

x_{11}	x_{12}	x_{13}	x_{14}	x_{21}	x_{22}	x_{23}	x_{24}	x_{31}	x_{32}	x_{33}	x_{34}	x_{41}	x_{42}	x_{43}	x_{44}			
	3	4				2	6				4							
0	0	1	0	0	0	0	0	0	0	0	1	0	0	0	0			8
	1	1														1	Eq	1
	1					-1	-1									0	Eq	0
		1				1					-1					0	Eq	0
							1				1					1	Eq	1

The corresponding LPP is given as:
Objective function is Minimize $Z = 3x_{12} + 4x_{13} + 2x_{23} + 6x_{24} + 4x_{34}$.
Subject to the constraints:

$$\text{For node-1: } x_{12} + x_{13} = 1$$

$$\text{For node-2: } x_{12} = x_{23} + x_{24}$$

$$\text{For node-3: } x_{13} + x_{23} = x_{34}$$

$$\text{For node-4: } x_{24} + x_{34} = 1$$

$$x_{ij} \in \{0, 1\}, \ \forall\,(i, j) \in E(G).$$

This formulated LPP can be solved in the routine manner. The optimal solution is given in table 2.18. The shortest path from 1 to 4 is 1–3–4 and the minimal cost is 8.

2.8 Dual simplex method

The other method for solving the LPP is the dual simplex method. In this method, we start with an optimal solution and move towards feasibility. Note that this method fails to produce an optimal feasible solution for some problems.

Steps involved in the dual simplex method:

1. Convert the objective function of the given problem into minimization, if the objective function is of maximization.
2. All constraints must be of $\leqslant$ type, if not convert.
3. Set up the initial table. Identify the leaving variable by choosing the most negative value in the solution column of the initial table and the corresponding variable is the leaving variable.
4. Construct the ratios between z-row values and the leaving variable row. Note that this ratio is constructed only for non-basic variables and the denominator must be negative.
5. Choose the minimal ratios and the corresponding variable (column head) is the entering variable.
6. Construct the next table as in the simplex method for the next iteration.
7. Check for optimality, i.e. (a) all values in the z-row must be $\leqslant 0$ and (b) all values in the solution column must be non-negative.

Note that if we cannot construct the ratio after identifying the leaving variable, that is, all values in the leaving variable row are non-negative, then the problem has an infeasible solution.

2.9 Software packages to solve LPP

LPP is intended for modeling real-world problems and hence one can find two types of software packages, one type represents the modeling perspective and the other type is solvers.

The most popular commercially available software packages (solvers) are OSL, MATLAB, LINDO, CPLEX, LOOQ, EXCEL, etc. The popular ones from the modeling perspective are GAMS, AMPL, MPL, Operation Research in NCS, etc. The statistical package R also provides packages to solve LPP.

Further reading

[1] Dantzig G B 1963 *Linear Programming and Extensions* (Princeton, NJ: Princeton University Press)

[2] Dantzig G B and Thapa M N 1997 *Linear Programming* (New York: Springer)

[3] Bazarra M S, Jarvis J J and Sharali H D 2005 *Linear Programming and Network Flows* (New York: Wiley Blackwell)

[4] Hiller F and Liberman G J 2005 *Introduction to Operations Research* 8th edn (New York: McGraw Hill)

[5] Winston W L and Albright S C 2007 *Practical Management Science* 3rd edn (Boston, MA: Cengage South Western)

[6] Winston W L 1996 The teachers' forum: management science for MBA at Indiana University *INFORMS J. Appl. Anal.* **26** 105–11

[7] Anderson D R, Sweeny D J, Williams T A, Camm J D and Cochran J J 2016 *An Introduction to Management Science: Quantitative Approaches to Decision Making* 14th edn (Boston, MA: Cengage Learning)

[8] Murthy K G, Kabadi S N and Chandrasekaran R 2000 Infeasibility analysis for linear systems: a survey *Arab. J. Sci. Eng.* **25** 3–18
[9] Greenbreg H J 1993 How to analyze the result of linear programs—part 1: preliminaries *INFORMS J. Appl. Anal.* **23** 56–67
[10] Greenbreg H J 1993 How to analyze the result of linear programs—part 2: price interpretation *INFORMS J. Appl. Anal.* **23** 97–114
[11] Greenbreg H J 1993 How to analyze the result of linear programs—part 3: infeasibility diagnosis *INFORMS J. Appl. Anal.* **23** 120–39

IOP Publishing

Modern Optimization Methods for Science, Engineering and Technology

G R Sinha

Chapter 3

Multivariable optimization methods for risk assessment of the business processes of manufacturing enterprises

Vladimir Gorbunov

The rapid rate of development of science and technology opens up broad prospects for the development of innovative businesses. However, opening a new business is associated with investment costs for the organization and development of this business. Before you meet these costs, you should make sure that the business idea in question is realistic. This can be achieved using special programs that allow you to create a mathematical model of the business and evaluate its possible future outcomes. However, business is associated with many random sources of data. The business model must be able to work with random variables as input data. This chapter discusses the possibility of forming such a model using the example of the program 'E-Project'. This program allows one, on the basis of the input source data, to determine the cash flow of the project during the entire period of its implementation and business performance indicators, and to build charts that reflect the internal and external data flows of the project. The program allows you to perform risk accounting and calculation of business process parameters taking into account these factors. The project manager has the opportunity to analyze the impact of anti-risk measures on the model and decide on the feasibility of their implementation. The decision is made on the basis of the project performance indicators, quantitative assessment of risk factors and the value of anti-risk costs.

3.1 Introduction

Optimization is the mathematical problem of maximizing or minimizing some function of one or more variables under constraints. Optimization is the basis of economic analysis of production processes. Typically, the value to be optimized is

doi:10.1088/978-0-7503-2404-5ch3

related to the performance of the project or facility in question. The optimized version of the object should be evaluated by a quantitative measure—the criterion of optimality. The criterion of optimality is a quantitative assessment of the optimized quality of the object.

On the basis of the chosen criterion of optimality, a function is made of the target that represents the dependence of the optimality criterion on the parameters that affect its value. The types of optimality criterion or target criterion functions are defined by a specific optimization task. Thus, the task optimization is reduced to finding the extremum of the objective function.

The most general statement of the optimal problem is the expression of the optimality criterion in the form of an economic indicator (profit, productivity, cost of production, profitability). The criterion of optimality in the general case should reflect the most significant aspects of the process, to yield a numeric measure and easily explain the physical meaning.

When setting specific optimization problems, it is desirable to write the optimality criterion in the form of an analytical expression.In the case where the random disturbances are small and their impact on the object can be ignored, the optimality criterion can be presented as a function of input (X), output (Y) and control (U) parameters:

$$L = F(X_1, X_2, \ldots, X_n, Y_1, Y_2, \ldots, Y_n, U_1, U_2, \ldots, U_n).$$

Since $Y = F(U)$, it is possible to write for fixed X

$$L = F(U).$$

In this case, any change in the values of the control parameters affects the change in the value of L, since the control parameters are directly included in the expression of the optimization criterion and thus change the output parameters of the process, which depend on the control ones.

If the random perturbations are large enough and they must be taken into account, then experimental and statistical methods should be used to obtain a model of the object in the form of a function that is fair only for the studied local area, and the optimality criterion will take the form:

$$L = F(X, U).$$

In optimization problems there are simple and complex optimization criteria. An optimality criterion is called simple if you want to define the extremum of the objective function without setting conditions on any other values. Such criteria are usually used in solving specific optimization problems (for example, determining minimum fuel consumption, optimal product processing time, etc). If you know the functional relationship between the target function and the control variable, you can directly calculate the value of the target function for a fixed value of the control variable.

If the objective function $F(x_1, x_2, \ldots, X_P)$ has continuous partial derivatives in its arguments, then putting the partial derivatives of F in x equal to zero and solving n equations together,

$$(dF/dx_i) = 0, \quad i = 1, 2, \ldots, n,$$

provides the value of the objective function.

One-parameter optimization provides a search of the extrema of the functions of one variable. The limiting condition is the one-modality of the objective function on the studied interval. The search for the minimum (or maximum) of the objective function is performed by different methods, which differ in the speed of finding the desired value. One such method is the dichotomy method. It uses the values of the objective function when searching for its extremum.

For example, take the function $F(x)$ (figure 3.1). It is necessary to find $\bar{x}$, delivering the minimum (or maximum) of the function $F(x)$ on the interval $[a, b]$ with a given accuracy ε, i.e. to find

$$\bar{x} = \arg\min F(x), \quad \bar{x} \in [a, b].$$

At each step of the search process, divide the segment $[a, b]$ in half, $x = (a + b)/2$—the coordinate of the middle of the segment $[a, b]$. Calculate the value of the function $F(x)$ in the neighborhood $\pm\varepsilon$ of the computed point x:

$$F_1 = F(x - \varepsilon),$$
$$F_2 = F(x + \varepsilon).$$

Compare F_1 and F_2 and discard one of the halves of the segment $[a, b]$ (figure 3.1).

When searching for a minimum:

- If $F_1 < F_2$, then discard the segment $[x, b]$, then $b = x$.
- Otherwise, drop the segment $[a, x]$, then $a = x$.

The division of the segment $[a, b]$ continues until its length is less than the specified accuracy $|b - a| \leqslant \varepsilon$. The algorithm of this method is presented in figure 3.2.

In the output, x is the coordinate of the point at which the function $F(x)$ has a minimum (or maximum), FM is the value of the function $F(x)$ at that point.

The optimality criterion is called complex if it is necessary to set the extremum of the objective function for several changing variables. The procedure for solving the optimization problem usually includes additional restrictions on these changing

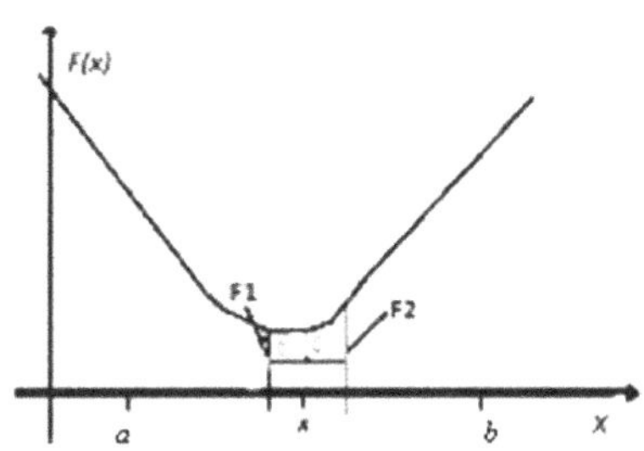

Figure 3.1. Determination of the minimum by dichotomy.

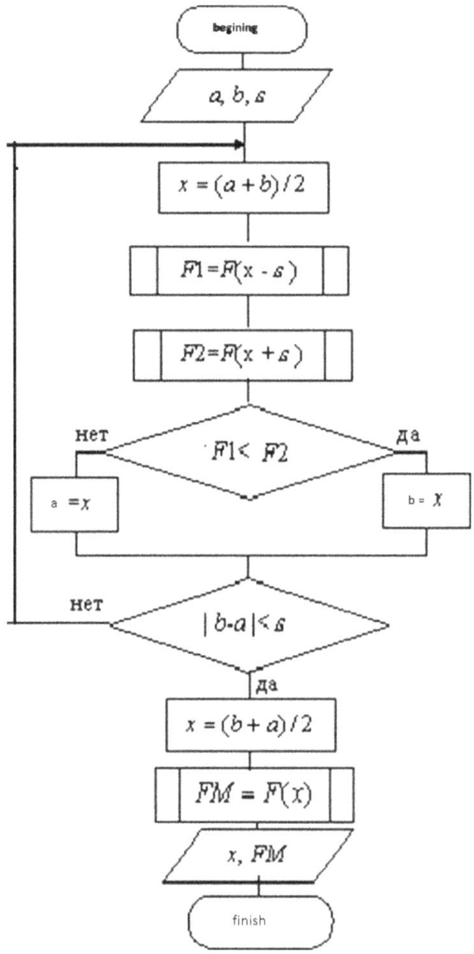

Figure 3.2. The scheme of the algorithm of the method of dichotomy.

parameters. In this case, the solution of the optimization problem is reduced to the sequential execution of the following operations: drawing up a mathematical model of the optimization object; selection of the optimality criterion and determination of the objective function; the establishment of possible restrictions on the variables; and the choice of the optimization method, which will find the extreme values of the required quantities.

The methods of multidimensional optimization are usually distinguished by the type of information they need in the process:

- Direct search methods (zero-order methods) that need only the values of the target function.
- Gradient methods (first order methods) which additionally require first order partial derivatives of the objective function.
- Newtonian methods (second order methods) that use second order partial derivatives.

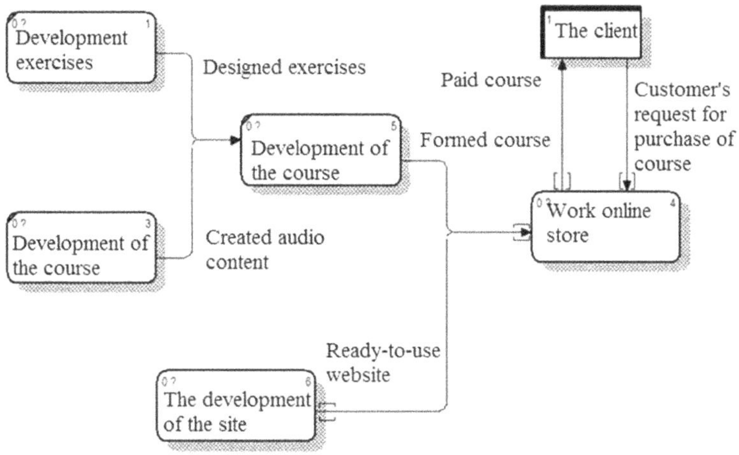

Figure 3.3. IDEF0 business model.

The recurrent scheme of most multivariate optimization methods is based on the expression $Xk + 1 = Xk + \lambda ksk$. They differ from each other by the way of choosing the step length λk and the search direction sk. The first is a scalar and the second is a vector of unit length. Methods of multidimensional optimization are reduced, in one form or another, to methods of one-dimensional optimization.

Modern production processes consist of a set of interrelated sub-processes, each of which has its own limitations and optimization conditions. Optimization of such processes is performed by several performance indicators in compliance with a set of limiting factors. Special computer programs are used to analyze and optimize such processes. For the features of multi-dimensional optimization of production business processes taking into account risk factors, consider the example of using the program 'E-Project'.

3.2 A mathematical model of a business process

Risks in manufacturing enterprises are associated with the appearance of random factors, which ultimately lead to changes in planned results. A financial plan allows you to consider and take into account the manifestation of potential risks.

The financial plan of the project involves the construction of business models, which show all the actions performed with the reflection of used material values, labor costs and the duration of the execution of all intermediate operations. Special modeling languages, e.g. the unified modeling language, are used to build such models [1]. This language describes the simple components of the procedure of a business process and reveals the relationship between the internal sub-processes and external data streams. Figure 3.3 shows an example view model of the company's activities (IDEF0 methodology). This company is planning the development of a training course to sell through an online shop.

The main parts of the methodology are given in the diagram. The diagram shows the functions of the system in geometrical rectangles, and also the existing relations between the functions and the external environment. The rectangles represent specific processes, functions, works or tasks that have a purpose and lead to a marked result. Arrows indicate the interaction processes between them and the external environment.

The model includes three types of documents (graphical charts, glossary, text), which refer to each other. Information about the system is displayed in the graphical diagrams with blocks and arrows, and their connections. The blocks represent the basic functions of the model elements. These functions can be broken down (decomposed) into their component parts and presented in the form of more detailed charts. The decomposition process continues until the subject is described at the level of detail necessary to achieve the objectives of a specific project. The glossary is created and maintained by a set of definitions, key words and explanations for each element of the chart and describes the essence of each element. The text gives an additional description of the operation of the system.

Often there are cases where some arrows do not make sense to continue considering in the child diagrams below a certain level in the hierarchy, or vice versa—individual blocks have no practical value above a certain level. On the other hand, sometimes you need to get rid of the separate 'conceptual' arrows and stop using them deeper than a certain point.

Standard IDEF0 introduces the concept of tunneling to solve such problems. The designation 'tunnel' in the form of two parentheses around the beginning of the arrow indicates that this arrow is not inherited from the parent functional block and appears (from the 'tunnel') only on this chart. In turn, the same designation around the end of the arrow in close proximity to the block receiver refers to the fact that the arrow is a child of this block diagram, and if displayed will not be considered in the diagram.

The business process model allows financial models to determine the cost characteristics of processes and their interaction in time. The financial plan of the project should reflect all costs associated with its preparation, take into account the cost of manufactured goods or services, and determine income from the sale of goods or services. The projects differ in the time interval start-up of the business, its development and completion. During the project the prices of goods, raw materials and debt capital can change and, in financial terms, these changes must be taken into account.

3.3 The market and specific risks, the features of their account

A financial plan is an important document containing detailed information about cash flow for current operations as well as the investment and financing activities of the enterprise. The manager may obtain information from this plan about:

- The sources of funds and directions of their use.
- The excess cash in the accounts of the enterprise. The extent to which the enterprise extension is provided at the expense of their own and borrowed funds.
- Any additional loans.

When making a report on cash flows for previous periods the entrepreneur uses the actual data available in the accounting records. The preparation of the forecast for the coming period requires more detailed analysis of the current situation and the trends of its change. Often entrepreneurs plan the distribution of funds for the best and worst cases and also for the most real situation.

A small enterprise can take into account the probabilistic nature of the factors influencing its activity. In this case, the calculation of the indicators for the future period is not reduced to the calculation of the three variants of development. In order to obtain correct results, we have to calculate a single process, which mathematically defines the probability of achieving the possible result of the activity of the enterprise depending on the identified characteristics of the involved processes. Such forecasting will enable the entrepreneur to choose the right development strategy for the enterprise, ensuring the timely payment of the accepted debt.

Values such as the price of the traded goods or services, the cost of materials or components, labor costs, etc, are connected through mathematical dependencies with the performance of the company. The variation of the initial indicators requires reproducing the calculations. In this regard, there is a need for an automated means of calculation for established procedures to instantly output when the source data change.

To help developers speed up the preparation of the business plan and provide the necessary information for the investor to know the level of quality of the design documents, the entrepreneur uses specialized programs for financial planning. Traditionally, such systems are developed versions of the document templates. The most popular systems provide capabilities similar to the program Excel, and all of their value for the user lies in a well-chosen list of topics. Filling it out, he or she will obtain a more or less acceptable financial plan. For example, the most popular system in this group is the program for Business Plan Pro, with hundreds of thousands of users.

The ability of some of the financial planning programs to account for risk factors is important because of the acceleration of the rates of economic development in every industry. An example of such a program is 'E-Project' [2, 3]. The program has the following features:

- Incorporates a probabilistic calculation of the business processes.
- Uses a widespread and reliable software package.
- The ability to navigate freely in the calculation methodology, adapted to specific user requests by creating their own forms of source data and calculation algorithms.
- The ability to enter data in the form of arbitrary shapes, and the results of calculations in the form of required reports. There are good editors for the formation of forms and reports.
- Performs an advanced analysis of the credit-worthiness of the project, i.e. the dependence of the results of calculations and changes in loan terms.
- A probabilistic calculation of the financial risk of the project depending on the probability characteristics of the source data.
- The results of the calculations are in the form of tables and diagrams.
- Has the ability to convert data into HTML format.

The software package 'AE-Project' consists of two related programs: a program for the calculation of economic indicators, realized in the program Excel, and a textual blank business plan that is implemented in Word. Using an input form in the analysis module, one enters all of the source digital data for the project. It then calculates the results and determines the course of economic processes associated with the implementation of the project.

Working with a design file starts with the opening title page of the project, where the author writes the project name, author name, chosen project start date, the interval calculation of the project (month, quarter) and the duration of the calculated period. In this form, you select the currency format in which to engage in financial calculations (RUB, thousand RUB, USD, thousand USD, etc).

After filling in and confirming the passport data, the program displays a main form, consisting of ten control buttons (figure 3.4).

The first eight control buttons provide data input on the project. The form of 'Investment costs' is a list of all activities that must be performed to implement the project. For each of these events the data are entered on its cost (with options for average, minimum and maximum values), data at the beginning of the activity and its duration, the availability of the plants and equipment associated with the event, and the rate of depreciation for each fixed asset group. The data input forms will be used to determine the financial flows required to support the production of products or the provision of services, to chart and plan the schedule of works for project implementation, and for the computation of the aggregate depreciation at any point in your project.

The main characteristics of the form 'Constant expenses' is the accounting of all project costs not related to production or services. Such expenses include costs associated with building maintenance, and payments for utilities, communications, transportation, security, etc. The data entry form 'Products' is associated with the data describing the income and expenses for each planned type of product or service. The form 'Production' is the total amount spent on marketing research, it also

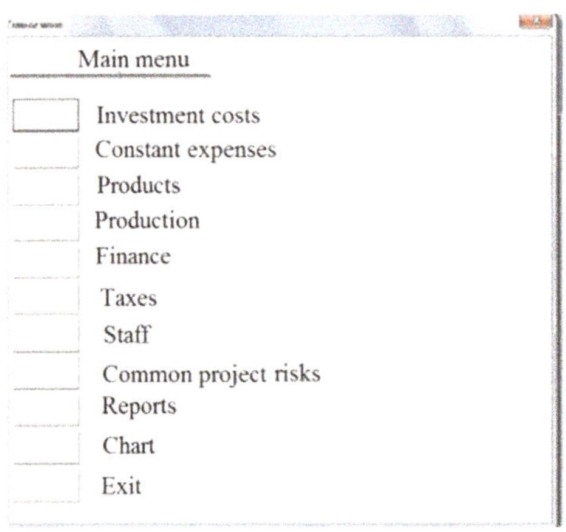

Figure 3.4. The main menu of the project.

answers the questions of how much and when the manufactured products listed in the form 'Products' will be implemented. The sales in this form will be submitted as the number of products, the characteristics of which are presented in the form 'Products'.

The form 'Finance' is used to enter the initial data on the financial activities and considers such indicators as equity, loans, repayment of loans, interest on loans, grants and government funding, and the payment of dividends.

The input form 'Taxes' specifies the source data for the calculation of tax planning. The procedure of payment and the amount of taxes depends on the legal form of the enterprise, and the accepted the forms of accounting and tax reporting. In this form, the authors of the draft introduce the adopted tax rate for the primary taxable base indicators: payroll, income, imputed income, profit and property.

The form 'Staff' is used to determine the financial flows associated with staff wages, receiving a set salary. This form indicates the period during which the employee will work on the project, his or her position and set salary. You can plan roles without names when you do not know the names of the professionals who will be involved in the project.

To calculate the impact of risks the software uses the input form 'Common project risks'. All risks of the project can be divided into two categories: market and special. Market risks form due to fluctuations in price parameters, i.e. the instability of the market. These indicators can change from their average value both upwards and downwards. Such risks must be characterized by the standard parameters for random distributions (expectation, variance), which are determined by the entered data and reviewed by the input forms. The second type (special risks) is associated with the specific situation in the project, which can occur with some probability and this will cause appropriate financial changes.

To assess the impact of special risks the following characteristics are considered: the period of manifestation of the risk factor, the probability of a risk situation R and the financial cost upon the occurrence of a risk situation Q. For the construction of the types of risks one of the accepted categories should be allocated: technical, organizational, financial, environmental, technological, etc. The program includes measures to reduce the impact of risks. The effect of these events will change the parameters P and Q on P_1 and Q_1. These events are recorded in the form of the investment measures the cost of these activities Q_2 and their execution time. Table 3.1 reflects the approach to risk assessment and events. Boolean variable K can take the value 0 or 1. When $K = 1$ the event risks are characterized by the parameters of P_1Q_1, if $K = 0$ the event fails and the risk is perceived with the parameters PQ.

The introductory data in the form end with the formation of the financial model of the project. The results for the generated models can be observed in the output reports that are in accordance with the selected algorithm and the results are given as tables (a button on the main menu 'Reports') or graphics (click main menu 'Chart'). Figures 3.5 and 3.6 show examples of output graphs.

Modern enterprises should take into account in their plans all tangible and intangible factors that can affect their condition in the future. The result of such

Table 3.1. Risks and measures for compensation of financial risks.

No.	Risk (finance)	PQ	P_1Q_1	Event	Q_2	K
1	Does not increase the return on investment.	10	4	Prepare to present information and analytical materials for the attraction of investments.	20	0
2	To continue the high production costs per unit of output.	10	15	To introduce an automation system to reduce the cost of production.	60	0
3	Will not decrease the average collection time of accounts receivable.	20	6	To automate the system of collection receivables.	30	1
4	Will not increase the net income of a separate contract.	60	12	Change contract.	30	1
5	Reduced sales growth.	50	15	To perform a range of promotion to increase sales.	10	1
6	Reduced labor productivity.	30	20	To introduce a machine to the assembly area.	60	0

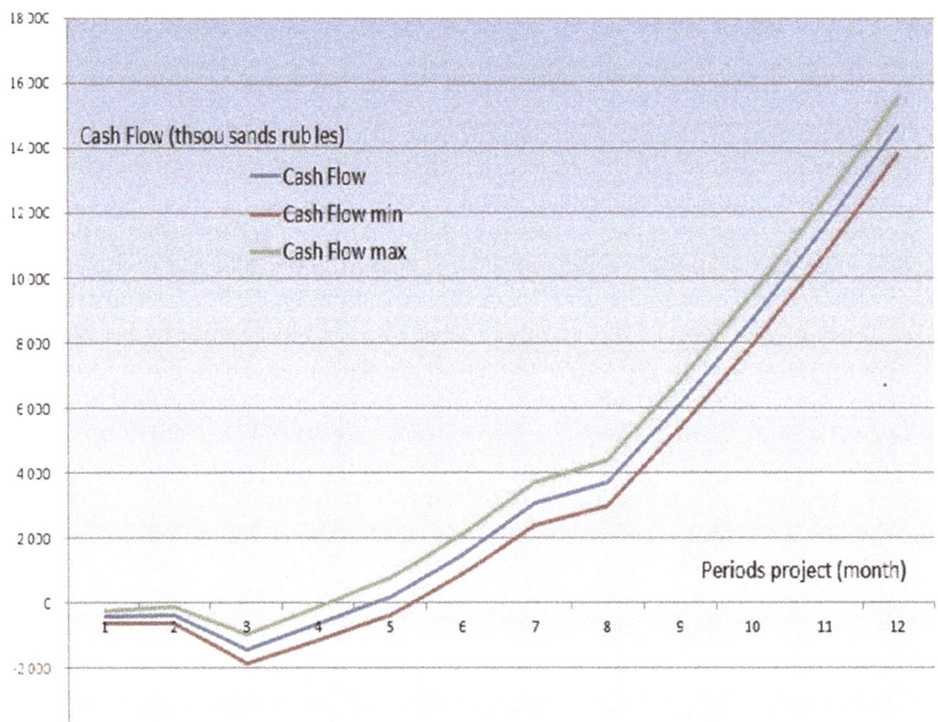

Figure 3.5. Example of chart 'The cash flow project'.

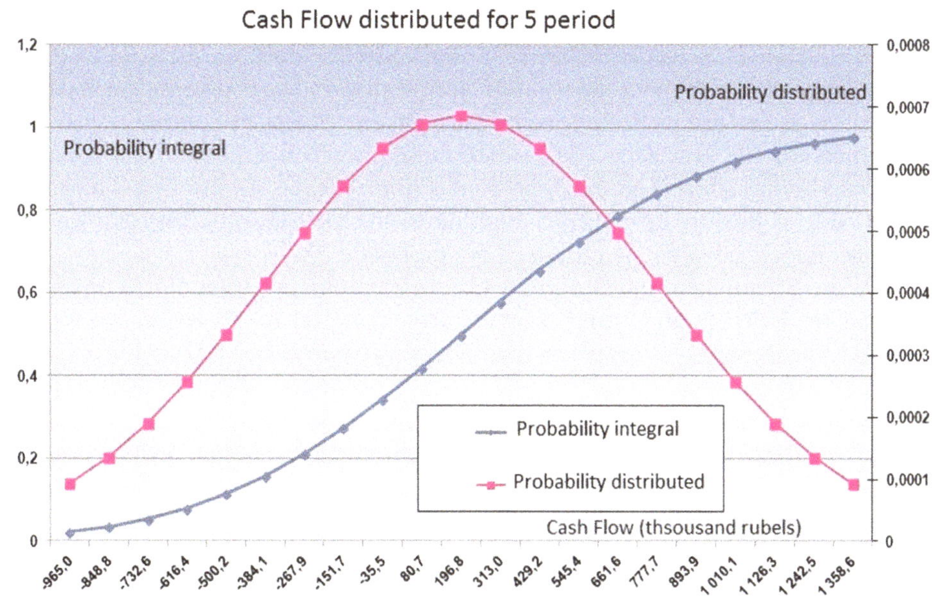

Figure 3.6. Example of chart 'Cash flow distribution'.

planning is a balanced scorecard. This metric is different from financial planning because it harmoniously takes into account financial and nonfinancial indicators that affect the activities of the enterprise. The office of strategic planning is usually done through risk management. In the formation of a balanced, strategic development plan, it is recommended that all risk indicators are split into four groups. Financial indicators occupy the main place in the system of strategic planning. Nonfinancial indicators include data for the company's customers and markets, characteristics of internal business processes, internal staff development and the growth potential of the company. These four areas of activity of the enterprise form the basis of the system of strategic planning. For individual companies other specific areas of activity can be dealt with, reflecting the peculiarities of their production. All of the indicator systems for the strategic planning of the united cause-and-effect relationships are such that changes in either direction indirectly or directly affect the performance in other directions.

A component of the strategic plan related to customer service shows how the organization looks in the eyes of customers, i.e. reflects the competitive capabilities of the company. The component for work with customers is important for the overall strategy of the organization because it clearly defines the queries with key customers, choosing a market position and the dynamics of the client groups on which it focuses. The assessment tool can serve in examining risk factors that determine the state of the client component of the strategic map. Possible deviations from the intended plans will be due to the risks of working with external clients. The results of detection of risk factors for the category 'Customers' produced a set of measures to reduce the impact of these factors.

A section of the strategic plan relating to the internal business processes necessary to achieve planned results defines the main innovations. The events section is aimed at the development of new products and improvement of the competitiveness of the enterprise. The performance section characterizes the processes contributing to the achievement of the efficiency of the enterprise, the receipt of planned financial results, improving internal technology, production process and meeting the changing demands of buyers. Insufficient attention to the internal processes can lead to activation associated with the risks of these processes.

The term development implies a structure that the organization must implement to ensure their development and growth in the long term. To ensure long-term success and prosperity requires a constant updating of processes taking into account new technical developments. The successful development of the organization involves its human resources, information systems and organizational procedures. To support their activities in the market, the company should invest in staff development, information technology, systems and procedures. Insufficient attention to these factors can lead to the emergence of risk factors associated with the personnel of the company and the growth of the company.

3.4 Measurement of the risk of using the discount rate, expert assessments and indicators of sensitivity

Using the discount rate is widespread to account for risk through the introduction of special allowances for the risk-free discount rate. The formula for the discount rate is

$$\mathrm{NVP} = \sum_{t=1}^{T} \frac{CF_t}{(1+i)^t} + \frac{CF_{(T+1)}}{i-g} * \frac{1}{(1+i)^T}, \tag{3.1}$$

where NPV is the net present value of future cash flows, T is the number of settlement periods within the planning horizon, CF_t is cash flow over the t period, $CF_{(T+1)}$ is the cash flow terminal of the first period, i is the value of the discount rate and g is the growth rate of cash flow in the post-planning period (percent per annum) [4].

In the general case, calculating the NPV considers the cash flow generated by the project lifecycle (during the planning horizon). Sometimes cash flow for the project in the post-planned period is omitted in the calculation of the NPV to outside investors (with the aim of increasing the reliability of the calculations). In formula (3.1) the first term characterizes the cash flow from the project during the planning horizon, the second does so in the post-planned period. Calculation of the cost cash flow in the post-planned period is carried out using the Gordon model [5]. The discount rate represents the average return that an investor received when investing in the project alternative under consideration.

Calculation (optional) of the discount rate is determined based on:

- Ways of taking into account inflation when calculating cash flow.
- The project participant for which NPV is calculated.
- Information about project components.

The project over time inevitably changes the factors that determine the value of the bet. For example, in the start-up phase the business may experience a permanent decline in risk through risk reduction 'non-realization of the project'. After the payback period, the risk of investors associated with a possible 'return' of funds is also reduced to zero. But at the same time, other changes can lead to an increased exposure to risk factors and this is equivalent to an increase in the discount rate. Therefore, the discount rate assumption is the basic assumption of the calculation of equation (3.1)—usually the adoption of constant values of the discount rate for the entire lifecycle of the constant bet throughout the project lifecycle in the preparation of preliminary calculations.

The simplified model developed by the magazine *Business Valuation Review* [7] for emerging markets may be used to build value models of the company and its elements. This model is based on the assumption that the yield on government Eurobonds reflects the risks associated with investing in the share capital of 'ideal companies', i.e. companies with no flaws. The disadvantages of a real company are equal to the risks specific to the company and the specific business. These flaws are marked as premium to the discount rate due to these risk factors (see table 3.2).

For practical application of this model for risk management, it is necessary to increase the risk-free rate in accordance with table 3.2. Each allowance should be determined taking into account the possibility of its appearance in the project and the consequences after the considered risk.

The method of expert estimations consists in the possibility of using the experience of experts in the analysis process of the project and considering the influence of diverse qualitative factors. The formal peer-review process often comes down to the following. The management of the project (firm) develops the list of evaluation criteria in the form of expert (polling) sheets containing the questions. Corresponding weighting coefficients are assigned for each criterion (at least are calculated), and are not disclosed to the experts. The answers are compiled for each criterion, the weight of which is not known to the experts. The experts should have information about the evaluation project and through examination should be able to analyze the questions and mark the chosen answer. Next, the completed expert

Table 3.2. Estimates of allowances for 'deviation of the ideal' corporate risks.

View corporate risk	Interval values
Quality control	0%–8%
Company size	0%–5%
Financial structure	0%–10%
Commodity/territorial diversification	0%–5%
Client diversification	0%–4%
Profits: rules and retrospective predictability	0%–5%
Other risks	0%–5%

sheets and the output or results of the examination are processed accordingly on the basis of well-known computer packages for the processing of statistical information.

In practice, risk analysis and decision-making is often not worth the task of obtaining quantitative characteristics. What is important is comparative analysis, which experts use to assess the occurrence of risk events on a simplified scale of gradations. For example, the place each of the events on the graph in the axes of 'impact' is 'probably'. The diagram consists of nine cells, each of which corresponds to a single set of estimates (figure 3.7). For example, an event characterized by the estimated 'low impact, low probability' should be displayed in the lower left cell of the chart, and event assessments 'low impact, high likelihood' should be displayed in the lower right cell, etc.

The whole chart is divided into three approximately equal parts. The three cells of the diagram located at the bottom left are an area of insignificant risk. The three cells of the diagram in the upper right are an area of significant risk. The remaining part of the diagram (three cells) is an area of medium risk. Thus, the risk associated with event A is insignificant, the risk for event B is average and the risk of event C is substantial. The resulting diagrams, which in accordance with expert assessments apply to all risk events, are called risk maps. This map shows what risk events can take place, what the correlation between different types of risks is and how risks should be given maximum attention (in this example, risk events C). This approach is widespread in the practice of risk management for companies in the real world. Risk managers typically use three or five (rarely seven) grades for probability of exposure and materiality. The described chart is a convenient way to visualize risk. In practice, there are other ways of visualization, such as using a circular or a color chart.

Sensitivity analysis and scenario analysis are the sequential steps in quantitative risk analysis; the latter allows us to avoid some of the shortcomings of sensitivity analysis. However, the scenario method is the most effective and can be applied when the number of possible values for NPV of course. However, the expert faces an unlimited number of different scenarios in the risk analysis of the investment project. The actual method of assessing the individual risk of the project helps to solve this problem (simulation); the basis of this method is the probabilistic assessment of the occurrence of various circumstances. By using specialized software packages for the calculation of the economic efficiency of projects, the evaluation of the impact of risks is obtained in the form of output tables and graphs reflecting the impact of risk factors on the project output.

		The probability		
		Low	*Medium*	*High*
The Impact	*Strong*			Event C
	Average	Event A	Event B	
	Weak			

Figure 3.7. The risk map.

In the sensitivity analysis, analyze the sensitivity of the project to the main parameters and modify one of the input parameters of the project. By sequentially changing project parameters, the developer defines input variables that strongly affect the project result. For these variables, measures are developed to reduce their impact. The aim of the project sensitivity analysis is to determine the impact of varying factors on the financial result of the project. The most common method used for sensitivity analysis is simulation. NPV is used as the integrated indicator of the financial result of the project. The result of sensitivity analysis can be reduced to similar conclusions, e.g. the project allows the reduction of the price of sale by 16%, an increased sales volume by 11% and increased direct costs by 14%.

Figure 3.8 shows a graph of the impact on the net present value, changes in sales volume, sales price and direct costs.

The sensitivity analysis does not perform the measurement of risk, but only assesses the influence of various factors on the results of the financial activities of the enterprise. The analysis allows one to identify the factors that most affect the amount of profit and then consider activities that control these factors to find ways to organize activities to neutralize these factors.

Sensitivity analysis (determination of the critical point) can be produced without the use of special computer programs based on the conditions of break-even production. Consider this analysis as an example.

In the calculation of the predetermined period from the start of series production the total costs of production (P) are defined as

$$P = V * M + F$$

where F and V are, respectively, fixed and variable costs.

Turnover (O) is defined as the product of the price of the product (C) and the quantity (M)

$$O = C * M.$$

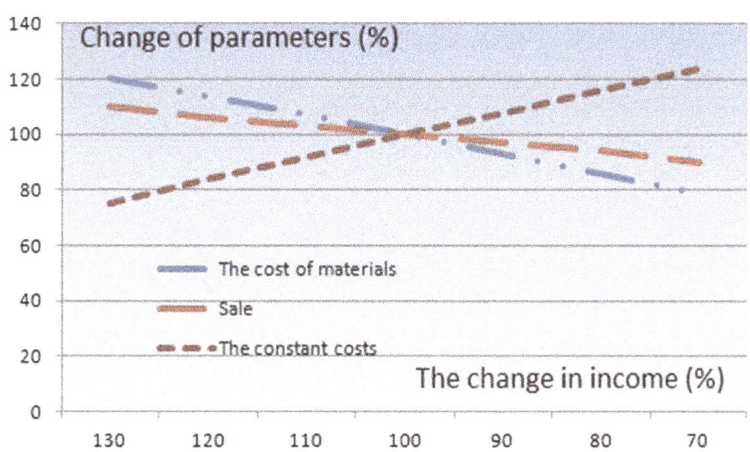

Figure 3.8. Graph of the evaluation of the sensitivity of the project to the sales, the cost of materials and constant costs.

The break-even condition describes equality:

$$P = O \text{ or } C * M = V * M + F.$$

Thus, the critical (break-even) volume of production is

$$M = F/(C - V).$$

Let us analyze the break-even conditional production period. For example, the company's revenue in the conditional period is planned to be the amount of 14.04 monetary units. It can be assumed that it is planned to implement a conditional piece of the 14.04 or 14.04 conditional sets of settings for the price of 1 monetary unit. The full current costs of this period are 12.26 currency units, including fixed costs of 2.38 and variable costs of 9.88.Variable costs per unit: $V = 9.88/14.04 = 0.704$.

The break-even threshold obtained from the break-even equation is

$$1 * x = 0.704 * x + 2.38 \quad \text{where } x = 8.04.$$

Break-even is reached when there is a sales volume of 8.04 units of money.

Critical values for fixed costs are calculated as follows: $14.04 = 0.704 * 14.04 + F$, where $F = 4.16$. Fixed costs can only increase up to 4.16 and can be 42% higher than planned.

The critical value of the variable costs is determined from the equations of momentum and costs, and is calculated as $V * 14.04 + 2.38$, where $V = 0.84$. Thus, variable costs per unit of output may increase to 0.84, i.e. the safety margin is 46% of that originally planned.

A critical value of the sales price P is determined from the conditions of equality, profit is zero. The price of conventional products (or the price of conventional sets) is defined as 1 unit of money. The price level at which profit is zero is defined by the equation of momentum and costs:

$$14.04 * Pr = 0.704 * 14.04 + 2.38; \quad Pr = 0.87.$$

This means that the selling price can only be 13% lower than planned if other conditions are constant.

The risk-category represents a probability, so in the process of assessing uncertainty and quantifying risk it is possible to use probabilistic calculations. The most important indicator for the measure of the financial risk of an enterprise is the level and probability of losses. This figure has a decisive impact on the level of profitability of the financial operations of the company. These two indicators are closely related and constitute a single system of 'profitability–risk'. The ratio of the level of risk and return is one of the main underlying concepts of financial risk management, in accordance with which the level of profitability of financial operations under other equal conditions is always accompanied by an increase in the level of their risk and vice versa. In addition, the level of financial risk is a main indicator in assessing the level of financial security of an enterprise, which characterizes the degree of protection of its financial activities from external and internal threats. Therefore, the assessment of the level of risk in the management of financial activities of the company is included in the preparation of almost all managerial decisions.

The level of financial risk characterizes the probability of its occurrence under the influence of a particular risk factor (or groups of factors) and possible financial loss upon the occurrence of a risk event. Concrete methodological tools for evaluation of risk level formed according this definition allow one to solve specific tasks of the financial management of the enterprise. Consider the key design and performance indicators.

The level of financial risk describes a common algorithm for estimating this level. It is represented by the following formula:

$$Ur = Pl * Fl,$$

where *Ur* is the appropriate level of financial risk, *Pl* is the probability of financial risk and *Fl* is the size of possible financial losses when implementing this risk.

In practice this algorithm calculates the size of possible financial losses usually expressed by an absolute amount and the probability of occurrence of financial risk—one of the coefficients is the measurement of this probability (the coefficient of variation, beta coefficient, etc). Accordingly, the level of financial risk when it is calculated according to the following algorithm will be expressed in an absolute measure, which greatly simplifies the comparison of project alternatives.

The variance characterizes the degree of variability of the studied indicator (in this case the expected income from the realization of financial operations) with respect to its average value. The more fluctuations—the greater the risk. The variance is calculated according to the following formula:

$$\sigma^2 = \sum_{i=1}^{n} (R_i - \bar{R})^2 * P_i,$$

where σ^2 is the variance, R_i is the specific value of the possible variants of the expected income for the financial transactions, $\bar{R}$ is the average expected value of income under financial transactions, p_i is the potential frequency (probability) of obtaining separate variants of the expected income on financial transactions and n is the number of observations.

Dispersion does not give a complete picture of deviations $\Delta X = X - \bar{R}$, which are more pronounced for the risk evaluation. However, dispersion allows establishing a link between linear and quadratic deviations using the well-known Chebyshev inequality.

The probability that a random variable X deviates from its expectation by more than a given tolerance $\varepsilon > 0$ does not exceed its variance divided by ε^2, i.e.

$$P(|X - \overline{R}| > \varepsilon) \leqslant \frac{D}{\varepsilon^2}.$$

This shows that a small risk of dispersion deviation corresponds to a small risk, and according to a linear deviation of point X are likely to be within ε, the neighborhood of the expected values.

The root mean square (standard) deviation is one of the most common in assessing the level of individual financial risk as the variable determines the degree of absolute variability and is calculated by the following formula:

$$\sigma = \sqrt{\sum_{i=1}^{n}(R_i - \bar{R})^2 * P_i}.$$

The standard deviation σ is a dimensional quantity and is specified in the same units in any measured variable characteristic. The advantage of standard deviation is that when you reach the observed distribution (e.g. distribution of investment income) to normal this parameter can be used to determine the boundaries in which, with a given probability, one should expect the value of a random variable.

The coefficient of variation (CV) lets you define the level of risk if the average expected income from financial operations differ [6]. The calculation of the coefficient of variation is carried out according to the following formula:

$$\mathrm{CV} = \pm\frac{\sigma}{\bar{R}} * 100\%.$$

The coefficient of variation is a dimensionless quantity. With it, you can even compare the variability of traits, expressed in different units of measurement. The coefficient of variation varies from 0% to 100%. The greater the ratio, the greater the variability. For different values of the coefficient of variation are established the following qualitative assessments: 10% is weak variability, 10%–25% is moderate variability and over 25% is high variability.

The beta coefficient (β) allows you to evaluate individual or portfolio systematic financial risk in relation to the risk level of the financial market as a whole. Usually, this indicator is used to assess the risk of investing in individual securities and is computed by the formula

$$\beta = \frac{K \times \sigma_i}{\sigma_p},$$

where β is the beta coefficient, K is the degree of correlation between the level of profitability on individual types of securities (or their portfolio) and the average level of profitability of the group equity instruments in the market as a whole, σ_i is the standard deviation of return on the individual securities (or on their portfolio as a whole), and σ_p is the standard deviation of return on the stock market as a whole. The level of financial risk of individual securities is based on the following values of beta coefficients: $\beta = 1$ is average; $\beta > 1$ is high level; and $\beta < 1$ is low level.

The risk not only of a particular transaction but also the business of the company as a whole can be assessed using the probabilistic evaluation method (analyzing the dynamics of its income) for a certain period of time. The choice of specific assessment methods is determined by the availability of the necessary information base and the skill level of management personnel.

A new methodology has been used and developed in the last decade for evaluating the measures of financial risk with the indicator 'cost of risk' or 'value at risk'

(VAR). The value at risk is a measure of statistical estimation expressed in monetary form of the largest possible size of financial losses in the prescribed form of the probability distribution of the factors influencing the value of the assets (tools) and a given level of the probability of occurrence of these losses over the estimated period of time.

From the above definition, it is clear that the methodology of the calculation of VAR consists of three main elements. One of these elements is to establish the characteristics of the original factors. The risk manager establishes the form of the probability distribution of the risk factors affecting the value of the assets (tools) or their total portfolio. It can be the normal distribution, Laplace distribution, *t*-test, etc. One of the qualities of VAR is to set the risk-manager level of probability that the maximum possible size of financial losses will not exceed the estimated value of this indicator. The specific value of VAR is chosen by the risk manager based on his/her risk preferences. In the modern practice of financial risk management this level is usually in the range of 90%–99%. A visual representation of the formation of VAR gives the plot shown in figure 3.6.

As can be seen from the graph in figure 3.9, the revenue curve illustrates the normal probability profit distribution on the financial instrument in a predetermined billing period of time. The field in this graph between -2σ and $+3\sigma$ corresponds to the chosen confidence level (90% of the area under the curve), and between -3σ and -2σ characterizes the value of possible losses beyond the confidence level (10%). On the chart, the VAR determined is the amount of −732.6 thousand rubles. This corresponds to a maximum size of possible financial losses on the financial instrument under the given confidence level and estimated valuation period; the value of VAR in the diagram separates the value of income beyond the limits of the confidence interval (10%).

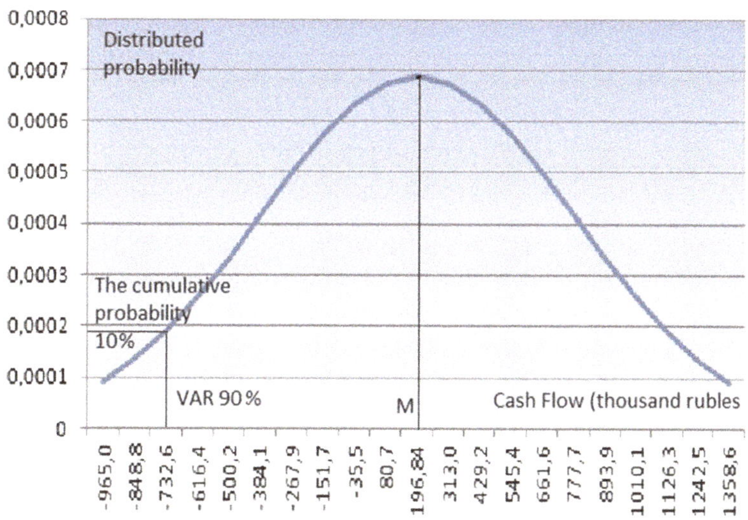

Figure 3.9. Graphical method of determining the value indicator VAR.

In order to fully describe the risk using the VAR measure, you must first specify the probability (small enough to consider the event 'almost' impossible), or the level of confidence associated with this probability value. If the probability is set as 5%, it means a confidence level of 95% (100%–5%) and represents the result in the form VAR95% (pronounced 'VAR at the 95% level'). The level of 95% is rather arbitrary, each individual sets this level based on their relationship to the possible unlikely events, and an understanding of what is considered an 'almost' impossible event. Therefore, VAR can be used with other levels of confidence, e.g. 90% or 99% (when talking about VAR90% or VAR99%). In addition, the assessment or the calculation of VAR is in practice used in the time horizon of the game (financial operations). Therefore, when speaking about risk, VAR determines what is the minimum financial result that can be obtained with a certain confidence level during a certain period of time.

Here is an example. The statement 'evaluation of the VAR of the risk of lower returns during the next week is minus 2% at a confidence level of 95%' or briefly 'a week VAR95% = −2%' means that:

- With a probability of 95%, the yield of the planned operation will be at least −2% for the week.
- With a probability of 95%, the loss for the week will not exceed 2%.
- A weekly loss of over 2% is possible with a probability of 5%.

There is a strong correlation between the two measures of risk—variance and VAR in the case of normal distribution. Since the normal distribution is completely determined by two parameters M and σ, any part of this distribution (in particular, any quantile) is determined by these two parameters. This means that for the normal probability distribution the relationship between the variance and the VAR at any confidence level is unambiguous and has the form:

$$\text{VAR}i = M[X] - Z(1 - i),$$

where $Z\,(1 - i)$ is the quantile of order $(1 - i)$ standard normal distribution.

The values of the tabulated quantiles, we present several important special cases:

$$\begin{aligned} \text{VAR90\%} &= M[X] - 1.283 * \sigma; \\ \text{VAR95\%} &= M[X] - 1.645 * \sigma; \\ \text{VAR99\%} &= M[X] - 2.326 * \sigma. \end{aligned} \tag{3.2}$$

These formulas are of practical importance. In the vast majority of cases the probability distribution of the results of economic games is not known. However, it is often possible to estimate some characteristics of the unknown distribution, in particular, the expected outcome and variance. Then you can make the assumption that unknown to us the distribution is very similar to normal, and we can estimate VAR using equation (3.2). This assumption is close to the truth for games in the financial markets, as the prices of many important assets are determined by many random factors that are often inconsistent and conflicting. Even if the probability distribution of the results of each of these random factors is not normal, their joint distribution will tend to normal.

Analytical models of business processes require a description of the procedures used and the task in the analytical form of the input process parameters. The complexity of this approach lies in the analytical conversion of the input parameters in accordance with the ongoing business processes. By adopting some limitations on input actions, it is possible to simplify the analytical analysis. For example, in the E-Project [2] the parameters of the random input signals used only two characteristics: M—mathematical expectation and D—standard deviation. Operations with such variables are performed according to the rules of operations with random variables [3]:

- For any two random variables and the expectation of their sum $M\ (X + Y)$

$$M(X + Y) = M(X) + M(Y). \tag{3.3}$$

The variance of the sum of two random variables is equal to the sum of their variances plus twice the correlation time K_{xy}:

$$D(X + Y) = D(X) + D(Y) + 2Kxy. \tag{3.4}$$

- The mathematical expectation of the product of two random variables is equal to the product of their mathematical expectations plus the correlation time K_{xy}:

$$M(XY) = M(X) * M(Y) + Kxy. \tag{3.5}$$

- The variance of independent random variables

$$D(XY) = D(X) * D(Y) + M(X)2 * D(Y) + M(Y)2 * D(X). \tag{3.6}$$

Most of the results of typical business processes are built on the combination of the operations of addition, subtraction and multiplication that can be defined by equations (3.3–3.6).

In the E-Project package the normal distribution of the output is accepted. This assumption is possible in cases when the business process involved a large number of random variables, and the final result is a complex combination of these input effects. It is possible to rate the difference between the results of modeling the business process using the analytical model and the method of simulation. For example, for changing the law Triang input data (figure 3.10), the distribution of the output can be represented by a normal distribution (figure 3.11) [7, 8].

From this figure, it is seen that model business processes at various distributions of the input parameters allow the output process to describe the distribution of the normal law. This example shows a normal distribution and Welbull distribution. The standard deviation of all the distributions differs by a negligible amount. An approximation of these distributions to normal will be more accurate the more processes will affect the output parameters. The output distribution final results are built according to the dependences of equations (3.2)–(3.5) for non-symmetric distribution input data and will have less asymmetry than the input distribution.

Figure 3.12 presents the simulation result of a mixture of input data, distributed over six different laws. The result of the comparison of the obtained distribution

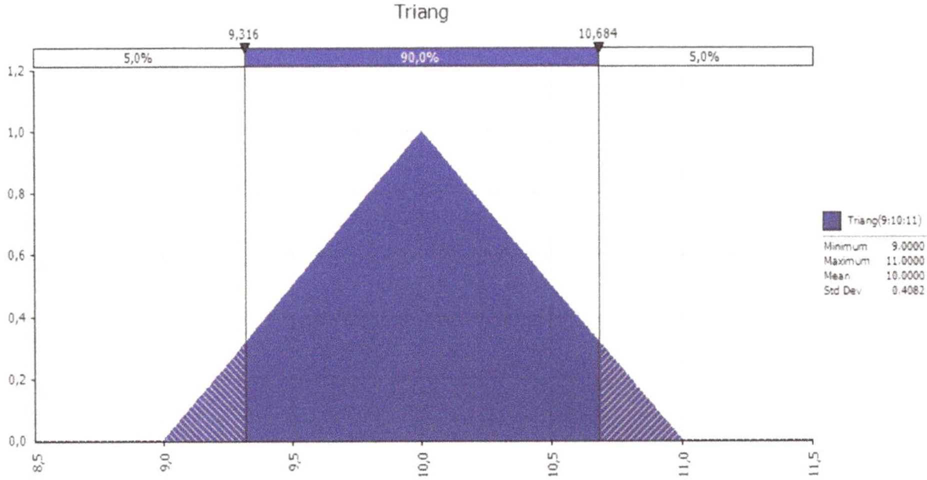

Figure 3.10. The distribution of the input parameter 'Sales'.

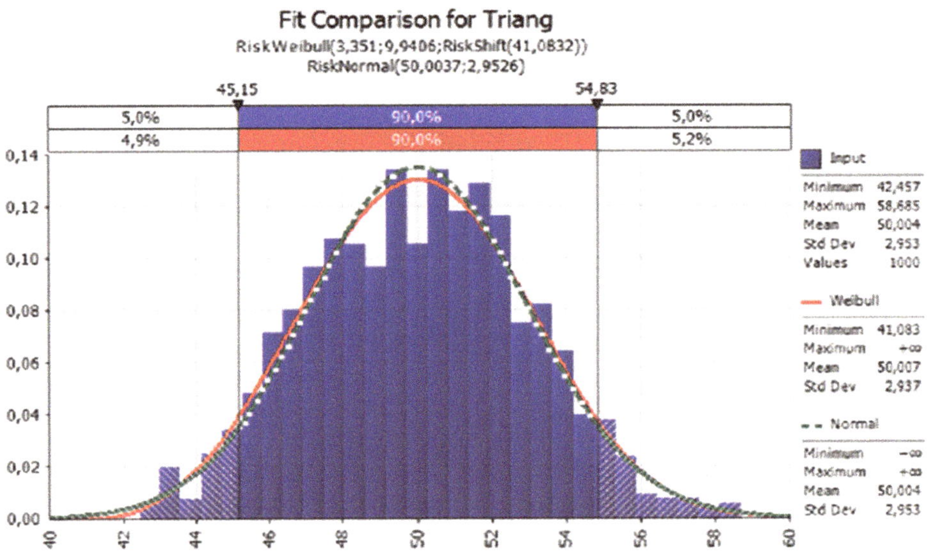

Figure 3.11. The distribution of the output parameter of the business process for Triang distribution input data.

shows that a normal distribution has a matching mathematical expectation and standard deviation with the result from the simulation values of the distribution of the output parameter. The simulation results show that the model of the business projects can be based on analytical relations (3.3–3.6), with simplified analytical calculations and process simulations.

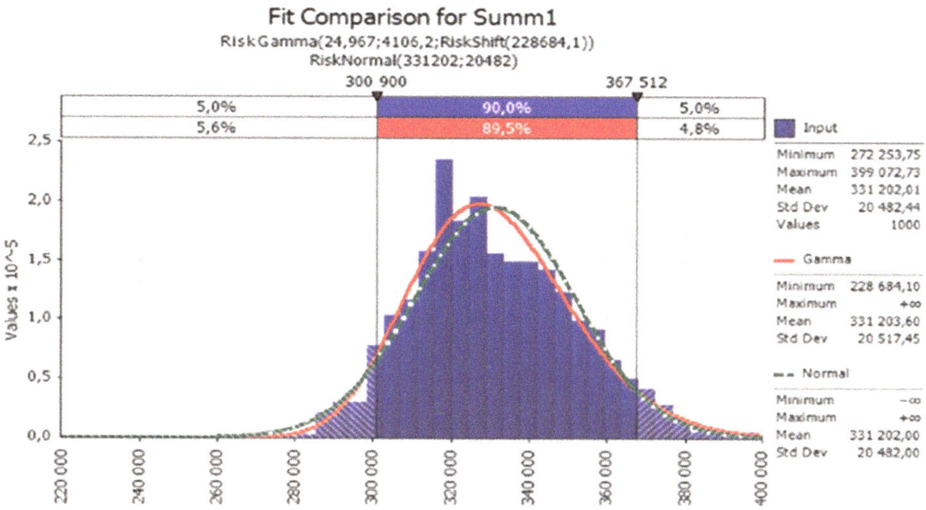

Figure 3.12. The distribution of the outputs of the business process while increasing aggregate sales from various distributions of input data.

The accuracy of the analytic calculation output parameters for business processes depends on the accuracy of determining the same statistical characteristics of the input parameters.

An important step of risk management is to reduce the cost of risks and anti-risk action when you run anti-risk measures to modify the characteristics of risk factors. Optimization of risks can be reduced to evaluating the cost of anti-risk measures and changing impact after the occurrence of the risk factors resulting from the application of anti-risk measures. With a limited budget, you cannot perform all anti-risk events. We select those that have the greatest effect. We can account for the feasibility of carrying out certain anti-risk activities based on an integrated assessment of the financial results of the project with different combinations of parameter *K* (table 3.1). E-Project uses a special add-in that enables the adopted criteria and conditions of restrictions to choose the optimal value of parameter *K* for all the risks involved.

First, activities are planned with the greatest efficiency [9]. Figure 3.13 shows the dependence of financial loss from the implementation of the 24 anti-risk measures and the costs of implementation of these activities.

Various algorithms can be proposed to automate the selection of anti-risk measures. For example, the set of events fits within the allocated budget or choice of activities, the effectiveness of which exceeds the threshold. The result of this optimization is presented in figure 3.14, with the financial losses from the original risk and the same rate for a project with risks.

The proposed method allows us to assess the impact of risk factors on the efficiency of the project, to assess the impact on financial performance anti-risk activities and to choose those which provide the greatest effect according to the chosen criterion of project evaluation.

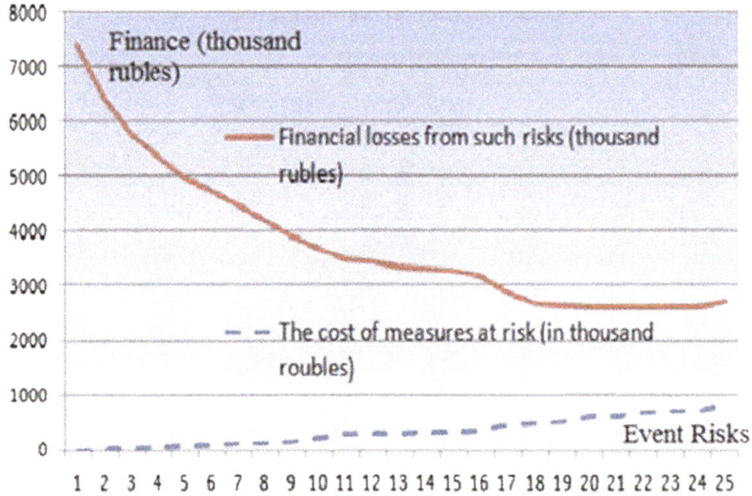

Figure 3.13. Losses from risks and the cost of measures from them.

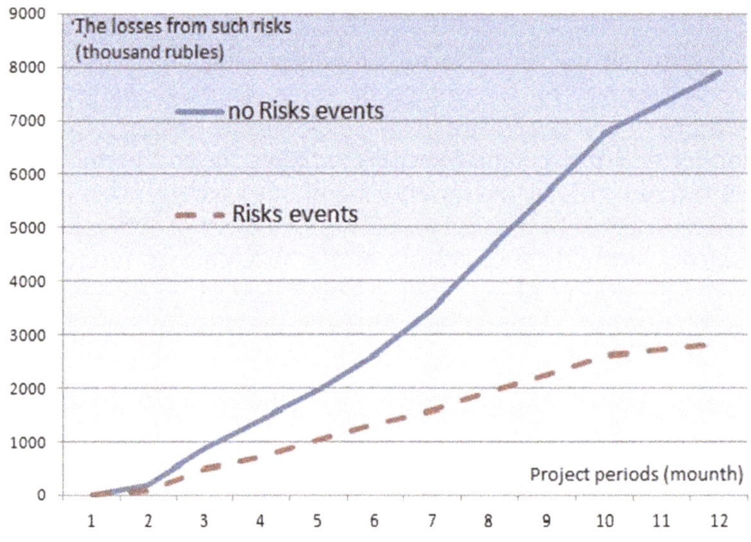

Figure 3.14. Losses from risk source and risk after special events.

3.5 Conclusion

The business process model based on risk allows you to more accurately plan the development of a business. It allows the opportunity to develop activities that improve the final result of the project with minimal cost before the start of the project.

References

[1] Sommerville I 2001 *Software Engineering* 6th edn (Reading, MA: Addison-Wesley), 693 p

[2] Gorbunov V L 2013 *Business Planning with Risk and Efficiency Assessment of Projects: Scientific-Practical Manual* (Moskow: RIOR: Infra-M.) p 248

[3] Gorbunov V L 2004 A database 'AE-Project' certificate of registration No. 2004629261
[4] Mikhailets V B 2002 The discount rate in the evaluation Russian Business Newspaper issue 90 pp 147–50
[5] Gregori A 2003 Strategic Valuation of Companies (Practical guide) (M. Kvinto-Consulting)
[6] Gracheva M V 2001 *Risk Analysis of the Investment Project* (Moskow: UNINI-DANA)
[7] Nersesian R L 2013 *Energy Risk Modeling* (New York: Palisade)
[8] Morris J R and Daley J P 2009 *Introduction to Financial Models for Management and Planning Textbook* (London: Chapman and Hall)
[9] Gorbunov V L 2014 Accounting balanced scorecard of the enterprise in the system of business planning *Proc. of the Int. Conf. Innovative Approaches to Solving Technical and Economic Problems* (Moscow: MIEE)

Chapter 4

Nonlinear optimization methods—overview and future scope

Somesh Kumar Dewangan, Siddharth Choubey, Jyotiprakash Patra and Abha Choubey

The technology for combinatorial optimization is changing quickly, and complexity of the fundamental technology is increasing with size expansion of extent of problems. The nonlinear programming problem related to the discrete lower bound farthest point examination problem is treated using methods for calculation where the need to linearize the yield criteria is avoided. The calculation is an interior-point strategy and is totally open, as no specific component discretization or yield basis is required. We classify current programming packages for processing the necessary nonlinear optimization problems. The packages include interior-point methods, progressive linear/quadratic programming methods, and increased Lagrangian methods. For each package the standard methodological components are discussed. The new methods are connected to the single-class client traffic equilibrium problem, the multi-class client traffic equilibrium problem under social marginal cost evaluation, and the stochastic transportation problem. In a constrained arrangement of computational tests the calculations end up being very efficient. Moreover, a feasible strategy with a multi-dimensional search for the stochastic transportation problem is created.

The traffic task problem is a nonlinear model which depicts how every traveler limits his/her own movement costs to achieve the optimal solution. In the administration of venture portfolios, the objective is to choose a combination of speculations in order to enhance returns while limiting risks. For understanding non-convex or large-scale optimization problems, deterministic methods may not be suitable for producing globally optimal results in a reasonable time due to the high complexity of the problems. Heuristic methods are therefore introduced to reduce the computational time for an optimization problem, however, the result obtained is

doi:10.1088/978-0-7503-2404-5ch4

not necessarily a possible or globally optimal solution. The two types of optimization methods have their advantages and disadvantages. Therefore, combining deterministic and heuristic techniques is suggested for handling large-scale optimization problems to find a global optimum. This chapter attempts to provide a future research direction in the area of deterministic and heuristic methods to improve the computational effectiveness of finding a globally optimum answer for various real-world application issues. As we introduce new ideas into model design, model examination, model changes, global restriction handling and hybridization, the advantages of this new optimization method can be used without expert knowledge. With increased complex optimization problems, hybrid optimization calculations and programming language concepts become essential.

4.1 Introduction

This section discusses an overview of optimization, non-linear programming (NLP) and a few case studies.

4.1.1 Optimization

Optimization can be related to existing or specifically built scientific models. The idea is that one might want to locate an extremum result of a model by shifting a few parameters or factors [1]. The typical motivation for discovering suitable parameter values is used in selection choices in design optimization.

Optimization is the process of achieving the best outcome under given conditions. In design, development, support, etc, engineers need to make choices. The objective of every such choice is either to limit costs or to increase advantages [2].

The costs and advantages can normally be communicated as a component of certain design factors. Thus, optimization is the process of finding the conditions that can give the maximum or the minimum estimate of value based on the suitable conditions. It is also commonly established as a hidden guideline in the investigation of numerous complex choice or allocation problems. Utilizing optimization theory, one solves a complex choice problem, including the choice of qualities for a number of interrelated factors, by concentrating on a single target function, designed to evaluate the execution and measure the nature of the choice. This one function is maximized (or minimized, depending upon the details) subject to criteria that may confine one of the choice variable qualities. In this case a reasonable single component of a problem can be isolated and described by a function, be it a benefit or risk in a business setting, or speed or distance in a physical problem [3].

No single strategy is available to handle all optimization problems efficiently. Thus, various methods are used to address different kinds of problems.

Linear and nonlinear optimization methods address the following problem: finding numerical values for a given set of variables with the feasible goal satisfying certain criteria, called target functions. The solution with target function achieves its minimum value among all the combinations of possible variables. An example is the dietary regimen problem, for finding a combination of foods which fulfills the health requirements at minimum expense. In contrast to traditional problems of connected arithmetic, the majority of which start in materials science, linear and nonlinear optimization problems for the most part need solutions given by closed formulae, and must be understood through numerical methods, with calculations performed on PCs [4].

In numerous numerical programming applications, linear optimization assumptions or approximations may allow proper depiction of the problem over the range of variables being considered. In nonlinear optimization, nonlinear constraints

$$f(x_1, x_2, x_3, \dots, x_n)$$

of the decision variable are used. If the possible solution space is bounded by nonlinear constraints then the method used to find possible solution is called nonlinear programming (NLP).

Statement of an optimization problem

Current literature suggests that no advances were made until the twentieth century in the field of optimization, when computing power made the execution of optimization strategies conceivable and thus invigorated further research [5].

Significant improvements in the field of numerical strategies for unconstrained optimization are reported [1, 6–11]. These include the advancement of the simplex method [12], the principle of optimality [13], and the necessary and sufficient conditions of optimality [14]. Optimization in its broadest sense can be considered to tackle any design problem, for example:

- Minimum weight of flying machines.
- Optimal (minimum time) directions for space missions.
- Minimum load of electric networks.
- Optimal production management, resource allocation and scheduling.
- Finding shortest routes.
- Optimum pipeline networks.
- Minimum handling time in production frameworks.
- Optimal control.

An optimization, or a scientific programming problem, can be described as

$$x = (x_1, x_2, x_3, \dots, x_n)$$

which reduces to $f(x)$, subject to the limitations

$$\text{for } k = 1, \dots, m, \; g_k(x) \leqslant 0$$
$$\text{for } k = 1, \dots, p, \; l_k(x) = 0.$$

A design vector can be defined through variable x known as $f(x)$ with objective function $g_k(x)$.

4.1.2 NLP

NLP is similar to linear optimization due to the fact that it has a goal, general requirements and variable constraints. The thing that matters is that a nonlinear program incorporates in any event one nonlinear function, which could be the target function or a few or the majority of the constraints.

In nonlinear optimization, models are intrinsically significantly more difficult to improve. There are some reasons behind this, which are briefly described as follows:

1. In numerical methods for tackling nonlinear optimization we have restricted data about the problem. This implies that it is difficult to recognize a local optimum from a global optimum. In such a case, we have accessible data, we expect that one point x is itself the estimation of the target function at x. There is sufficient data to determine when you are near the minimum or maximum, yet there is no opportunity to determine whether there exists an alternative and better local maximum [10, 11].
2. In a linear optimization system there is a set number of points to search for the optimum solution in a linear program, by checking optimum points, or corner points, of the feasible polytope space. In contrast, in a nonlinear optimization system an optimum solution could be in any place—at an optimum point, along an edge of the feasible space, or in the interior of the feasible space [15].
3. In the event that nonlinear constraints are considered, there might be various diverse feasible regions, regardless of whether you can locate the optimum inside a specific feasible space.
4. In nonlinear problems, specific initial conditions may produce diverse final solutions, where there might be various distinctive minima, and beginning at some other point may produce an alternative final solution point and target function value.
5. In a linear optimization system one will either discover a point that fulfills every one of the requirements or it will specifically confirm that no feasible points exist anywhere. However, in nonlinear procedures there is no such certainty.
6. In linear optimization methods, the initial segment routine finds a solution that fulfills every one of the constraints and from there on these are never broken, however, in a nonlinear optimization strategy, finding a solution that fulfills the folding and curving conditions is difficult in itself, regardless of whether a solution is found sooner or later; the uniformity may again be violated when the calculation attempts to move to another point that has a superior estimation of the solution [16].
7. In a linear optimization strategy, we have only a few possible results: a globally optimum solution point, feasible yet unbounded, and infeasible.
8. There is a huge number of complex numerical hypotheses and various solution calculations. The reason why NLP is so much more difficult than LP is due to its being dependent on possible nonlinear capabilities.

There are also considerably more reasons why NLP is difficult, which depend on increasingly practical considerations.

9. It is difficult to decide the suitable conditions that result in obtaining the best possible solution.
10. Various calculations converge at certain conditions and provide the optimum solution.
11. Different conditions used by different users may produce distinctive solution.

In nonlinear optimization, a few things which were considered difficult in the past can now be generally achieved effectively:

- *Derivatives.* The derivative functions are regularly required by solvers (those involved in attempting optimum solution). There were two methods: numerical estimate by limited differences, or a modeling function which can provide code to the subsidiary.
- *Input groups.* At a certain point every solver had its own specific information organization to depict the model and if we found that one solver was not successful for the current problem it was a monotonous and mistake-laden process to recode the model into an alternative configuration and thus it could be attempted by different solver [17, 18].

4.1.3 Nonlinear optimization problem and models

This section highlights a few models of the non-linear optimization problem [18, 19].

Quadratic optimization models

This model suggests a linear structure

$$Ax = b,$$

which is over-selected (larger number of conditions than variables), and we can obtain least squares arrangement by dealing with the NLO problem [1]

$$\min||Ax - b||2 \text{ (quadratic optimiztion problem)}.$$

Further,

$$\min||Ax - b||2 = (Ax - b)T(Ax - b) = xT\ AT\ Ax - 2bTAx + bTb.$$

The concrete mixing problem can be considered as an example. The important characteristics for the strength of concrete are its sand and rock structure and these are used in particular proportion for an optimum design and architecture. For every type of concrete architects and engineers could determine the ideal combination of these characteristics.

Scientific model

This model suggests a plan that operates on different vectors $v = (v_1, v_2, v_3, \ldots, v_n)$ T, where $0 \leqslant v_i \leqslant 1$.

Portfolio examination model

In this model, we define different variables, arrangement preferences and distinction of hard and soft elements [20]. Numerically, x_i is the degree placed into the major mean-fluctuation problem:

$$\min\left\{\frac{1}{2}x\hat{}TV(x): f\hat{}Tx = \kappa,\ Dx = d,\ x \geqslant 0\right\}.$$

4.2 Convex analysis

Convex analysis is used in cases where the nonlinear optimization problem utilizes curved objective function [15].

4.2.1 Sets and functions

As per the principles of sets we have two situations, x^1 and x^2, in IR^n [21].

Explanation 4.1. Considering two points x^1, x^2, IR^n we obtain

$$x = \lambda x^1 + (1 - \lambda)x^2,$$

where the two points x^1, x^2 result in a convex combination.

In this scenario, $C \subset IR^n$ is convex if any two points $x^1, x^2 \in C$ and if every raised combination of $x^1, x^2 \in C$.

In other ways, we can say that the line partition interfacing two sensitive motivations behind a convex set is contained in the set [22]. A parabola $f(x) = ax^2 + bx + c$ with $a > 0$ can also be defined by the same function.

Explanation 4.2. This type of function $f: C \rightarrow R$ defined on a convex set C is called convex if for all $x^1, x^2 \in C$ and $0 \leqslant \lambda \leqslant 1$ one satisfies

$$f(\lambda x^1 + (1 + \lambda)x^2) \leqslant \lambda f(x^1) + (1 - \lambda)f(x^2).$$

Explanation 4.3. The function of $f: C \rightarrow R$, where $C \subset IR^n$ is defined as an $n + 1$-dimensional set:

$$\{(x, T): f(x) \leqslant T,\ x \in C,\ T \in IR\}.$$

Explanation 4.4. $f: C \rightarrow IR$ is also denoted as a one group value C becomes the convex function for all x^1, $x^2 \in C$ and for $0 < \lambda < 1$ we can write

$$f(\lambda x^1 + (1 + \lambda)x^2) < \lambda f(x^1) + (1 - \lambda)f(x^2).$$

Explanation 4.5. A function $f : \mathcal{C} \to IR$, characterized by the upper raised set is called concave if the capability profile of the function is curved.

4.2.2 Convex cone

Explanation 4.6. The set $\mathcal{C} \subset IR^n$ is defined as a convex cone for all $x \in \mathcal{C}$ and $0 \leqslant \lambda$ satisfying $\lambda x \in \mathcal{C}$.

A convex cone does not contain any subspace with the exception of the function.

4.2.3 Concave function

Concave functions are essentially the opposite of convex functions.

Explanation 4.7. A function $f(x)$ is called concave if, for every y and z and for $0 \leqslant \lambda \leqslant 1$,

$$f|[\lambda\gamma + (1 - \lambda)z].$$

4.2.4 Nonlinear optimization: the interior-point approach

This section discusses logarithmic boundary method for dealing with with curved optimization. We will consider the CO issue in the following equation,

$$(CO)\ \min\{f((x)) : x \in \mathcal{F}\},$$

where $\mathcal{F}$ means the feasible space, which is given by

$$\mathcal{F} := \{x \in IR^n : g_j(x) \leqslant 0,\ 1 \leqslant j \leqslant m\}.$$

The constraint function $g_j : IR^n \to IR (1 \leqslant j \leqslant m)$ and the objective function $f : IR^n \to IR$ are convex limits [23].

When managing calculations for solving CO functions f and $g_j (l \leqslant j \leqslant m)$ are taken into consideration and it is also assumed that $f(x)$ is linear, for example $f(x) = -c^T x f$ is defined as $c \in c \in IR^n$. In case this is not valid, one may present an extra factor x_{n+1}, an extra requirement $f(x) - x_{n+1} \leqslant 0 f(x) - x_{n+1} \leqslant 0$ and a limit x_{n+1} and then objective function converges as linear.

We can write

$$(\text{CPO}) \begin{cases} \min -c^T x \\ g_j(x) \leqslant 0,\ j = 1, \dots, m. \\ x \in IR^n \end{cases}$$

The Lagrange–Wolfe dual of CPO is expressed as

$$\begin{cases} \max - c^T x + \sum_{j=1}^{m} y_j g_j(x) \\ \sum_{j=1}^{m} y_j \nabla g_j(x) = c \\ y_j \geqslant 0, j = 1, \ldots, m \end{cases}.$$

The primary feasible region is

$$\mathcal{F}^0 := \{x \in IR^n : g_j(x) < 0, \ j = 1, \ldots, m\}.$$

This will satisfy the IPC if $\mathcal{F}^0$ is nonempty [24].

4.2.4.1 Genetic algorithms

Genetic algorithms (GAs) are based on natural selection, both constrained and unconstrained types of optimization. Biological evolution is considered as a basis for GA to work as an optimization method [3].

4.2.4.2 Simulated annealing

Simulated annealing (SA) is an effective and general form of optimization. It is useful in finding global optima in the presence of large numbers of local optima. 'Annealing' refers to an analogy with thermodynamics, specifically with the way that metals cool and anneal.

The 'finding that every specie has obtained is exemplified' [25, 26].

Genetic algorithm procedural steps are:

1. Start with different variables s and q.
2. Set initial value of q as 0.
3. Evaluate the value of function $D(q)$.
4. With the help of incremental operators q ++, take the function from $D(q)$ to $D(q-1)$.
5. Combine the functions of $(D(q))$.
6. Return to step 3 and repeat.

4.2.4.3 Scatter search

This strategy joins favored subsets of reference centers to create a fundamental point by weighted linear combinations, and selects the best members to transform into the source of the new reference point. After introduction and assessment, the dissipate/tabu search calculation characterizes the population of arrangement $X_1 \ldots\ldots X_{\text{pop\_size}}$ into a few sets [27].

Procedural steps are:

1. Start with different variable s and q.
2. Set initial value of q as 0.
3. Calculate and classify the value of function $D(i)$.
4. With the help of incremental operators q ++, create the function E(q).
5. Calculate the function $E(q)$.

6. Take the function from $D(q)$ to $D(q-1)$ and $E(q)$.
7. Calculate the function $(D(q))$.
8. Return to step 4.

This was derived from true mechanics for discovering least cost answers for immense upgrade issues [7]. It aggregates up the hillclimbing systems and discards their standard damage: dependence of the course of action on the starting stage, and accurately certifications to pass on a perfect plan. This is cultivated by displaying a probability ρ of affirmation $\rho = 1$.

The procedure is described as [28]:
1. Start with 0 to q.
2. Define the temperature T.
3. Arbitrarily define the variable $\overline{X_c}$.
4. Calculate $\overline{X_c}$.
5. Go to step 2.
6. Go to step 3.
7. Create a new variable $\overline{X_n}$ with $\overline{X_c}$. If $f(\overline{X_c}) < f(\overline{X_n})$. Then $\overline{X_c} \leftarrow \overline{X_n}$.
8. Temperature T is assigned to function $D(T, q)$.
9. q is incremented by 1 and repeat the steps from Step 2.

4.2.4.4 Evolutionary strategies

Evolutionary strategies are methods used for parameter optimization problems. These are advances of development techniques having many members in the field of optimization problem solving approaches. This can be written as [29]

$$(\mu + \lambda) - ES_s \text{ and } (\mu, \lambda) - ES_s.$$

Evolutionary procedure uses the following steps [30]:
1. The initial value of q is 0.
2. Evaluate the value of function $D(q)$.
3. Start the loop with condition of endless.
4. With the help of incremental operators $q++$, take the function from $D(q)$ to $D(q-1)$.
5. Combine the function of $(D(q))$.
6. Go to step 3.

4.2.4.5 Sequential unconstrained minimization technique (SUMT)

Reducing a nonlinear problem to linear constraints is not difficult when the requirements are nonlinear. The main reason is that it is difficult to move along the limit of a nonlinearly constrained region, while it is generally easy to move along the limit of a linearly constrained region. Rosen's anticipated inclination technique of nonlinear programming with linear constraints provides a decent methodology for moving along the limit of linearly constrained regions [6].

It is advantageous to compose the nonlinear programming problem with non-linear imbalance and uniformity requirements in the following way. Choose q to minimize $f(q)$ subject to

$$g_i(q) \geqslant 0 \text{ the value of } l \text{ is denoted as 1 to } l$$
$$g_i(q) = 0 \text{ the value of } l \text{ is denoted as 1 to } m'$$

where point z is such that $g_i(z) > 0$ $i = 1, \ldots, l$.

Define the function (called the P function) [9]

$$P(q, r_1) = f(q) + r_1 \sum_{i=1}^{i} \frac{1}{g_i(q)} + r_1^{-1/2} \sum_{i=i+1}^{m} g_i^2(q),$$

where r_1 is a positive number.

As a starting point determine z^n such that $g_i(z^n) > 0$ $i = 1, \ldots, l$, proceed from z^n to a point of $z(r_1)$ that approximates the minimum of $P(z, r_1)$ in the set of points [15],

$$P(q, r_2) \equiv f(q) + r_2 \sum_{i=1}^{i} \frac{1}{g_i(q)} + r_2^{-1/2} \sum_{i=i+1}^{m} g_i^2(q),$$

i.e. $r_1 > r_2 > 0$ starting from $z(r_1)$ approximates the minimum of $P(z, r_2)$.

4.3 Applications of nonlinear optimizations techniques

Here, we will discuss a few applications of non-linear optimization techniques.

4.3.1 LOQO: an interior-point code for NLP

Interior-point systems are a well-known augmentation to nonlinear programming compared to the ordinary expansion of the simplex methodology [31]. The methodologies perform very well when interior-point systems are used as procedures for compelled curved nonlinear optimization that minimizes $f(x)$

$$\text{subject to } b \leqslant q(x) \leqslant b + s,$$

where variables b, h and r take values in $\mathbb{R}^m$ and l, z, and u take values in $\mathbb{R}^m$.

4.3.2 Digital audio filter

In digital audio filter applications, the presence of noise and other similar signals is analyzed and their effect is minimized by optimization methods [32]. Hence, the condition of the tweeters is noteworthy yet the woofer can be put anywhere within hearing range. For sound systems it is not even essential to have two woofers as the low repeat signal can be joined into one. The speakers would then have the alternative to be proposed to work ideally in a smaller range of frequencies [2]. With computerized measurement one has significantly more control over how the split is accomplished than can be achieved with a basic signal.

4.4 Future research scope

The best local and global search heuristics structures in this theory are mainly based on calculations utilizing different physical and numerical parameters. These parameters center around a few calculations for randomization and creating neighborhood arrangements. The basic aspects of actualizing these calculations efficiently and successfully depend on exploiting huge databases and on maintaining suitable potential solutions for updating possible accessible moves. Similarly to interior-point techniques for direct programming, the number of cycles is affected only a little by the problem estimate. The new techniques are connected to the single-class client traffic balance problem, the multi-class client traffic harmony problem under socially negligible cost estimation, and the stochastic transportation problem. In a constrained arrangement of computational tests the calculations end up being very effective. Moreover, an achievable direct technique with a multi-dimensional search is created for the stochastic transportation problem.

In the last three decades, the field of linear programming has grown rapidly due to the pioneering contributions of the active researchers in the field. However, as a tool for multi-objective decision analysis, the field of fuzzy target programming is relatively new in the area of fuzzy multi-objective decision making. Interval programming is even newer than fuzzy programming, so there remains a huge open area for new research and development. There is also wide scope for new research in different application areas for the development of the field of stochastic programming. The implementation of genetic algorithm methods to various multi-objective decision making problems is still at its initial stage. There is a vast scope for future research in this field for the development of a new generation of information technology applications. The potential which the genetic algorithm techniques offers over existing techniques is enormous.

References

[1] Klafszky E 1976 *Geometric Programming, Seminar Notes 11.1976* (Budapest: Hungarian Committee for Systems Analysis)

[2] Moore C 1993 Braids in classical gravity *Phys. Rev. Lett.* **70** 3675–9

[3] Chenciner A and Montgomery R 2000 A remarkable periodic solution of the threebody problem in the case of equal masses *Ann. Math* **152** 881–901

[4] Coleman J O 1998 Systematic mapping of quadratic constraints on embedded fir filters to linear matrix inequalities *Proc. of 1998 Conf. on Information Sciences and Systems*

[5] Spergel D N 2000 A new pupil for detecting extrasolar planets, arXiv: astro-ph/0101142

[6] Ho J K 1975 Optimal design of multi-stage structures: a nested decomposition approach *Comput. Struct.* **5** 249–55

[7] Karmarkar N K 1984 A new polynomial–time algorithm for linear programming *Combinatorica* **4** 373–95

[8] Anstreicher K M 1990 A standard form variant, and safeguarded linesearch, for the modified Karmarkar algorithm *Math. Program.* **47** 337–51

[9] Jarre F 1990 Interior-point methods for classes of convex programs *Technical report* SOL 90-16 Systems Optimization Laboratory, Department of Operations Research, Stanford University, CA

[10] Han C-G, Pardalos P M and Ye Y 1991 *On interior-point algorithms for some entropy optimization problems* Working paper Computer Science Department, Pennsylvania State University, University Park, PA

[11] Kortanek K O and No H 1992 A second order affine scaling algorithm for the geometric programming dual with logarithmic barrier *Optimization* **23** 501–7

[12] Dantzig G B 1987 Origins of the simplex method *Technical report* SOL 87–5 https://apps.dtic.mil/dtic/tr/fulltext/u2/a182708.pdf

[13] Lin Y 2002 Bellman's principle of optimality and its generalizations *General Systems Theory: A Mathematical Approach* (Kluwer), pp 135–61

[14] http://web.math.ku.dk/~moller/undervisning/MASO2010/notes/LKKT.pdf

[15] Vanderbei R J 1999 LOQO user's manual—version 3.10 *Optimiz. Methods Soft.* **12** 485–514

[16] National Electrical Manufacturers Association (NEMA) 2018 Volt/VAR optimization improves grid efficiency *Technical report* http://assets.fiercemarkets.net/public/sites/energy/reports/voltvarreport.pdf

[17] den Hertog D, Roos C and Terlaky T 1993 The linear complementarity problem, sufficient matrices and the criss–cross method *Linear Algebra Appl.* **187** 1–14

[18] Jarre F 1994 Interior-point methods via self-concordance or relative Lipschitz condition (Würzburg: Habilitationsschrift)

[19] Lustig I J, Marsten R E and Shanno D F 1994 Interior point methods for linear programming: computational state of the art *Oper. Res. Soc. Am. J. Comput.* **6** 1–14

[20] Baxter J and Bartlett P L 2001 Infinite-horizon policy-gradient estimation *J. Artif. Intell. Res.* **15** 319–50

[21] Coleman J O and Scholnik D P 1999 Design of nonlinear-phase FIR filters with second-order cone programming *Proc. of 1999 Midwest Sym. on Circuits and Systems*

[22] Bertsekas D P 1995 *Nonlinear Programming* (Belmont, MA: Athena Scientific)

[23] Jarre F 1996 Interior-point methods for convex programming ed T Terlaky *Interior-Point Methods for Mathematical Programming* (Dordrecht: Kluwer), pp 255–96

[24] Chenciner A, Gerver J, Montgomery R and Simó C 2002 Simple choreographic motions of N bodies: a preliminary study *Geometry, Mechanics, and Dynamics* ed P Newton, P Holmes and A Weinstein (New York: Springer)

[25] Broucke R 2003 New orbits for the *n*-body problem *Proc. of Conf. on New Trends in Astrodynamics and Applications*

[26] Vanderbei R J, Spergel D N and Kasdin N J 2003 Circularly symmetric apodization via star-shaped masks *Astrophys. J.* **599** 686–94

[27] Vanderbei R J, Spergel D N and Kasdin N J 2003 Spiderweb masks for high contrast imaging *Astrophys. J.* **590** 593–603

[28] Kasdin N J, Vanderbei R J, Spergel D N and Littman M G 2003 Extrasolar planet finding via optimal apodized and shaped pupil coronagraphs *Astrophys. J.* **582** 1147–61

[29] Duan Y, Chen X, Houthooft R, Schulman J and Abbeel P 2016 Benchmarking deep reinforcement learning for continuous control *Int. Conf. on Machine Learning (ICML)* pp 1329–38

[30] Sen P K and Lee K 2014 Conservation voltage reduction technique: an application guideline for smarter grid *IEEE Trans. Indus.* **52** 2122–8

[31] Lobo M S, Vandenberghe L, Boyd S and Lebret H 1998 Applications of second-order cone programming *Technical report* Electrical Engineering Department, Stanford University http://rutcor.rutgers.edu/~alizadeh/CLASSES/12fallSDP/Papers/socp.pdf
[32] Bendsøe M P, Ben-Tal A and Zowe J 1994 Optimization methods for truss geometry and topology design *Struct. Optimiz.* **7** 141–59

Chapter 5

Implementing the traveling salesman problem using a modified ant colony optimization algorithm

Zar Chi Su Su Hlaing, G R Sinha and Myo Khaing

In this chapter, the main modifications induced by ant colony optimization (ACO) are presented. In fact, the traveling salesman problem (TSP) could easily be solved by using an improved version of the ant colony optimization method. The same is attempted in this chapter. There are two main ideas in the proposed algorithm for the modification of the ant algorithm. The first phase involves defining the candidate set which is applied to the construction of the ant algorithm. The solution construction phase includes defining the value of exploitation parameter q_0. The second phase focuses on the variation of pheromone information that is used to adapt the heuristic parameter automatically throughout the algorithm runs. Additionally, a local search algorithm is applied to all the solutions that are produced by ACO and thus the performance of ACO is improved for TSP related applications.

5.1 ACO and candidate list

ACO is a very popular and much researched area in the metaheuristic domain. It improves the convergence speed and escaping from local optimal solutions under the condition of guaranteeing solution quality. The ACO metaheuristics are population based constructive metaheuristics. At each step of the algorithm, each ant first searches for an element (the next city for the TSP) to add it to the ant's solution. When the size of the problem is increased in any application, then the difficulty increases and thus ACO is said to have this problem as bottleneck. It is also difficult when applied to large-sized problems in the solution construction process. As ACO is a constructive metaheuristic in nature, the ants do consider the whole set of

doi:10.1088/978-0-7503-2404-5ch5

elements which are possible at each and every step, and then the current solution is added just before choosing one element. It can therefore be seen that in most algorithms utilizing the ant concept, the evaluation of utilities is done based on reachable elements, otherwise the techniques employing ACO can perform adversely due to the large value of the runtime, because if the construction of appropriate subsets is not done appropriately then this problem is likely to occur.

At the solution construction phase, there are possible problems that cause slow convergence due to the fact that the ants start scanning the set of all possible state elements (cities) before choosing a particular one and the probability is very small to visit the same state for most ants. Then, the computational time can be large for each scanning step of the algorithm. When the TSP size is large scale, such a situation can occur. This is suggested even though we are aware that the system performance can be improved as well as the efficiency in a number of real world applications of ACO and when large TSPs are taken into consideration. So, the major problems of ACO include slow convergence or speed, long runtimes, long execution times, trapped local optima, etc.

To consider the above mentioned TSP, there is a limited number of elements (cities), called the candidate list (CL), which can reduce the scanning set of possible elements. The CL is a limited set of possible neighborhood elements from the current element. CLs are created statically or dynamically using available knowledge of the problems. Using a CL, ACO algorithms can reduce the search space of problems in the solution construction phase.

The local search procedure employs CL strategies that help to improve the ACO generated solutions and outcomes. However, the implementation of optimization has its own constraints in strategies related to 2-Opt and 3-Opt local search heuristics, because the use of the construction phase in ACO becomes not very appropriate. These strategies aim to improve the ant colony system (ACS) so that candidate set based strategies can successfully be applied in the construction phase.

Now, there are two important approaches which are used for solving general purpose problems in each of the elements with a strong relationship between the elements. These approaches are the Delaunay graph candidate set approach and nearest neighbor approach, which provide relationships between elements such as in the TSP. A static candidate set is used in all such strategies as an *a priori* candidate which is derived and does not receive any updates or modifications while being applied in the process of the ACO metaheuristic. On the other hand, throughout the search process, dynamic candidate set strategies are required.

5.2 Description of candidate lists

A CL restricts the number of possible choices that need to be considered at each construction step. In the case of the TSP, the CL is chosen for any city i and contains a fixed number nn of cities having the smallest distance to city i. These cities are called nearest neighbors. In order to determine them, all cities are sorted by their distance to city i in increasing order and the first nn cities are inserted in city i's CL. For the tour construction rule (equation (5.2) or, respectively, equation (5.1)) the biased search is performed only for cities in the CL. If all cities from this list have already been visited, one of the remaining cities is chosen.

5.3 Reasons for the tuning parameter

There are many disadvantages in the basic form of ant system (AS) implementation, a few of them are trapping, premature convergence and stagnation in the behavior of the system. If all ants start producing similar or equal solutions again and again then the behavior is said to be stagnated or in stagnation. This demands new ways or modified and adaptable paths, which is a very difficult task but can result in a situation of obtaining the best possible local solution. This may arise when a large amount of pheromone trail on the edges of solutions is seen and some advanced phases of search process are also observed, which actually happens as a consequence of improper tuning of ACO parameters such as α, β, ρ.

The ACS has this problem of stagnation which can be observed if the parameters are not tuned properly. The individual candidates with the highest and lowest values for q_0 affect the construction phase of ACO, which is not very surprising, but the individuals try to exploit the values. Ranks which are low valued are far from pheromone information and these ranks with low values affect the algorithm. If the value goes down too low the search becomes unsupervised and erratic even though the pheromone trail is fine. On the other hand, a high value of rank improves the solution and helps in obtaining the local best possible option which is not based on a probability concept but is obtained on the basis of a high value of rank. As an effect, short tours are created rather than long tours and therefore the candidates supporting such exploration are considered as less fit and consequently their corresponding genes start vanishing. Sometimes, short tours result which usually support the exploration and the solution becomes better, but the problem of premature convergence still remains to some extent.

The problem of stagnation and premature convergence can easily be handled and avoided using a random number q that is generated and compared with pheromone trails on edges. For high values of trails τ, the random number is smaller and vice versa. So, the algorithm does not succeed if a suitable value of trail is not chosen, which means that it should not be too high, otherwise the selection of the best path and point cannot be made. If the value of the trail is very high, then the algorithm tends to edges and therefore the updating of these values is required in a dynamic manner.

5.4 The improved ACO algorithm

At this stage, a modified version of the ACS is presented. The modifications include a flexible state transition rule, dynamic candidate set strategy and heuristic parameter updating rule [1–4, 6]. The system introduces a mechanism for escaping from tapping in a local optimal solution and increases the efficiency of the algorithm. The complete IACO algorithm is described in detail below.

Transition probability

Ants choose the next city from the current city using the state transition rule which is the same as for the ACS. The rules are taken from the ACS of Dorigo and Stutzle. In this stage, an ant will iteratively select from city to city. Some of the time an ant will select a city using the random proportional action choice rule as before. Then the ant chooses only the 'best' city based on the visibility and the pheromone trail.

Formally, an ant k positioned on city x chooses a city y to move to according to

$$y = \begin{cases} \arg\max_{u \in J_k}(x)\left\{[\tau_{(x,u)}]^{\alpha}.\ [\eta_{(x,u)}]^{\beta}\right\}, & \text{if } q \leqslant q_0 \\ Y, & \text{otherwise} \end{cases}, \tag{5.1}$$

where q is a number which has been randomly generated in the range between 0 and 1, q_0 is a constant, $J_k(x)$ is the set of cities which are not visited by the kth ant and Y is a random city according to

$$P_k(x,\ y) = \begin{cases} \dfrac{[\tau_{(x,y)}]^{\alpha}.\ [\eta_{(x,y)}]^{\beta}}{\sum_{u \in J_k(x)}[\tau_{(x,u)}]^{\alpha}.\ [\eta_{(x,u)}]^{\beta}}, & \text{if } y \in J_k(x) \\ 0, & \text{otherwise} \end{cases}, \tag{5.2}$$

where $P_k(x, y)$ is the probability of choosing city y from city x, $J_k(x)$ is the set of cities which are not unvisited by ant k, $\tau(x, y)$ is the pheromone value for moving from city x to city y, $\eta(x, y)$ is the local heuristic (distance) for moving from city r to city s which is $1/d(x, y)$, α determines the significance of pheromone information and β determine the significance of heuristic visibility.

Exploration and exploitation

A suitable adaptation is required to be carried out in the algorithm, which is between the exploration and exploitation of the search space. The experience gained so far is considered as exploitation, which is gained from visibility and trail amount, whereas the exploration is counted as the number of unvisited or relatively unexplored or relatively less explored possible set of elements in search space regions. Exploitation is the action of taking the path with maximum probability. Exploration is the process of calculating the probability where the ants decide which path to take based on this probability. Hence, a path with higher probabilities would have a chance of being more favorable to choose. So exploitation or exploration is an important factor of the algorithm. The decision of the exploitation or exploration factor is defined by a factor q_0, the exploitation factor, which has a point value between 0 and 1 (a floating point value).

The tours are constructed in the AS by the ants which differ to a large extent from the best possible values of tours. In the ACS of [8], the parameters q_0, β and ρ are taken as fixed values or constants whose values do not vary when the construction of the solution is attempted for the ACS algorithm. In Dorigo *et al* [8], the value for q_0 is chosen as 0.9, which implies that tours will be constructed by ants after exploring

10% of the total time and exploiting 90% of the time. This further suggests that the relative significance of exploitation can be determined by q_0 in comparison to biased exploration, on the other hand, the desirability of the edges is determined by the values of β and ρ.

In the algorithm suggested in this chapter, the next city is chosen by the ants with more exploration at the beginning and the ants tend to have more exploitation at the end of the algorithm. The information which is accumulated by exploitation values is utilized properly. The values of q_0 and β are permitted to vary and the appropriate exploitation can be achieved. A new value for q_0 is updated after every iteration of the suggested scheme, which can be understood through

$$q_0 = \frac{\text{IterationCounter}}{\text{MaxIteration}} \tag{5.3}$$

$$q_0 = \begin{cases} 0.1 & \text{if } q_0 < 0.1 \\ 0.9 & \text{if } q_0 > 0.9, \\ q_0 & \text{otherwise} \end{cases} \tag{5.4}$$

where IterationCounter is an important number that indicates the current iteration cycle, and maximum number of iterations is MaxIteration.

When we increase the value of q_0 then we will obtain the exploitation more frequently and with greater value. The greater value of exploitation enables the ants to use the edges of tours, which are globally best with a greater frequency. In such cases, a local optimum may occur, which can be avoided by finding the same tour iterating again and again, keeping in mind that the global best tour is not improved after completing a number of iterations. This condition is not advisable because such exploitation for a best global tour can erase memory and disturb the chosen edges for the best global tour. Here, we also have to perform dynamic updating of heuristic parameter so that the global best tour is not affected and proper utilization of memory is also maintained.

Global pheromone update

In the process of global pheromone updates, the pheromone values are allowed to be updated by the global best ant once the best tour is detected [10–13]. After the completion of an iteration, the global pheromone updating rule is applied to obtain the pheromone update and this is actually an important step in the updating process. If there is no updating of the pheromone value, each ant will repeatedly find the same probability on all moves and find the same solution again and again. The evaporation and depositing of pheromone is only added to the best route found when the beginning of the trial is detected or observed. The global updating rule is

$$\tau(x, y) = \begin{cases} (1 - \rho)\tau(x, y) + \rho(L_{gb})^{-1}, & \text{if } (x, y) \in \text{global\_best\_tour} \\ \tau(x, y), & \text{otherwise} \end{cases}, \tag{5.5}$$

where ρ is the decay parameter of pheromone and L_{gb} is the length of the global best tour. As the effect of this rule, the ants' searching concentrates around the best known tour and also increases exploitation.

Local pheromone update

The local updating rule is used whenever we find the ant moving from one city to the next city, and on movement from a city to the next city only the local updates are applied. At the end of each iteration, while the global pheromone update is applied to all edges once, the local pheromone update is applied to many. In the case where local updates are not done, the ants will become non-collaborative in various iterations. The ants will then only make use of the beginning of the pheromone trail. The updates by ants on the pheromone trail over the edges are achieved by the following rule:

$$\tau(x, y) = (1 - \rho)\tau(x, y) + \rho\tau_0, \tag{5.6}$$

where ρ is a constant whose value is found in the range (0, 1) and τ_0 is the initial value of the trail level. This initial value for the trail is calculated as

$$\tau_0 = \frac{1}{n \cdot L_{nn}}, \tag{5.7}$$

where n is the number of cities, and L_{nn} is tour length covered by the nearest neighbor search.

Using the above strategy, i.e. local updating, there is a decrease in the pheromone concentration upon the traversal of edges. Subsequently, encouragement is given to ants choosing other edges so that different solutions are produced. This avoids the possibility of obtaining the same solutions by the ants during one iteration. If we look at the original ACS, the ants are found to choose the next city or next element and, based on the set of possible cities, the ants move to non-visiting cities. However, this is a time-consuming process as ants, if they do so, will consume more time. If the set of elements is restricted, then the efficiency of IACO can be improved considering k nearest neighbors using a dynamic CL.

5.4.1 Dynamic candidate set based on nearest neighbors

The CL uses a strategy to attempt to obtain better performance of the algorithm used by the ants. The strategies are defined and interpreted nicely in Gambardella and Dorigo [7, 9] which involves the searching process of ACS applied over a large amount of data. If a fixed candidate size is used, which is not flexible option, then various data sizes cannot be addressed, but in order to improve the performance of the algorithm the system requires the application of different CLs that are dynamic in nature. A dynamic candidate list (DCL) can take on an appropriate number of nodes depending on the total number of nodes [14, 15, 17, 18]. The DCL also determines which sets are to be calculated and calculated again (recalculated) throughout the search procedure. The static data structure has a limited number of closed cities of preference which can be visited, as when we increase the number of cities with the increase in the distance of cities. While constructing the solution the next city is chosen by the ants, then the probability of movement from city i to city j needs to be computed and, based on the probability of transfers, the next city is chosen. If a city is not found by an ant, then that city is moved out of the total number of the CLs, but this limits the selection capability of the ant and the list subsequently becomes smaller. While moving, all different cities which are found by the ants are added to the list. When the algorithm uses the candidate set strategy then a few factors become very important:

- The number of edges in the global optimal solution in a set of candidates.
- Restriction of candidates to the scope of choosing or selection.

The improvement results in the performance of the algorithm on application of a dynamic CL in the process of construction of the solution space. The DCL is based on the theory of the nearest neighbor approach. The cities which are outside the DCL are not allowed to be covered by the ants. This results in the following:

- A more limited number of cities than the actual number n.
- A limited value for the size of the candidate.

For instance, computation of the CL is determined and size is calculated for Oliver30 data through the following procedure:

Procedure: Candidate list selection/determination

---

1. Initialize an empty node *Node* and *MaxLength*.
2. Set DCL = $n/4$ /*candidate list size*/.
3. **if** DCL > *MaxLength* DCL = *MaxLength*.
4. Get the cities which have not yet been visited.
5. **repeat**
1. **for** $i = 1$ to n.
2. Find the unvisited city j within the nearest neighborhood of $k(j \in N(k))$.
3. Calculate the distance between city j and city r.
4. **if** distance < distance of previous city j.
5. *Node* ← *city j* (move city j to *NL*).
6. **end if**
7. **end for**
8. *DCL* ← *Node*.
9. **until** list of candidates (*DCL*) is full.

The above procedure proved high speed in the performance of the algorithm using the DCL strategy, which further helps in obtaining an improved computational time and improved quality of solution. Experimental studies have shown that this proposed method results in an improvement of solution quality and a significant performance gain.

5.4.2 Heuristic parameter updating

We studied a number of ACO strategies and observed that in the theoretical background dealing with ACO parameters, α and β have not been attempted much and therefore the amount of research on this specific domain is limited. These parameters are very important as they allow ACO for controlling the relative weights of pheromone trails and also the information of the heuristic search. The use of parameters in the analysis of the possibility of the use of heuristic information is pertinent since the construction of probable solutions results in exploitation of specific knowledge for the problem chosen. The value referred to as the heuristic value can obtain a prior value or, also in the runtime of all TSP problems, during the initialization step the heuristic information η is calculated and once it is computed it should not change throughout the operation of the algorithm. Another parameter β actually describes the relative influence between the level of pheromone and the information of the heuristic search, and hence the heuristic desirability η or the information can be controlled by β. The larger the value of β, the faster convergence becomes and the speed of the algorithm becomes very high, which subsequently results in local best solutions in a more efficient manner. If the value of $\beta = 0$, then only the information in the pheromone is taken into consideration, which leads to stagnation in the algorithm. A stagnation situation in the algorithm means that the ants will obtain the same solution rather than a sub-optimal solution. Thus, the value of β needs to be investigated properly so that an appropriate adjustment of this parameter can obtain the best local solution without any stagnation. High quality tours also need appropriate heuristic information for the initial search stages. Initially, the pheromone trail amount was kept as equal for all the edges, but this did not allow a favorable path for the ants in the construction of the solution. This situation demands a larger value of parameter β.

Later, when implementing the algorithm, the heuristic parameters may require certain changes, but not much, due to the fact that the pheromone trails are gathering enough information as per the requirement and the information may misdirect the search process. So, we have to be very aware of what value is required for the heuristic parameter, sometimes a constant value is also desired and accordingly we have to provide this, particularly in conventional ACO algorithms. High values of parameters are required in achieving better quality tours and the influence of the pheromone is also reduced by a considerable extent. Under such situations, ants can search their paths effectively in the process of construction of feasible solutions. In the initial process, a small value can help, but performance improvement always requires high values of parameters [14, 15]. The values of parameters need to be adaptively changed as per the requirement so that performance is also satisfactory and no stagnation or other such problems prevail. This study suggests that dynamic updates are required in the values of heuristic parameters for ACO towards improved search performance.

Now, when the algorithm begins to be executed then the pheromone information for all paths is equal to each other and the value of entropy is highest at this point of time. After every iteration, updates are done on the pheromone and it is also enhanced on the path chosen so that the entropy value starts slowly decreasing. If the entropy is allowed to decrease, then it may also reach zero and that may not lead to a global best tour and the situation will be termed as premature convergence. In order to deal with this type of difficulty, which occurs due to behavioral defects, some complex optimization problems require colony optimization which can handle the issues related to entropy. The discussion of entropy issues is required for dynamic updates of heuristic parameters, which can also control the value of entropy. Shannon's theory of entropy as information theory is very popular in this area, also known as Shannon's information theorem [3]. This was introduced in 1948 and is usually defined as a measure of uncertainty which is concerned with events having disorder in a system. The entropy represents that information which is associated with the probability of the occurrence of the event. Why this concept is relevant is an important question. It is because of the fact that the ACO algorithm has a path selection procedure and the selection of the path is also not certain, which means there is uncertainty. Therefore, we suggest the entropy information to be estimated in ACO as variation of the pheromone matrix. Each trail in this case would be a random variable in the pheromone matrix. Such entropy is defined as

$$H(X) = -\sum_{i=1}^{r} P_i \log P_i, \tag{5.8}$$

where $_{pi}$ is the probability of occurrence of each trail in the pheromone matrix. For a symmetric n city in the TSP, there are $n(n-1)/2$ distinct pheromone trails and $r = n(n-1)/2$. It is obvious that the probability of each trail is the same, and H will be the maximum ($H_{\max}$), given as

$$H_{\max} = -\sum_{i=1}^{r} P_i \log P_i = -\sum_{i=1}^{r} \frac{1}{r} \log \frac{1}{r} = \log r. \tag{5.9}$$

We are suggesting using entropy as an index or degree of how much information is trained by the pheromone trails, and updates in the heuristic parameter are required accordingly. It is worth noting here that the heuristic parameter β value has to be considered as an integer value because this can avoid complicated computation processes. The value of β is also used as a power in equations (5.1) and (5.2). The suggested system behaves adaptively and the following rules are proposed with regard to β:

$$\beta = \begin{cases} 5 & 0.9 < H' \leqslant 1 \\ 4 & 0.7 < H' \leqslant 0.9 \\ 3 & 0.5 < H' \leqslant 0.7 \\ 2 & 0 < H' \leqslant 0.5 \end{cases} \tag{5.10}$$

$$H' = 1 - \frac{H_{\max} - H_{\text{current}}}{H_{\max}}, \tag{5.11}$$

where H' is the value of entropy of the pheromone matrix that is currently being considered.

5.5 Improvement strategy

After the success of the proposed algorithm, there needs to be a lot of effort to improve the performance of the algorithm. The solution can be improved using a local search, keeping in mind the fact that we have to look in its neighborhood as well for better ones. The ants have generated the tours and local search methods are now responsible for improving the performance. In the case of the TSP, this involves improving a given tour with small local changes. Having found a tour, the order of the cities is slightly changed into another valid tour by applying one of the following methods.

5.5.1 2-Opt local search

The simplest way to change a TSP tour is to exchange two edges. The 2-Opt algorithm is basically used to eliminate two edges from the tour, and then start rebuilding the two chosen paths. This is done by a reconnection process of two other edges in some different way so that we can obtain a new tour. This process is generally called a 2-Opt move. The way in which reconnection can work is shown in figure 5.1(a), where we can see [18] that two paths are reconnecting and generating a new and valid tour. This can be obtained for shorter tours only. The process of removing and reconnecting tours continues until there remains on further scope of improvement in Opt-2 and now the resulting tour is 2-optimal.

For example, to realize this, two non-adjacent edges, call them *ab* and *cd*, are chosen (note that *ab* means that city *a* is visited before city *b* on the tour). These edges are deleted and new edges *ac* and *bd* are inserted and hence form a new TSP tour (see in figure 5.1(b)). Local search in 3-Opt starts working in a similar manner to 2-Opt, but rather than removing two edges we need to remove three, which means that the exchange of three edges in the tour is considered to compute the final best tour. There are two possible ways in which we can reconnect the three paths into a valid tour.

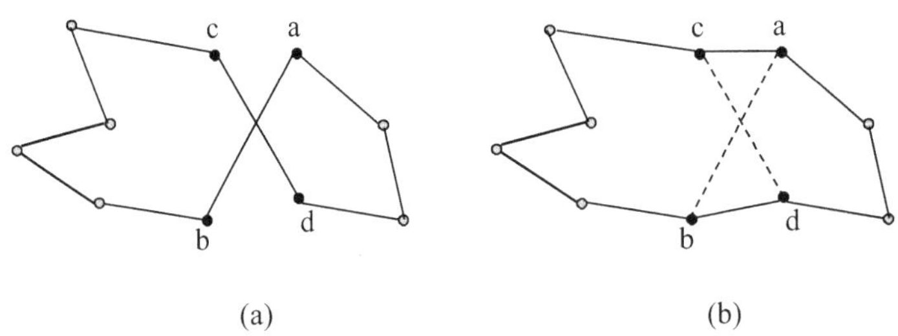

Figure 5.1. A 2-Opt local search. Edges *ab* and *cd* from the graph illustrated in (a) were deleted and the tour was completed again by inserting edges *ac* and *bd*, thus resulting in the valid tour shown in the graph (b).

5.6 Procedure of IACO

This section briefly describes the proposed algorithm step by step.

Algorithm. Improved ACO (IACO) algorithm for TSP.

---

1. Initialize
        Find pheromone trails τ_0 using a nearest neighbor heuristic
        for every *edge*(x, y)
            Set $\tau(x, y) = \tau_0$
        end for
2. Calculate the maximum entropy
    globalBestTour $\leftarrow \phi$
    globalBestTourLength $\leftarrow \infty$
    determine candidate_list strategy
3. Place the n ants randomly on the starting city
    for $k = 1$ to n ants do
        Place and sum with the starting city of the kth ant to its tabular list
    end for
        localBestTour $\leftarrow \phi$
        localBestTourLength $\leftarrow \infty$
    Define the value of exploitation parameter q_0
4. **for** $k = 1$ to n do /*n is the number of ants */
        $q \leftarrow random$
            repeat
                Find the next unvisited node j for kth ant from kth ant's current node i by using state transition rule with CL
                Append j into the kth ant's tabu list
                Perform local pheromone update
            until tour has been completed by ant k
        Apply local search (2-Opt or 2.5 Opt) to improve tour
        Compute *tourLength* of kth ant
        if *tourLength* < *localBestTourLength* then
            localBestTourLength $\leftarrow$ *tourLength*
            localBestTour $\leftarrow$ best tour found
        end if
    end for
        if *localBestTourLength* < *globalBestTourLength* then
            globalBestTourLength $\leftarrow$ *localBestTourLength*
            globalBestTour $\leftarrow$ *localBestTour*
        end if
5. **for** each *edge* (r, s) belonging to the global best tour
            Perform global pheromone update
    end for
6. **for** every pheromone $\tau(r, s)$
            Compute value of entropy for current pheromone trails

```
    end for
    Update the heuristic parameter
7. // Check end condition
      if (end_condition = true or MaxIteration)
         print global best tour
         exit
      else
         go to 3.
      End if
```

5.7 Flow of IACO

The flow diagram of the suggested IACO algorithm is shown in figure 5.2. This begins from initializing the parameters and terminates by obtaining the best tour results. The steps and procedures shown in the flow diagram are self-explanatory.

5.8 IACO for solving the TSP

This section explains how the IACO executes, which is applied to solve a considerable problem for TSP and the candidate set is defined as a set of cities. In the following example, there are eight cities P to W and two ants (ant A_1 and ant A_2). Assume that you place the starting city of ant A_1 and ant A_2 on city P and Q, respectively. When an ant k needs to move from current city x to next city y, it first searches in predefined set of CLs. Then, it also adds its current start city to a short-term set of lists called the tabu list and applies equation (5.3), where it produces a value of the q_0 parameter and also produces a randomly generated value for parameter q, and compares the two parameter. When a random parameter $q \leqslant q_0$, the state transition rule, equation (5.1), is applied to choose the next city. An ant k exploits the knowledge available about the problem and moves to city (y) which has the highest product of the amount of pherormone trail on the edge (x, y) and the shortest distance between the two cities. When $q > q_0$, the random proportional rule, equation (5.2), is applied in which the ant k uses it to explore new solutions using the candiate list. Each city has a candidate list (cl); the number of cities listed is the length of the CL. In this example, assume that the length of the CL, $cl = 2$, where the candidate list of city (P) includes cities (Q) and (W) and they will be explored by ant A_1 before the other cities. It is assumed that ant A_1 goes from city (P) to city (Q), then city (Q) will be added to the tabu list to avoid being visited again by the same ant. After moving from city (P) to city (Q), ant A_1 updates the pheromone on the edge between the two cities using the local pheromone updating rule, equation (5.6). For the next step, ant A_1 again computes the possibilities of moving from its current city (Q) to those other cities that are not in its tabu list (R to W) using the same equation (5.1) and so on until ant A_1 visits all the other cities as shown in figure 5.3. Ant A_2 also proceeds in such a way to find its solution.

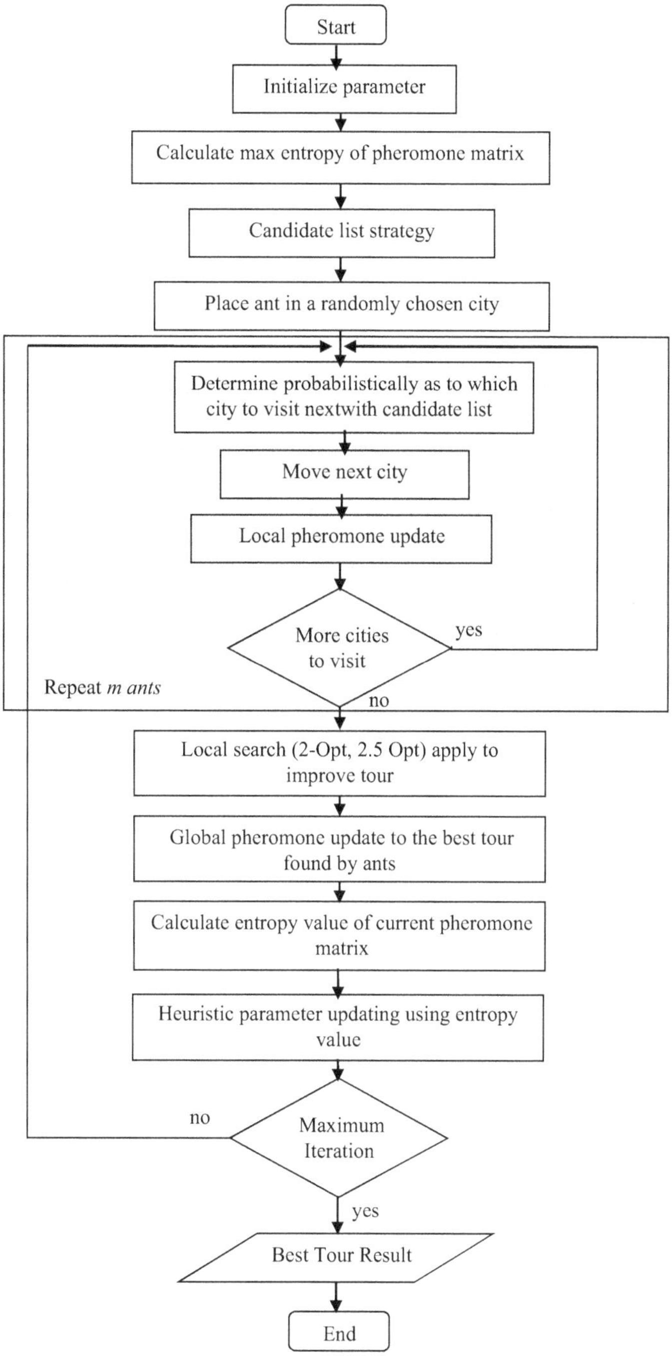

Figure 5.2. Flow of improved ACO algorithm.

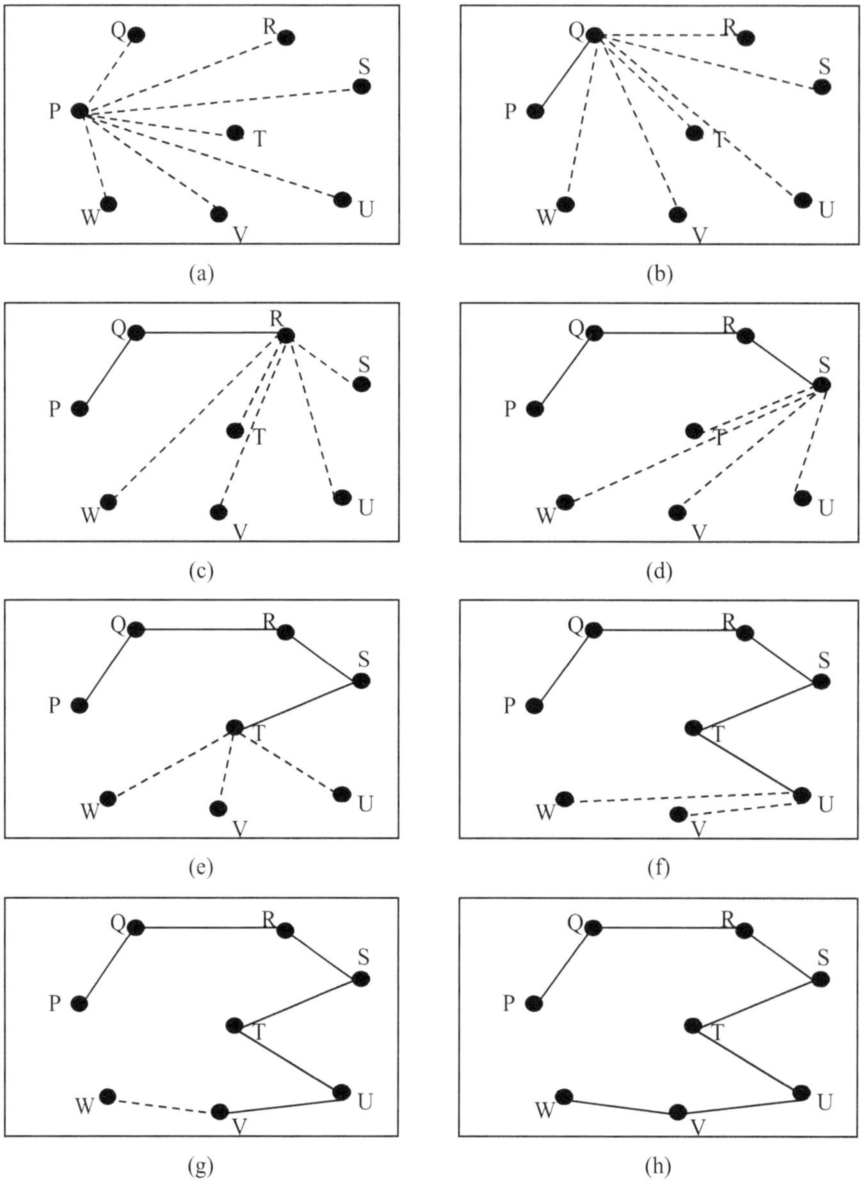

Figure 5.3. IACO for a simple TSP.

The tour length generated by the ant will be calculated by adding the length of the edge between each two cities from the tour. The process will be accomplished by each ant and at the end of the iteration there will be five tours generated by five ants. The local search improvement is applied to improve the tour consturcted by the ants.

The shortest of these tours will be selected as the best tour and the edges that form this tour will be updated using the global update formula in equation (5.5). Then, we calculate the current entropy by analyzing the pheromone information to update the heuristic parameter β. Then, the ants are again placed randomly for a second iteration and redefine the exploitation parameter q_0. The algorithm goes on until the maximum number of iterations has finally found the global best tour.

5.9 Implementing the IACO algorithm

This section describes in detail the steps for implementing an improved ACO algorithm for the TSP. The basic considerations for the implentation of the IACO algorithm are similar to the ACS and the necessary changes for implementation are presented. The problem can be represented as a construction graph used by the construction procedure, and then it is only necessary to:

1. Define the pheromone trail variables to the construction graph.
2. Define the number of aritificial ants to be used for constructing solutions according to equation (5.1).
3. A randomized version of the construction procedure.

This section describes an implementation in a pseudo-code description.

Procedure: IACO for TSP

```
begin
    InitializeData
    Calculate MaxEntropy
    while (not termination) do
        ConstuctSolutions
        ApplyLocalSearch
        UpdateStatistics
        UpdatePheromoneTrails
        ComputeCurrentEntropy
        UpdateHeuristicParameter
    end-while
end
```

In data initialization: (i) read the TSP instance; (ii) calculate the distance matrix of the read TSP instance; (iii) define and compute the the CLs for all cities; (iv) the ants randomly place the their starting cities; (v) the algorithm's parameters must be initialized; and (vi) some variables that keep track of statistical information, such as number of iterations, or best solution found (best tour length), and best tour, need to be included.

Procedure: InitializeData

---

begin
    ReadTSPInstance
    ComputeDistances
    ComputeCandidateLists
    InitializeAnts
    InitializeParameters
    InitializeStatistics
end

The following two constuction steps are repeated until the tour is finished by all the ants. When exploiting the procedure the *ConstructExploitDecisionRule* needs to be adapted. If not, the procedure *ConstructExploreDecisionRule* needs to be computed. In *ConstructExploitDecisionRule*, a major change that can be seen is that while choosing the next city, we need to find the city which is not visited from the entire list of candidates from the current city. Another change which can be seen is the necessity for dealing with the situations where all cities have been covered in the CL by any ant k. Under this condition and taking the changes into account, the variable *node* maintains its initial value as −1 and the city beyond the CL is chosen. We are required to choose the maximum product of the pheromone value and heuristic information $[\tau_{ij}]^{\alpha}[\eta_{ij}]^{\beta}$ for moving to the next city. In *ConstructExploreDecisionRule* there are two changes as in the *ConstructExploitDecisionRule*. The exploring procedure helps in choosing the next city which is unvisited as per the acton choice rule in equation (5.2).

Procedure: ConstructSolutions

---

begin
    curNode ← *startNode*
    q_0 ← *IterCounter*/*Iterations*
    for k = 1 to m ants **do**
      repeat
        q ← *random number*
        if($q < q_0$) **then**
          newNode ← ConstructExploitDecisionRule(k, *curNode*)
        else newNode ← ConstructExploreDecisionRule(k, *curNode*)
        end if
        Add *newNode* to ant k's tour
        LocalUpdatingRule(*curNode*, *newNode*)
        curNode ← *newNode*
      until ant k completes tour
    end for
end

Procedure: ConstructExploitDecisionRule(*k*, *curNode*)

```
begin
    sum_probability ← 0.0        // CandidateListConstructionRule
    node ← −1
    for j = 1 to DCL do
      if kth ant's node j is not visited in candidate list then
        selection_probability ← value of transition probability
        node ← j                  /* city with maximal τ^α η^β */
      end if
    end for
    if (node == −1) then          // city outside candidate list
      for j = 1 to n do
        if kth ant's node j is not visited outside the candidate list then
        selection_probability ← value of transition probability
          node ← j   /* city with maximal τ^α η^β */
      end if
    end for
  end if
    return node
end
```

Procedure: ConstructExploreDecisionRule(*k*, *curNode*)

```
begin
    sum_probability ← 0.0     // CandidateListConstructionRule
node ← −1
for j = 1 to DCL do
    if kth ant's next node j is not visited in candidate list then
      partial_product ← pheromone_value*exp(1/distance,β) /* city with τ^α η^β */
      sum_probability ← sum_probability + partial_ product
    end if
end for
for j = 1 to DCL do
    if kth ant's next node j is not visited in candidate list then
      selection_probability ← partial_ product / sum_probability
      rno ← random number
      if selection_probability >= rno then
        node ← j
      break
      end if
    end if
end for
      if node = −1 then         // city outside candidate list
    for j = 1 to n do
        if kth ant's next node j is not visited outside the candidate list then
```

```
            partial_product ← pheromone_value*exp(1/distance, β) /*city with τ^α η^β*/
               sum_probability ← sum_probability + partial_ product
            end if
         end for
         for j = 1 to n do
         if kth ant's next node j is not visited outside the candidate list then
            selection_probability ← partial_product / sum_probability
            rno ← random number
         if selection_probability >= rno then
               node ← j
            break
            end if
         end if
      end for
      return node
end
```

It is evident that the use of CL and computation time are very important factors for the ants in the construction of solutions, in particular if lower values of these factors are expected. It can be seen also that the values are considerably reduced when ants are choosing the cities.

The next step is about local pheromone updating which acts as a trigger after the ants have moved to the next city.

Procedure: LocalUpdatingRule(*curNode*, *newNode*)

```
begin
   value ← (1 – ρ)*pheromone[curNode][newNode] + ρ* pheromone[curNode][newNode]
   pheromone[curNode][newNode] ← value
   pheromone[newNode][ curNode] ← value
end
```

Once the solutions are constructed, the generated tours may be improved by a local search procedure (for example 2-Opt or 2.5-Opt). The next step in an iteration of the algorithm is the pheromone update (global pheromone updating). This is implemented by the procedure of *UpdatePheromoneTrails*, which makes up two pheromone update phases: pheromone evaporation and pheromone deposit. Pheromone evaporation decreases the value of the pheromone trails on only the best path by a constant pheromone decay factor ρ. The pheromone deposit adds pheromone to the edges belonging to tours constructed by the ant's best path length. *ComputeCurrentEntropy* computes the pheromone information of the current pheromone matrix to be used in the next step. *UpdateHeuristicParameter* dynamically updates the heuristic parameter based on the entropy value of current pheromone information.

Procedure: UpdatePheromoneTrails

```
begin
  for i = 1 to n − 1 do
    tau ← 1/ bestLength
    evaporation ← (1 − ρ) * pheromone[bestPath[i]][bestPath[i + 1]]
    deposition ← ρ * tau;
    pheromone[bestPath[i]][bestPath[i + 1]] ← evaporation + deposition
    pheromone[bestPath[i + 1]bestPath[i]] ← pheromone[bestPath[i]][bestPath[i + 1]]
end for
end
```

Procedure: ComputeCurrentEntropy

```
begin
  sum ← 0.0
  current ← 0.0
  max_entropy ← Math.log(noOfNodes*(noOfNodes-1)/2)
  for i = 2 to n do
    for j = 1 to i do
      sum ← sum + pheromone[i][j]
    end for
  end for
  for i = 2 to n do
    for j = 1 to i do
      current ← current + (−((pheromone[i][j]/sum)*
  (Math.log(pheromone[i][j])/sum)))
    end for
  end for
  current ← 1−((max_entropy-current)/max_entropy)
end
```

5.10 Experiment and performance evaluation

This section reports on the experiments performed to test the proposed system and compares the results with the ACS ant algorithm. An experimental study is presented which focuses on comparing the convergence speed of the proposed system with that of ACS and also the analysis results. To obtain meaningful performance comparison results, a relative error (degree of approximation) calculation is conducted. The measurement of the performance of the experiments is described and the tour length results are also presented.

The TSP is a classical optimization problem which has been used for the evaluation of the performance of ant algorithms. This chapter presents experiments on TSPs comparing the performance of ant colony optimization and the modified ant colony optimization approach with CL strategy based on entropy. The implementations use the ACS version of the ant algorithm. The proposed method includes a candidate set which is applied to the construction phase and one extension to the update phase of the algorithm's heuristic parameter based on entropy.

5.10.1 Evaluation criteria

Numerous studies in the field of TSPs and applications using ACO systems suggest the metrics used by the method in the implementation, such as standard deviation, best tour iteration and best possible tour. The best tour is interpreted as the optimum tour with the least computation time and the standard deviation indicates how much deviation is reported in the solution after each iteration. The degree of approximation, also referred to as relative error, is also computed, which determines how much deviation there is in solutions from the optimal solution. Among several methods reported in the literature, the TSP has found a good position as a traditional method to address various combinatorial problems of optimization.

5.10.2 Path evaluation model

The goal of the path evaluation model is to find an improvement space for each path in an optimal solution. Given a random sample (X_1, X_2, …, X_n) from an n-dimensional random variable, the sample mean is

$$\overline{X} = \frac{\sum_{i=1}^{n} x_i}{n}, \tag{5.12}$$

where X means the tour length, and n is number of the tour length. The sample variance σ^2 is the second sample central moment and is defined by

$$\sigma^2 = \frac{\sum_{i=1}^{n} (X_i - \overline{X})^2}{n}, \tag{5.13}$$

where σ is the standard deviation. The algorithm for optimization is better, with more robustness and better stability, if the value of the standard deviation of a performance criterion over a number of simulation runs is small. Another similar factor σ^2 is known as variance in the evaluation of performance criterion. The smaller the value of σ^2, the smaller is the range of performance values fitting into stability like a swarm.

5.10.3 Evaluation of solution quality

In optimization search problems, the algorithm generates some solution but the quality of the solution needs to be assessed carefully. Suitable measures are required for the assessment and there are four popular measures for evaluating the quality of the solution: approximation, accuracy, completeness and belief.

a. *Approximation*. The approximation ε describes the size of the deviation of s from s^*. s^* means the optimal solution value and s is the value of the feasible solution:

$$\varepsilon = \frac{s - s_*}{s_*}. \tag{5.14}$$

Note that $0 \leqslant \varepsilon < \infty$. If the value of approximation is equal to 0, it is an optimal solution.

b. *Accuracy*. The accuracy, also called the degree of accuracy, α, indicates how close the feasible solution s is to the optimal solution s^* and is defined as

$$\alpha = \frac{\min\{s_*, s\}}{\max\{s_*, s\}}. \tag{5.15}$$

Note that $0 \leqslant \alpha \leqslant 1$. The optimal solution has an accuracy equal to 1.

c. *Completeness*. This is the measure of retrieving the complete solution. If the search ceases, then the solution may not be complete. Adequate resources are to be used so that no premature convergence takes place. The value of this measure is between 0 and 1, and for a value of 1, the solution is said to be complete.

d. *Belief*. The degree of belief indicates heuristically or statistically the belief that the solution obtained is the one desired. This is useful in searching in a probabilistic environment. The degree of belief is a value between 0 and 1. A totally confident solution has a degree of belief equal to 1.

In this system, we are interested in metaheuristic solutions; therefore, completeness and belief are irrelevant. Approximation and accuracy are duals because one can be determined from the other. In this system, the approximation is taken because it is commonly used in approximation searches and also presents the degree of accuracy of both algorithms' results.

5.11 TSPLIB and experimental results

The experiments in this approach use problems exclusively from TSPLIB, a widely known library of TSPs [16]. TSPLIB is an online library, developed by the University of Heidelberg in Germany, that contains several samples of TSPs and similar related problems in a list of different files. The optimal solutions of the problems are recorded within the library, which is a convenient feature of TSPLIB. Due to this being a prominent source for TSPs, many other research papers explore the TSP instances defined within TSPLIB. TSPLIB is thus used to benchmark solutions. The benchmark instances are given with varying complexity and difficulty. Only strongly connected instances of TSP were chosen from the TSPLIB, and through symmetric Euclidian TSPs were analyzed, referenced as EUC_2D, meaning that distances (or weights) between nodes are expressed as an Euclidean 2D coordinate system. The coordinates are decimal numbers (or doubles).

5.11.1 Experiment 1 (analysis of tour length results)

First, the proposed algorithm was tested to see how it performed; the eil51 case was chosen as the test case. The results were shown over 20 trials with 20 iterations per trial (run) for eil51. The parameters used in this test case are described in table 5.1.

The results show how large the deviation from the optimal distance is when running the proposed system several times, both with and without optimization including local search optimization.

As seen in figure 5.4, the proposed algorithm obtains an average tour length of 426.6 when using optimization with local search, which is 0.6 higher than the optimal of 426. The deviation of the proposed algorithm's result (degree of approximation) is

Table 5.1. General parameters used in the test case.

Parameters	Value
α	1
β	2–5
ρ	0.1
No. of ants	10
No. of iterations	20

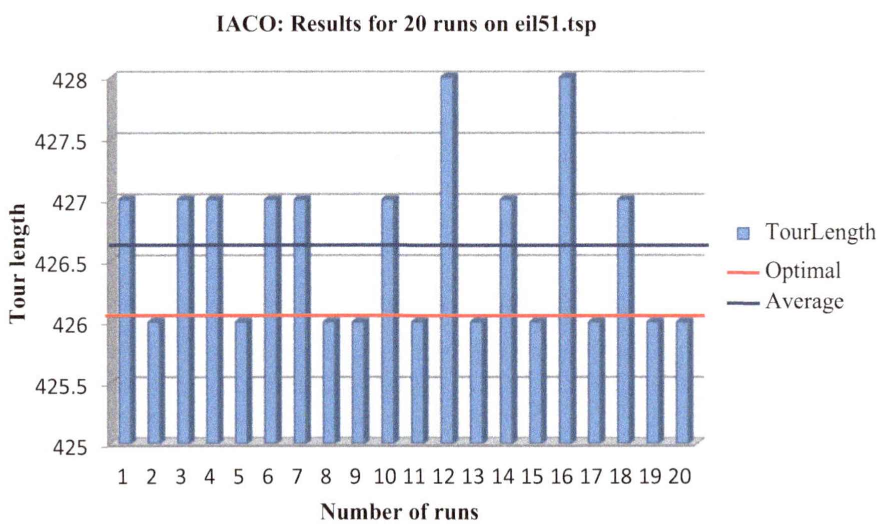

Figure 5.4. Results for the eil51 instance using local search (2-Opt).

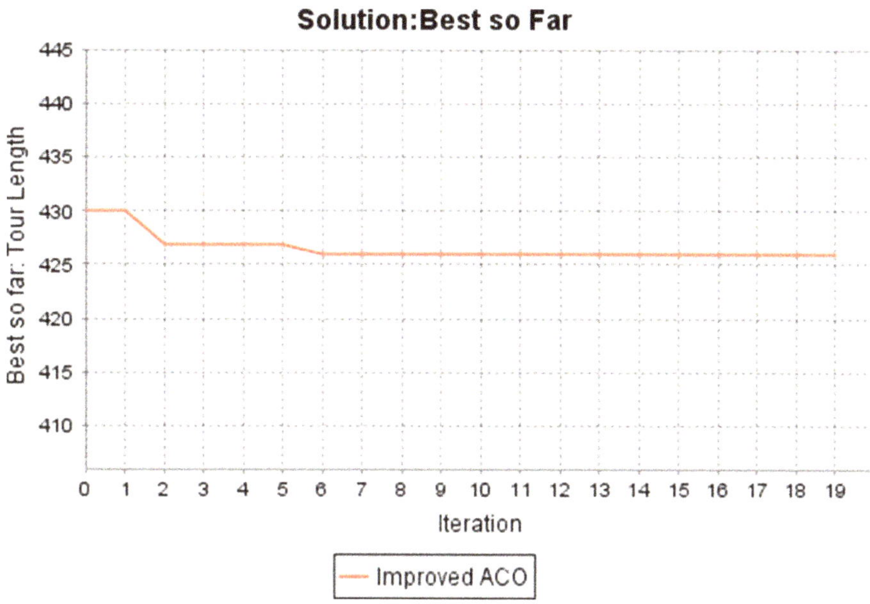

Figure 5.5. The best-so-far solution of the tour length of the eil51 instance.

Figure 5.6. The tour best result of each iteration for the eil51 instance.

0.14% from the optimal distance. The proposed algorithm's results vary from 426 (0% deviation) to 428 (0.47% deviation) and considering half of the running results are converged to the optimal and average length of the runs are much closed to the optimal. Figures 5.5–5.7 show the analysis results for best tour-so-far solution, tour best and standard deviation of tour length, respectively.

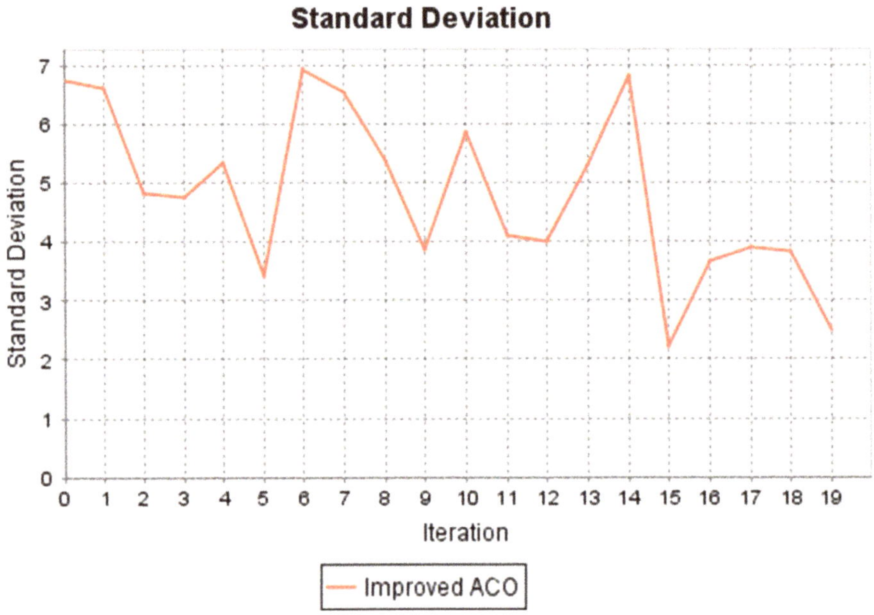

Figure 5.7. Standard deviation of tour length for the eil51 instance.

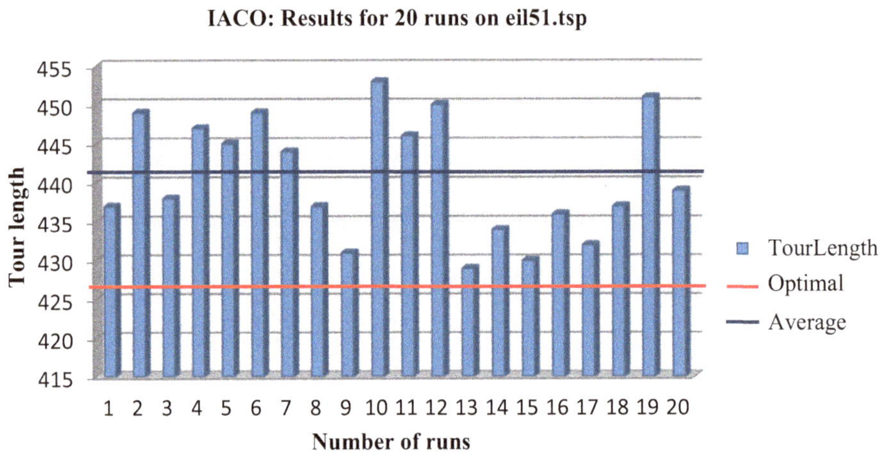

Figure 5.8. Results on the eil51 instance without using local search.

In figure 5.8, when the proposed algorithm does not use the local search optimization to optimize the tour, the results change dramatically. The test case was changing the number of iterations to 100 and the remaining parameter values were kept as in table 5.1. The best solution found is at 429 for 20 trials. It is 0.7% deviated from the optimal solution 426. The results vary between 429 and 453, with an average tour length of 440.7 (0.7%, 6.34% and 3.45% deviation, respectively).

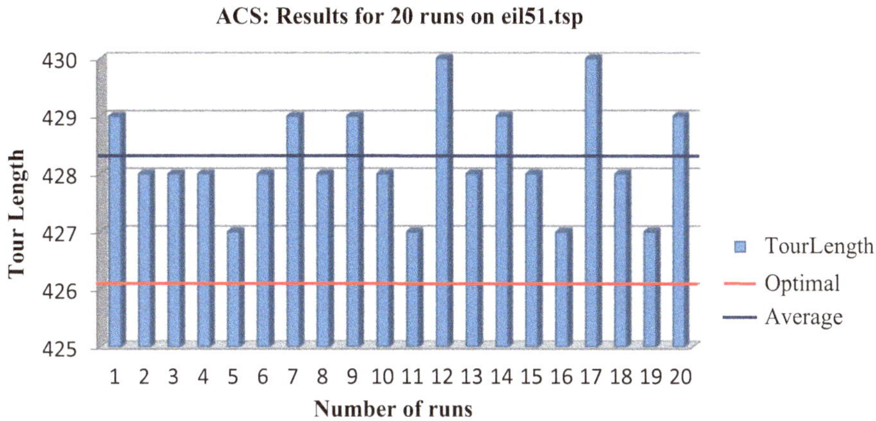

Figure 5.9. Results on the eil51 instance with local search (2-Opt).

Table 5.2. Comparison of the final best solution and convergence number between IACO and DSMACS.

TSP problem	Best length of IACO	Best length of DSMACS	Convergence number of IACO	Convergence number of DSMACS
eil51	426	426	4	5
berlin52	7542	7542	3	4
st70	675	N/A	3	N/A

The analysis results of the ACS are also described in figure 5.9. The ACS algorithm runs the eil51 test case with the parameter value of $\beta = 5$ and the other parameter values as in table 5.1. ACS obtains an average tour length of 428.25 when using optimization with local search, which is 2.25 higher (making it a deviation (degree of approximation) of 0.53%) than the optimal of 426. The ACS algorithm results vary from 427 (0.24% deviation) to 430 (0.94% deviation) and more than half of the runs are below the average length of the runs.

5.11.2 Experiment 2 (comparison of convergence speed)

In the second experiment, the proposed approach is compared with DSMACS [5] in terms of convergence speed. In order to compare the proposed system with DSMACS, some instances of TSP that are the same as those used in DSMACS are chosen. A comparison of the final best solution and the convergence number between the proposed algorithm (improved ACO) and DSMACS are shown in table 5.2 (the results of DSMACS are directly taken from [5]). It should be noted that the performance of the proposed IACO algorithm is excellent. It not only discovers the global optimal solutions for the following TSP instances but also has very quick convergence speed. The speed of convergence is also shown graphically as figure 5.10 for the proposed algorithm.

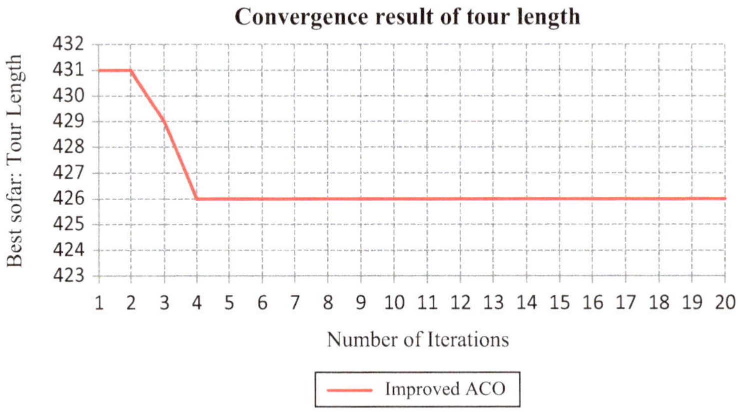

(a) eil51

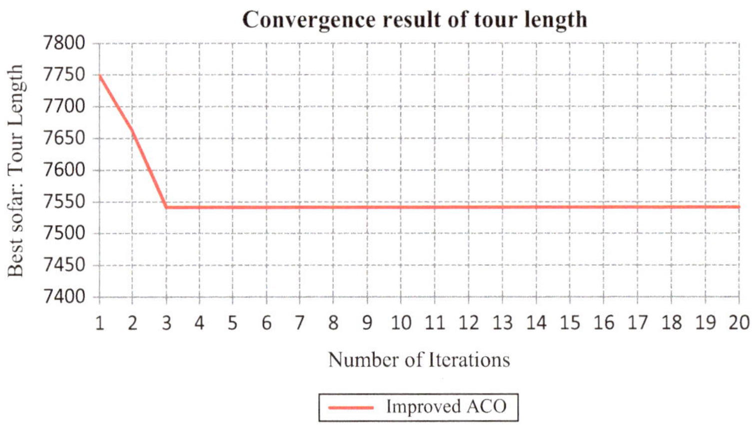

(b) berlin52

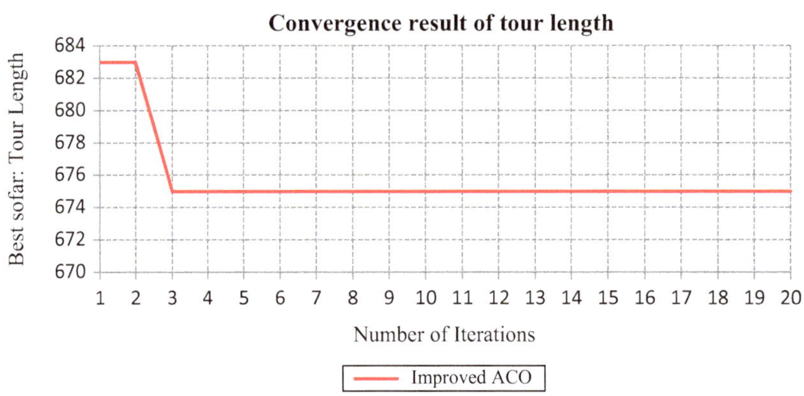

(c) st70

Figure 5.10. Convergence speed of tour length for (a) eil51, (b) berlin52 and (c) st70 TSPs.

5.12 Comparison experiment

For the purpose of demonstrating the efficiency of the improved ACO algorithm proposed in this system, we have constructed a simulation and applied it to problems from the TSPLIB library: oliver30, eil51, eil76, eil101, berlin52, st70, rat99, kroA100, lin105 and pr144. In this study the proposed algorithm results are compared to the results of the ACS algorithm in the aspects of algorithm convergence and experiment results. The ACS algorithm combines with local search.

In tables 5.3–5.12, the first column is the best distance from the beginning of the trial (run), as compared to the benchmark distance. The second column shows the

Table 5.3. Comparison results of oliver30.

oliver30 (20 runs, 20 iterations/run)		Best solution	Average solution	Relative error	Standard deviation
ACO	$\beta = 2$	420	420	0%	0
	$\beta = 3$	420	420	0%	0
	$\beta = 4$	420	420	0%	0
	$\beta = 5$	420	420	0%	0
IACO algorithm		420	420	0%	0
Optimum			420		

Table 5.4. Comparison results of eil51.

eil51 (20 runs, 20 iterations/run)		Best solution	Average solution	Relative error	Standard deviation
ACO	$\beta = 2$	427	429.05	0.23%	1.62
	$\beta = 3$	427	429.3	0.23%	1.63
	$\beta = 4$	427	429.05	0.23%	1.63
	$\beta = 5$	427	428.21	0.23%	0.92
IACO algorithm		426	426.6	0%	0.68
Optimum			426		

Table 5.5. Comparison results of eil76.

eil76 (20 runs, 20 iterations/run)		Best solution	Average solution	Relative error	Standard deviation
ACO	$\beta = 2$	541	546.8	0.56%	3.53
	$\beta = 3$	538	546.1	0%	3.46
	$\beta = 4$	541	545.75	0.56%	4.27
	$\beta = 5$	538	543.95	0%	2.14
IACO algorithm		538	539.9	0%	1.33
Optimum			538		

Table 5.6. Comparison results of eil101.

eil101 (20 runs, 20 iterations/run)		Best solution	Average solution	Relative error	Standard deviation
ACO	$\beta = 2$	637	647.35	1.27%	3.47
	$\beta = 3$	636	644.9	1.11%	3.95
	$\beta = 4$	638	644.45	1.43%	3.66
	$\beta = 5$	639	643.05	1.59%	3.10
IACO algorithm		629	635.55	0%	3.03
Optimum			629		

Table 5.7. Comparison results of berlin52.

berlin52 (20 runs, 20 iterations/run)		Best solution	Average solution	Relative error	Standard deviation
ACO	$\beta = 2$	7542	7561.85	0%	53.11
	$\beta = 3$	7542	7581.25	0%	54.95
	$\beta = 4$	7542	7599.65	0%	66.56
	$\beta = 5$	7542	7556.2	0%	36.51
IACO algorithm		7542	7542	0%	0
Optimum			7542		

Table 5.8. Comparison results of kroA100.

kroA100 (20 runs, 20 iterations/run)		Best solution	Average solution	Relative error	Standard deviation
ACO	$\beta = 2$	21 282	21 420.85	0%	90.91
	$\beta = 3$	21 282	21 383.75	0%	82.06
	$\beta = 4$	21 292	21 375.75	0%	59.75
	$\beta = 5$	21 282	21 353.15	0%	63.03
IACO algorithm		21 282	21 297.95	0%	24.31
Optimum			21 282		

average distance of the trials. The third column shows the relative errors, also called degree of approximation ((best solution-optimum)/optimum) which is how much the deviation of the optimal distance from the best solution, and the fourth column is the standard deviation.

In all cases, the proposed algorithm shows better performance than the ACS algorithm. The experiment shows that the improved ant colony algorithm proposed

Table 5.9. Comparison results of lin105.

lin105 (20 runs, 20 iterations/ run)		Best solution	Average solution	Relative error	Standard deviation
ACO	$\beta = 2$	14 397	14 439.35	0%	41.71
	$\beta = 3$	14 397	14 416.3	0%	32.51
	$\beta = 4$	14 397	14 417.85	0%	47.83
	$\beta = 5$	14 397	14 404.65	0%	33.24
IACO algorithm		14 397	14 391.35	0%	16.72
Optimum		14 397			

Table 5.10. Comparison results of pr144.

pr144 (20 runs, 20 iterations/run)		Best solution	Average solution	Relative error	Standard deviation
ACO	$\beta = 2$	58 537	58 583.5	0%	35.46
	$\beta = 3$	58 537	58 592	0%	51.48
	$\beta = 4$	58 537	58 589.65	0%	54.33
	$\beta = 5$	58 537	58 571.4	0%	27.03
IACO algorithm		58 537	58 537	0%	0
Optimum		58 537			

Table 5.11. Comparison results of rat99.

rat99 (20 runs, 20 iterations/run)		Best solution	Average solution	Relative error	Standard deviation
ACO	$\beta = 2$	1217	1233.2	0.5%	8.02
	$\beta = 3$	1219	1234.85	0.66%	10.07
	$\beta = 4$	1215	1232.45	0.33%	9.33
	$\beta = 5$	1213	1228.5	0.17%	10.14
IACO algorithm		1211	1216.4	0%	4.87
Optimum		1211			

in this approach achieved better results for TSPs, and its efficiency of solutions is better than that of ant colony algorithm.

Looking at the above tables the deviation (relative error) from the optimal solutions can be seen, and the differences in the deviations between the IACO algorithm and the ACS are compared. Moreover, the average distance of the IACO

Table 5.12. Comparison results of st70.

st70 (20 runs, 20 iterations/run)		Best solution	Average solution	Relative error	Standard deviation
ACO	$\beta = 2$	676	680.7	0.15%	2.83
	$\beta = 3$	675	680.2	0%	3.17
	$\beta = 4$	675	680.5	0%	2.55
	$\beta = 5$	675	680	0%	2.55
IACO algorithm		675	676.8	0%	1.91
Optimum			675		

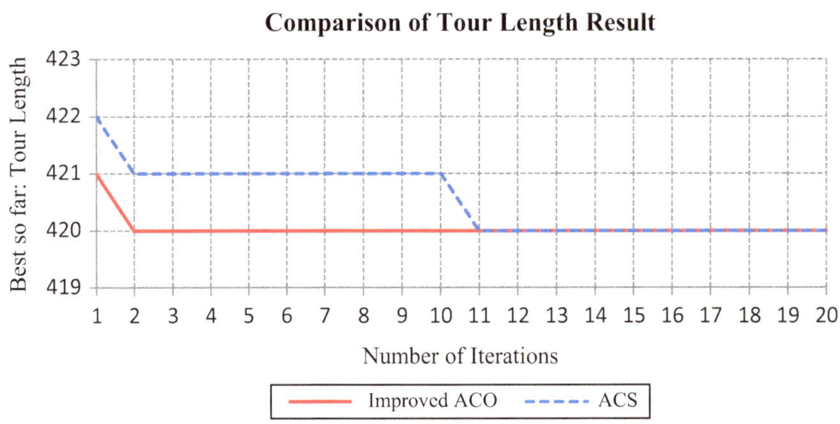

Figure 5.11. Comparison of the tour length result of oliver30.

and ACS are more obvious. Figures 5.11–5.20 present the comparison of convergence results of the tour length to obtain the best results for oliver30, eil51,eil76, eil101, berlin52, kroA100, lin105, pr144, rat99 and st70, respectively. The tested TSP datasets were executed for 20 trials, with the same number of iterations for each dataset. The best values obtained from a trial are given below for each dataset as in the figures. These figures show the best solution found since the start of the algorithm run, or the iteration number at which the best solution was found. The distance which is illustrated in the graph is the best distance found so far in the iterations.

In figure 5.21 the performance and optimal solution can be seen for IACO and ACS with the index 100 (also called degree of accuracy). The IACO algorithm obtains better results than the ACS and all of the results are strikingly close to the optimal solution.

The analysis results for the larger TSP instances can be seen in table 5.13. These problems were tested with 20 iterations for 30 trials. The number of ants used for all these test cases is 10. The parameters are as follows $\alpha = 1$, β is a dynamic value of the algorithm and the pheromone decay parameter ρ is 0.1. However, $\beta = 5$ is used for the ACS and other parameters are used as the proposed algorithm. Both of these

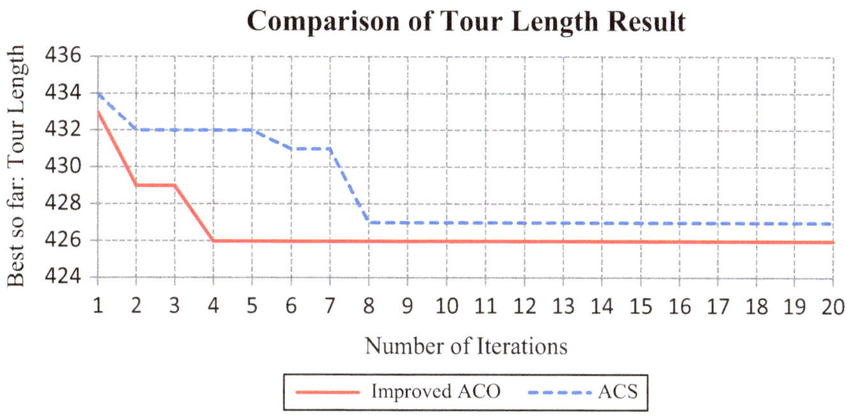

Figure 5.12. Comparison of the tour length result of eil51.

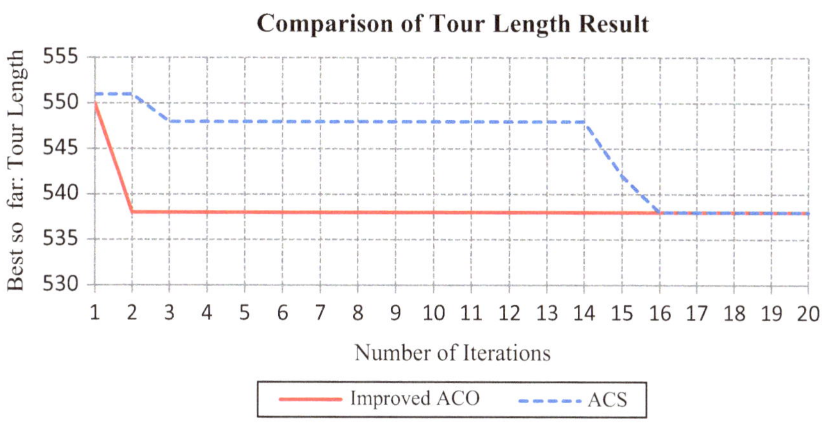

Figure 5.13. Comparison of the tour length result of eil76.

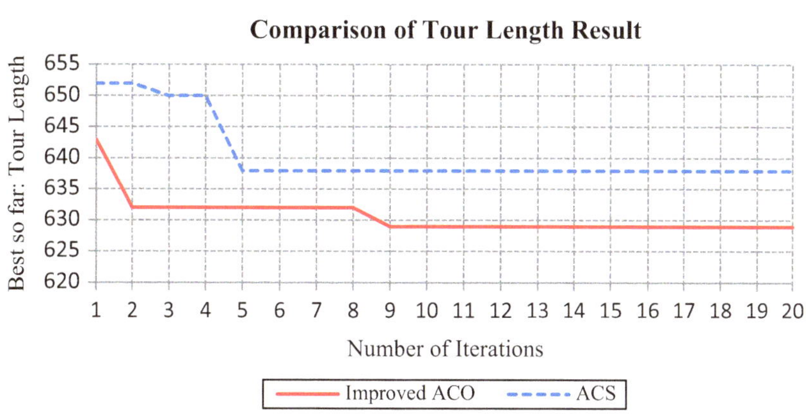

Figure 5.14. Comparison of the tour length result of eil101.

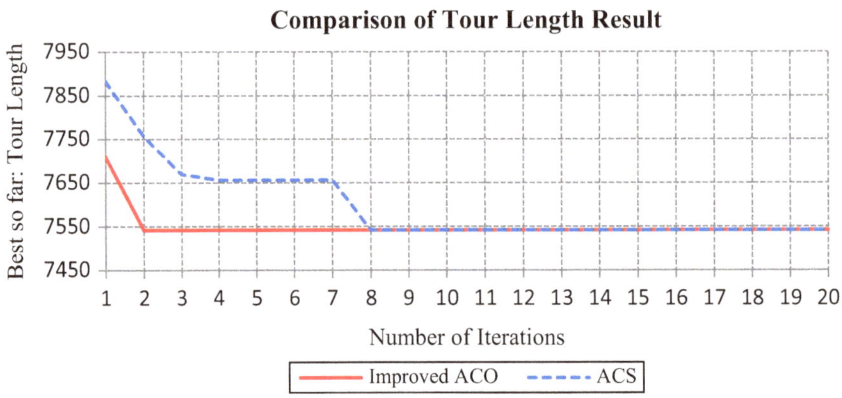

Figure 5.15. Comparison of the tour length result of berlin52.

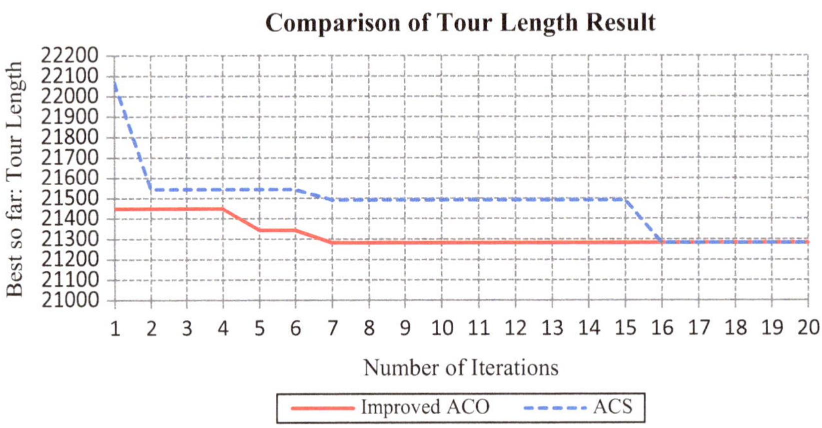

Figure 5.16. Comparison of the tour length result of kroA100.

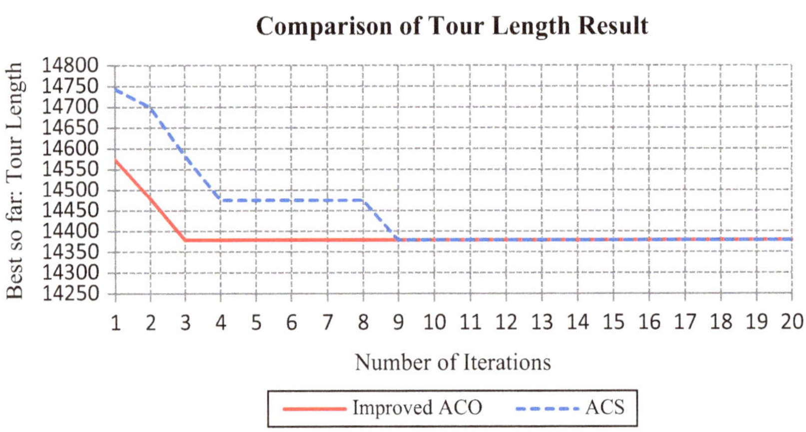

Figure 5.17. Comparison of the tour length result of lin105.

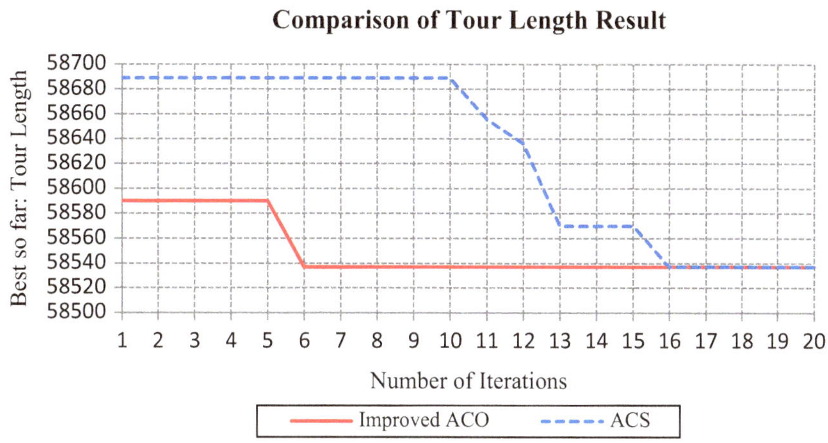

Figure 5.18. Comparison of the tour length result of pr144.

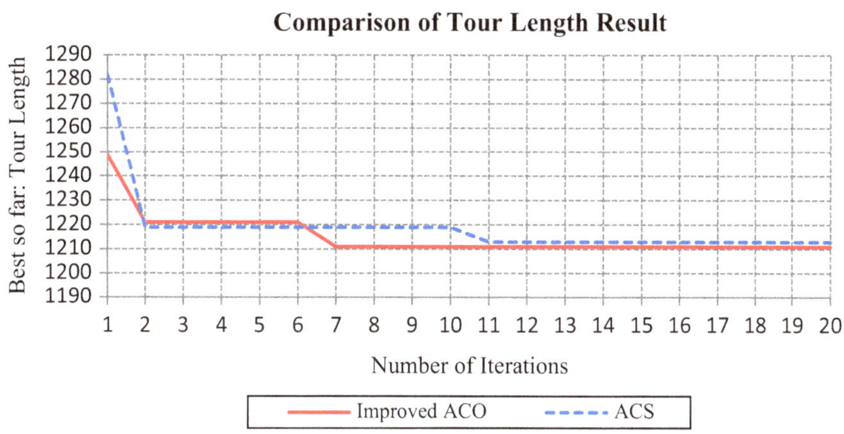

Figure 5.19. Comparison of the tour length result of rat99.

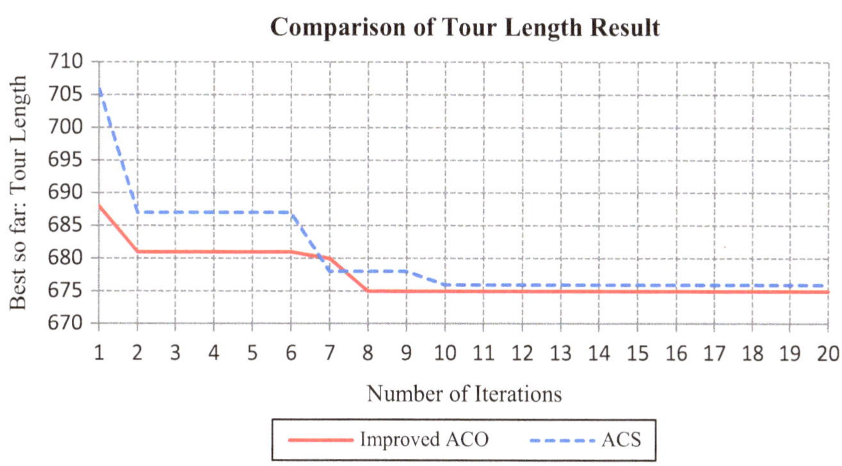

Figure 5.20. Comparison of the tour length result of st70.

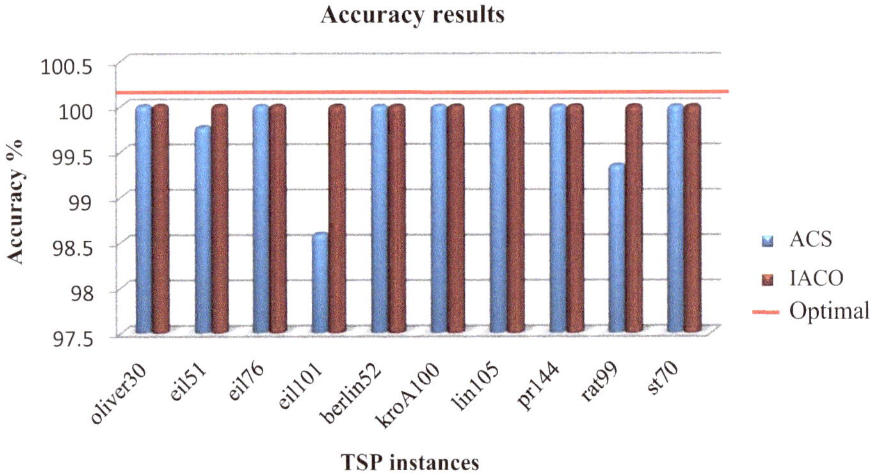

Figure 5.21. The accuracy of IACO and ACO performance compared to the optimal solution.

algorithms tested all the problems with local search optimization (2-Opt). The results were shown as the best length, average tour length and relative errors of the algorithms. Most results of the IACO algorithm achieved the optimal solutions. Therefore, the results of the IACO algorithm are more satisfactory than the results of the ACS algorithm. Furthermore, the results of both algorithms are compared to the optimal solution with the degree of accuracy. In figure 5.22 it can be seen that most of the proposed algorithm results achieve the optimal solution and some results are less deviated from optimum. The log graphs are also presented.

5.13 Analysis on varying number of ants

The section presents the analysis results when the number of ants is varying. The analysis on the number of ants consists of two parts: analysis of the number of ants starting at different cities versus the same city, and analysis of an increasing number of ants versus an increasing number of iterations.

5.13.1 Analysis of ants starting at different cities versus the same city

If the TSP finds a good solution with a good tour length, then the algorithm is very effective when the ants start from different cities. Generally, cities are similar or identical in implementations of the TSP, and while maturity of the algorithm is obtained and at the beginning of the tour, ants can freely select any appropriate city for the next movement. So, ants may also use the cities with short distance and when they reach the end of the tour almost all cities might have been covered. Once the cities with small edges or distance are covered, the ants will also be required to reach the next cities with some bigger edges. The method of selection for a starting city is random in this work and the search space can therefore take multidirectional space. The possibility of obtaining better results along with handling short and long edges is greater.

Table 5.13. Tour length results and relative errors (deviation) on several TSP instances.

		IACO			ACS		
TSP problems	Optimum (1)	Best length (2)	Average tour length	Relative errors (deviation) ((2)−(1))/(1)	Best length (3)	Average tour length	Relative errors (deviation) ((3)−(1))/(1)
ch130	6110	6123	6175.3	0.21%	6144	6200.6	0.56%
ch150	6528	6528	6573.9	0%	6548	6600.83	0.31%
d198	15 780	15 815	15 894.03	0.22%	15 900	15 994.77	0.76%
kroB100	22 141	22 141	22 183.33	0%	22 146	22 278.5	0.02%
kroC100	20 749	20 749	20 789.6	0%	20 753	20 906.1	0.02%
kroD100	21 294	21 294	21 379.6	0%	21 309	21 584.7	0.07%
kroE100	22 068	22 068	22 135.77	0%	22 116	22 284.67	0.22%
kroA150	26 524	26 524	26 806.5	0%	26 820	27 189.47	1.08%
pr76	108 159	108 159	108 303.5	0%	108 304	108 723.4	0.13%
pr124	59 030	59 030	59 110.67	0%	59 076	59 228.23	0.08%
pr152	73 682	73 682	73 772.73	0%	73 818	74 243.67	0.18%
pr226	80 369	80 377	80 628.16	0.01%	80 524	80 959.67	0.19%
rat195	2323	2339	2356.17	0.69%	2352	2379.27	1.25%

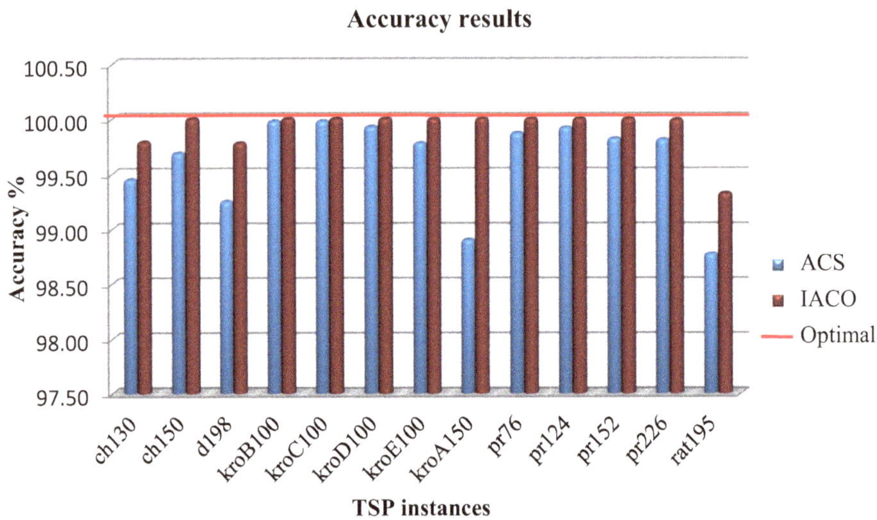

Figure 5.22. Accuracy of IACO and ACO performance compared to the optimal solution.

Table 5.14. The effect of the same starting city and random starting city.

		Same starting city			Random starting city		
TSP problem	Optimal (1)	Best length (2)	Average length	Relative error((2)−(1))/(1)	Best length (3)	Average length	Relative error((3)−(1))/(1)
eil51	426	430	441.6	0.94%	428	438.5	0.47%

The effects when all ants start from same city randomly can be seen in table 5.14. The eil51 (51-city) problem has been tested and accordingly we found the best possible path using 100 iterations in 20 trials. For this data, the best result was obtained when ants started from same city randomly. In the case of the best optimal solution, the distance is greater than the optimal distance by 2.0. The results for distances greater than 4.0 are also highlighted.

5.13.2 Analysis on an increasing number of ants versus number of iterations

One way to analyze the solution of tour length is using more iterations or more ants. This case was tested using the eil51 TSP instance without using local search with 20 trials. As seen in the figure, the results seem vary random between the two extreme values found in figure 5.23 (428 and 455) and in figure 5.24 (436 and 466). In figures 5.23 and 5.25, the former figure shows the result of testing with 10 ants in each iteration and the latter figure presents the testing result of using 100 ants in each iteration. It can also be seen that the tour length value is better when varying the number of iterations than the number of ants.

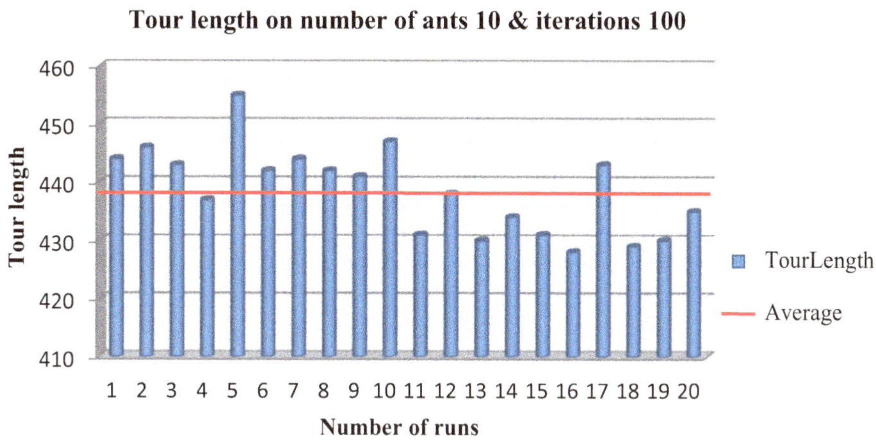

Figure 5.23. Tour length result for the eil51 instance with 10 ants and 100 iterations.

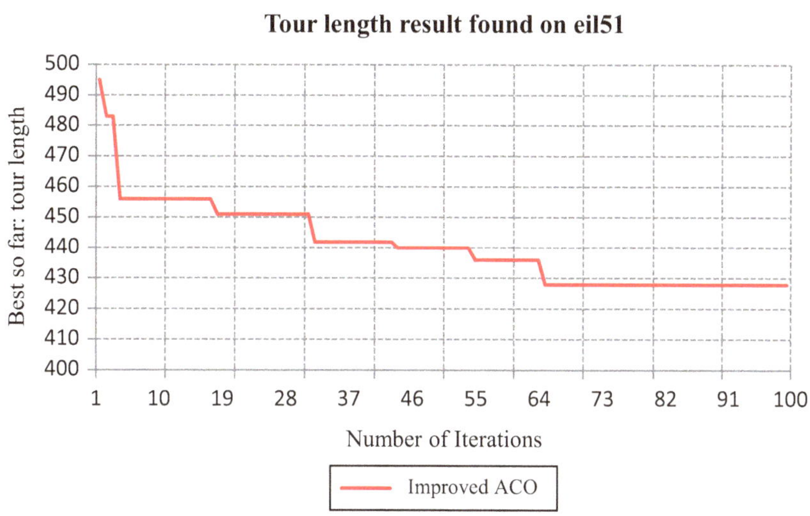

Figure 5.24. Tour length distance achieved for eil51 using 10 ants and 100 iterations.

The algorithm was run for 100 iterations, where each iteration used 10 ants and the best solution was found after 60 iterations. The best value was quite close to the optimum value (figure 5.26).

Summarized results for the eil51 TSP instance:

Best distance achieved:	428
Optimum solution:	426
Iteration which achieved best distance:	65

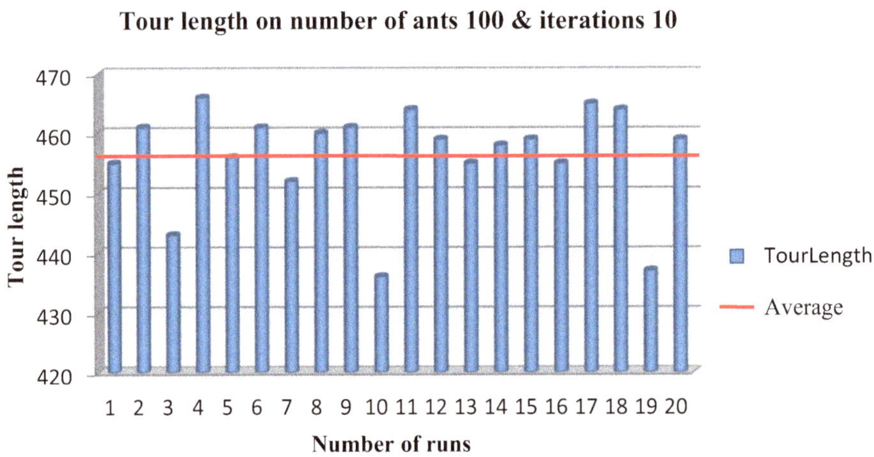

Figure 5.25. Tour length result on the eil51 instance with 100 ants and 10 iterations.

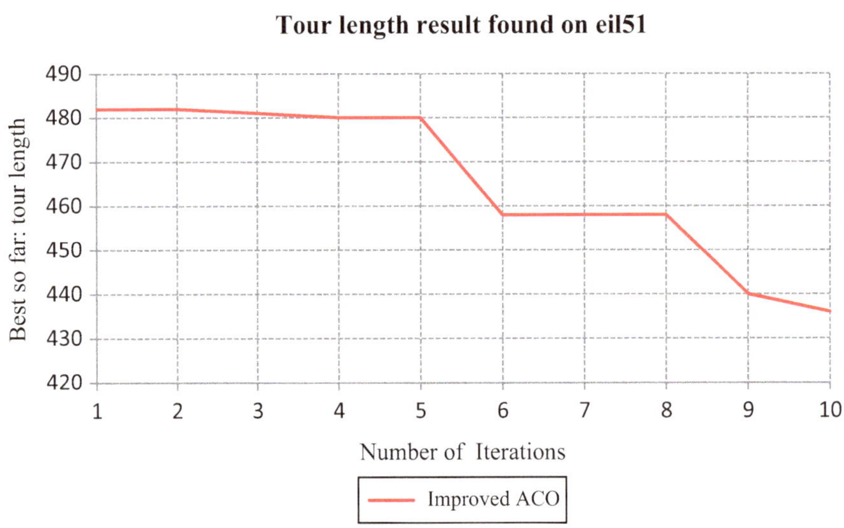

Figure 5.26. Tour length distance achieved for eil51 using 100 ants and 10 iterations.

The algorithm was run for 10 iterations, where each iteration used 100 ants and the best solution was found at 10 iterations. The best value was not far from the optimum value.

Summarized results for the eil51 TSP instance:

Best distance achieved:	436
Optimum solution:	426
Iteration which achieved best distance:	10

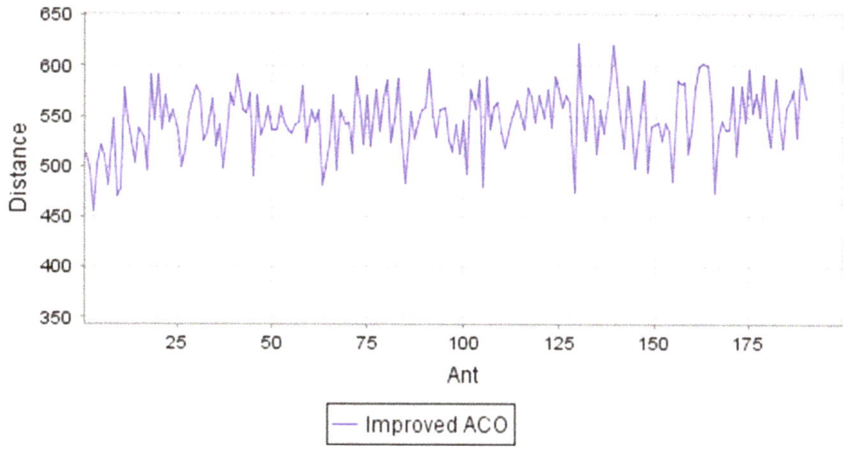

Figure 5.27. Tour length results for an increasing number of ants.

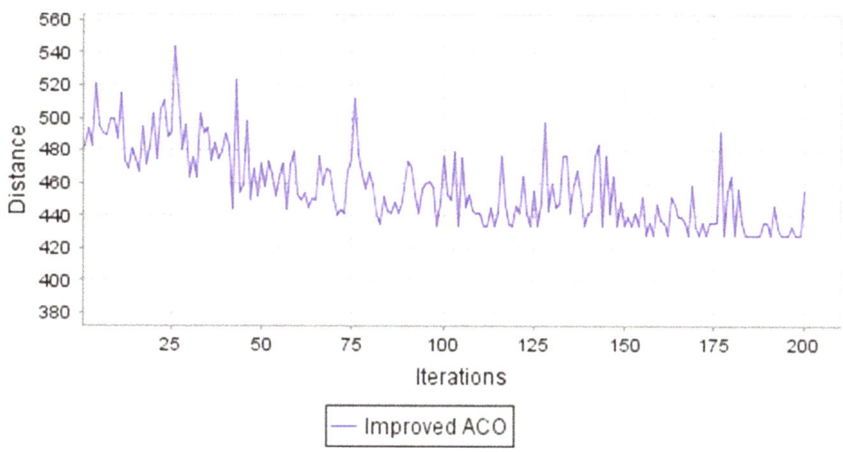

Figure 5.28. Tour length results for an increasing number of iterations.

It can also be seen that the solution result is better when increasing the number of iterations than the number of ants.

Another alternative way to analyze the varying number of ants and iterations is also presented in the following figures, which suggest that the chance of obtaining a good result would be higher. Figure 5.27 shows the analysis result for an increasing number of ants throughout the iterations (for example, the first iteration uses 1 ant, the second iterations use 21 ants, 41 ants, 61 ants and so on and the last iteration uses 181 ants). This test iterates 10 times. Figure 5.28 shows the result of increasing the number of iterations to 200 iterations. The individual iteration uses 10 ants. The average length of the increasing number of ants is 545.96 and that of the increasing number of iterations is 457.89. It can be seen that increasing the number of iterations obtain better quality results than increasing the amount of ants.

5.14 IACO comparison results

We discuss here the analysis of results obtained for the suggested algorithm, IACO, both using CL and without it. Three TSP instances were tested, i.e. oliver30, eil51 and berlin52, which are highlighted in table 5.15 in four different columns. The distance was found for 500 iterations among 10 trials in the first column, in comparison to the optimal distance, which is considered as a benchmark distance. The average length is reported in the second column and the best number of iterations is in the third column. The average time is taken as 10 trials in the fourth column where 500 iterations are there in each run or trial. Local search optimization is used and convergence was found to be satisfactory (table 5.15).

Table 5.15. IACO comparison results.

		IACO with candidate list				IACO without candidate list			
TSP problem	Optimal	Best length	Average length	Best iteration	Average time	Best length	Average length	Best iteration	Average time
oliver30	420	420	420.5	69	5.66	420	423.3	356	7.30
eil51	426	426	430.8	382	8.61	431	444.3	475	12.87
berlin52	7542	7542	7635.2	235	9.33	7542	7678.6	302	13.86

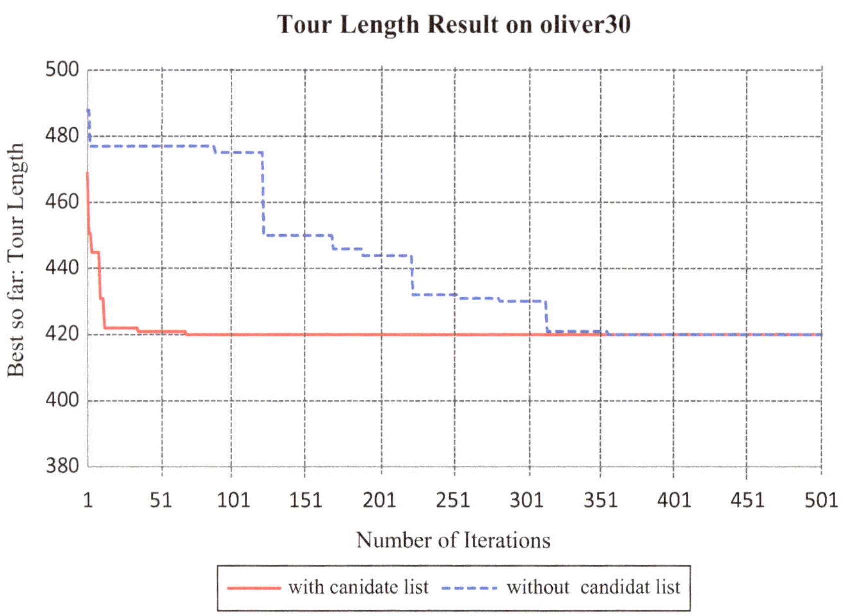

Figure 5.29. Comparison of the tour length result of IACO with the candidate list and without the candidate list for oliver30.

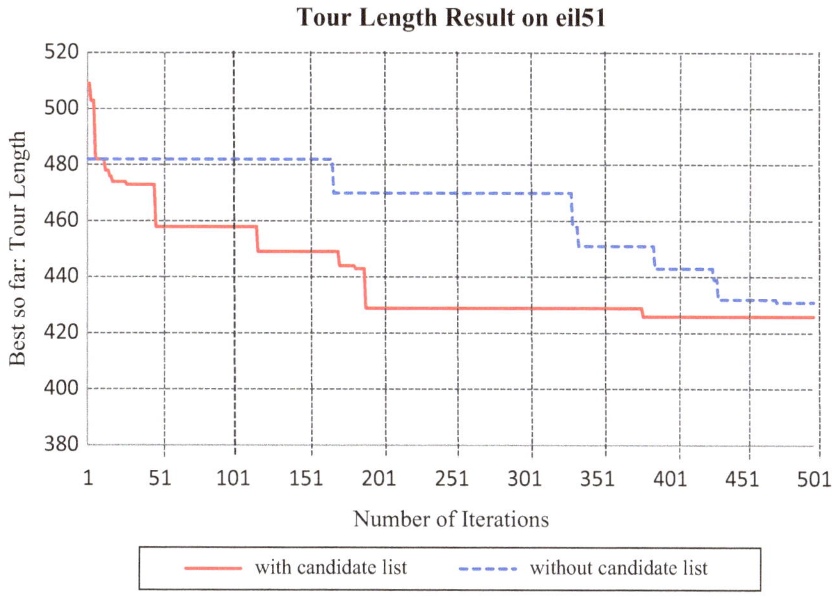

Figure 5.30. Comparison of the tour length result of IACO with the candidate list and without the candidate list for eil51.

Figures 5.29–5.31 show the results with CL and without CL, respectively. We have achieved considerable improvements in our searching ability, with the number of iterations reduced and less time taken.

5.15 Conclusions

Research studies suggest that ACO is a modern and emerging research field in the area of optimization and is used for a number of engineering problems, in particular in the age of artificial life and operations research. One of the most important applications is swarm intelligence, used in applications related to AI. The simulations using ant colonies and exploring their behavior foraging for the food resulted in the birth of ACO whose principal element is pheromone information. This chapter has discussed the basics of ACO, establishing a metaheuristic structure which provides a number of options for implementation in the design of algorithms. Various alternative approaches are explained briefly, highlighting successful algorithms such as MMAS and ACS. ACO algorithms are an emerging field and are considered as state-of-the-art algorithms in solving combinatorial problems of optimization.

One major contribution of the chapter is suggesting IACO as an improved ant colony optimization algorithm, that is designed for addressing combinatorial optimization problems such as TSPs. The search space, dynamic updates, computation time and best possible tours have been addressed well with a substantial amount of results and discussions. The heuristic parameter and selection of the values have been discussed in terms of how the different values affect the

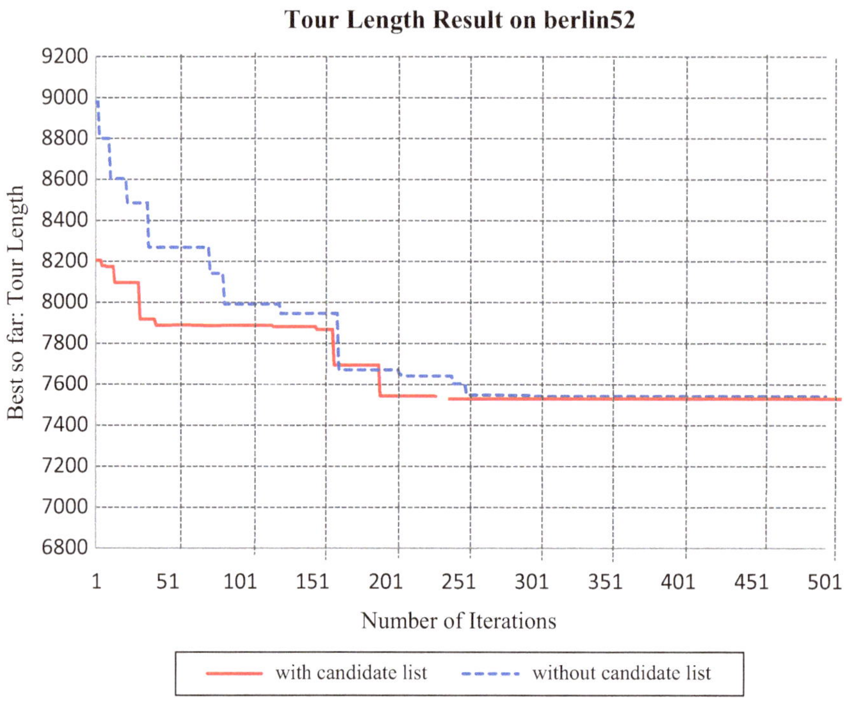

Figure 5.31. Comparison of the tour length result of IACO with the candidate list and without the candidate list for berlin52.

performance of the system. The suggested approach proved to produce much better results compared to a single approach which is used to solve TSPs. Optimal results are obtained and some solutions to the problems were found to be optimal. An interesting finding regarding the time taken was made that suggests the time required in finding the solution can be optimized if flexible and dynamic updates are properly applied.

References

[1] Colorni A, Dorigo M and Maniezzo V 1991 Distributed optimization by ant colonies *Proc. of Ecal91—Eurpoean Conf. on Artificial Life* (Paris, Amsterdam: Elsevier) pp 134–42

[2] Colorni A, Dorigo M and Maniezzo V 1992 An investigation of some properties of an ant algorithm *Proc. of the Parallel Problem Solving from Nature Conf. (PPSN 92)* (Brussels, Amsterdam: Elsevier) pp 509–20

[3] Shannon C E 1948 A mathematical theory of communication *Bell Syst. Tech. J.* **27** 379–423

[4] Pintea C-M and Dumitrescu D 2005 Improving ant system using a local updating rule *Proc. of the Seventh Int. Symp. and Numeric Algorithms for Scientific Computing (SYNASC'05)* (Piscataway, NJ: IEEE)

[5] Wang C-X, Cui D-W, Zhang Y-K and Wang Z-R 2006 A novel ant colony system based on Delauney triangulation and self-adaptive mutation for TSP *Int. J. Inform. Technol.* **12** 89–99

[6] Hung K S, Su S F and Lee S J 2007 Improving ant colony optimization for solving traveling salesman problem *J. Adv. Comput. Intell. Intell. Inform.* **11** 433–42
[7] Gambardella L M and Dorigo M 1995 Ant-Q: a reinforcement learning approach to the traveling salesman problem *Proc. of the Twelfth Int. Conf. on Machine Learning* (San Francisco, CA: Morgan Kaufmann) pp 252–60
[8] Dorigo M, Maniezzo V and Colorni A 1996 The ant system: optimization by a colony of cooperating agents *IEEE Trans. Syst. Man Cyber.* B **26** 29–41
[9] Dorigo M and Gambardella L M 1997 Ant colony system: a cooperative learning approach to the traveling salesman problem *IEEE Trans. Evolut. Comput.* **1** 1–24
[10] Dorigo M and Gamgardella L M 1997 Ant colonies for the traveling salesman problem *BioSystems* **43** 73–81
[11] Dorigo M, Birattari M and Stuzle T 2006 Ant colony optimization—artificial ants as a computational intelligence technique *IEEE Comput. Intell. Mag.* **1** 28–39
[12] Dorigo M and Stützle T 2004 Ant colony optimization (Cambridge, MA: MIT Press)
[13] Dorigo M and Stützle T 2002 The ant colony optimization metaheuristic: algorithms, applications, and advances *Handbook of Metaheuristics* ed F Glover and G Kochenberger (Amsterdam: Kluwer)
[14] Randall M and Montgomery J 2002 Candidate set strategies for ant colony optimisation *Ant Algorithms, 3rd International Workshop, ANTS 2002, Proceedings* ed M Dorigo, G di Caro and M Sampels (Lecture Notes in Computer Science (including subseries Lecture Notes in Artificial Intelligence and Lecture Notes in Bioinformatics) vol 2463) (London: Springer), pp 243–49
[15] Stutzle T and Hoos H H 1997 MAX–MIN ant system and local search for the traveling salesman problem *Proc. of the 1997 IEEE Int. Conf. on Evolutionary Computation (ICEC'97)* (Piscataway, NJ: IEEE), pp 309–14
[16] TSPLIB 2005 *TSPLIB: Library of Sample Instances for the TSP* University of Heilderberg, Department of Computer Science http://iwr.uniheidelberg.de/groups/comopt/software/TSPLIB95/tsp/
[17] Hlaing Z C S S and Khine M A 2011 An ant colony optimization for solving traveling salesman problem *Int. Conf. on Information Communication and Management (ICICM), IPCSIT* **vol 16** (Singapore: IACSIT)
[18] Hlaing Z C S S and Khine M A 2011 Solving traveling salesman problem by using improved ant colony optimization algorithm *Int. J. Inform. Educ. Technol.* **1** 404–49

Chapter 6

Application of a particle swarm optimization technique in a motor imagery classification problem

Rahul Kumar and Mridu Sahu

In many real-world problems one needs to find the best solution from all feasible solutions to solve any particular problem. Finding the best among all solutions is a basic theme of optimization. Minimization of the time and space to solve real problems is still a challenging task in many areas such as the biomedical, behavior and prediction sciences, etc. Optimization establishes relationships among problem objectives, predefined constraints on these and the targeted variables. Optimization techniques have been widely employed by researchers to improve the performance of computers in many cognitive detections. This chapter presents the application of particle swarm optimization (PSO) in motor imagery (MI). The dynamic state of the brain is simulated with the action of different body parts in MI. The brain–computer interface (BCI) provides a conduit between the brain and computer, and it performs classification of MI action, which is very helpful for those people who are paralyzed due to high-level spinal cord injury and are not able to perform any muscular activities. The quality of signals and their performance on classifiers are still crucial challenges in MI classification. For classification various steps are required, such as preprocessing, feature extraction, selection, etc. The presented chapter includes various feature extraction techniques and selections for the classification of MI. For the recording of brain activity electroencephalography (EEG) is normally used. This is the non-invasive mode of recording brain signals. The brain signal from the recording is noisy and complex data, and several preprocessing steps are involved to improve the quality of signals. The transformation of the signal from one domain to another is called signal transformation, which has been done in the current chapter through wavelet transformation; this also helps in extracting features from these

doi:10.1088/978-0-7503-2404-5ch6

signals. EEG signal nonlinearity is a serious problem in finding a solution for the detection of many diseases and disorders. Fuzzy logic is a powerful tool for handling nonlinearity. This chapter focuses on the classification of EEG signal using an adaptive neuro-fuzzy inference system (ANFIS). The discrete wavelet transform (DWT) has been used here for useful feature extraction.

Also, in this chapter, PSO has been applied for optimizing the network parameters of ANFIS. In standard ANFIS all parameters are updated using a gradient descent algorithm. The problem present in gradient descent is that when the search space is large, the complexity of gradient computations is high. The PSO method is inspired by the biological nature of bird flocking and fish schooling, which is applied here to tune the parameters of the network. Normally PSO finds an iterative candidate solution for the specified problem. It is a metaheuristic, which provides a set of rules that is high level and independent of the problem. PSO can optimize the functional module, tune the neuro and fuzzy systems as well as modify the rule set generated from various systems. In the presented study, PSO minimizes the mean square error (MSE) by tuning the parameters of membership functions of the fuzzy inference system (FIS) and increases the performance for classification of right hand and foot movement detections. PSO possesses a central advantage on the implementation side as it is easy and requires very few parameters. Two models have been proposed: one is based on PSO and the other is gradient descent for comparative analysis. The results confirm that the PSO based model gives better accuracy than the others.

6.1 Introduction

Optimization plays an important role in our day-to-day lives. In many disciplines, such as the scientific, social, economics and engineering fields, optimization is used to find a desirable solution to problems by adjusting their parameters. The main aim of optimization is finding the best solution among all feasible solutions that are available to us. A feasible solution satisfies all constraints in optimization problems. Currently the problems in optimization are multi-objective and multidisciplinary. To solve that kind of complex problem, not only is gradient descent based optimization used, but also the evolutionary algorithms such as genetic algorithms, and particle swarm optimization (PSO) are employed. Over the years, many methods have been developed for optimization. The PSO method is a popular method of optimization which is inspired by the social behavior of birds and fish. PSO is used in many fields, such as the medical field, engineering, social economics, etc. In medicine, biomedics is one of the important fields which focuses on the study of the brain and its behavior (how the brain works in different environments). In this field, the design of a stable and reliable brain–computer interface (BCI), which provides the basic connection between a brain and computer, is a major challenge. To design a BCI, various problems need to be overcome. To solve these problems PSO can be used. Channel selection is an important step in recording brain signals and also a major challenge in recording brain signals. PSO is widely used for selecting the minimum number of channels. During the recording of signals, a lot of noise and artifacts are generated that affect the performance of the model. To remove those artifacts and noise, PSO based filters are used. Feature extraction methods are used to obtain the best possible characteristics from brain signal for dimensionality reduction. Feature selection is a method which is applied after extraction to filter out irrelevant and redundant features. PSO based feature selection is a widely used technique to reduce the dimensionality.

Human motor activity is described by notable adaptability; a human can perform a number of undertakings, for example, walk forward and backward, run, dance, shuffle and produce a wide range of activities. We appear to have the option to produce a practically limitless stream of movements to achieve objectives in the surrounding environment [1]. The motor system is based on brain activity. However, due to spinal cord injury and other neurological diseases the motor ability can be damaged [2]. People may be paralyzed and they are not able to perform any muscular activity. To restore the damaged motor function, a concept called MI, which is fundamentally a mental task in which patients perform the motor task in their mind without performing any physical task. This concept is used with a brain–computer interface (BCI), which provides a connection between the brain and external environment (figure 6.1) [3]. An MI based BCI performs the classification of MI action such as movement of the hands and feet and other motor related action. It provides one-way communication for a patient who is suffering from a motor disability. BCI is categorized into two types: invasive and non-invasive. In invasive BCI, electrodes are placed in the head using cranial surgery in order to compute brain activity. This method provides a higher signal-to-noise ratio (SNR), but its downsides are a higher infection rate and other side effects. To counter the negative effects of the above method, non-invasive BCI was introduced, but it has a lower SNR. This calls for the development of more precise, non-invasive BCI in order to accurately measure brain activity.

In the BCI, people produce a different brain activity pattern by performing the MI action and the pattern is identified and converted into a command and control

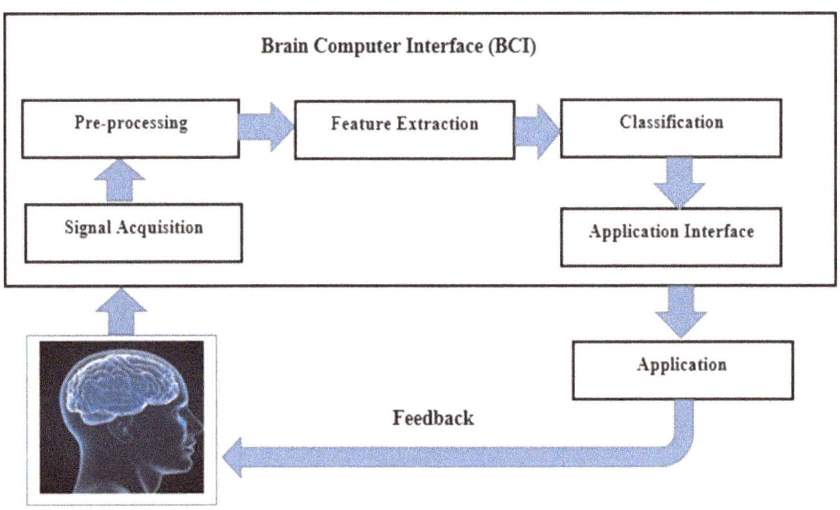

Figure 6.1. Brain–computer interface system.

signal by the system. Identification depends on the classification algorithms. The performance of a BCI depends on the features and classification algorithm. To select the most suitable classifier for designing the BCI, it is important to understand which types of features we have to use. Various features are extracted in order to design the BCI, such as band power (BP) [4], power spectral density value [5, 6], amplitude values of EEG signals, autoregressive (AR) and adaptive autoregressive parameters [7], time-frequency features and inverse model-based features [8].

In the BCI classifier there are two basic problems: first, the curse of dimensionality and, second, the bias–variance trade-off [3]. Solving these problems is the main challenges for classification algorithms. This chapter focuses on the classification of EEG signal using ANFIS and DWT, used here for the extraction of useful features.

Also, in this chapter, PSO has been applied for optimizing the network parameters of ANFIS. In standard ANFIS all parameters are updated using a gradient descent algorithm. The problem present in gradient descent is that when the search space is large, the complexity of gradient computation is high. The PSO method is inspired by the biological nature of bird flocking and fish schooling, which is applied here to tune the parameter of the network. Normally PSO finds an iterative candidate solution for a specified problem. It is a metaheuristic, which provides a set of rules that is high level and independent from the problem. PSO is also able to optimize the functional module, tune the neuro and fuzzy systems, as well as modify the rule set generated from various systems.

6.1.1 Literature review

Previously, a lot of work has been done on MI and BCI application. Channel selection for the classification of brain signals is crucial in designing a reliable and stable BCI. PSO plays an important role in channel selection for recording signals, filtering the signal with optimum parameters, and PSO is also used by some researchers to find an optimum number of features, which is more relevant and efficient in classification. A number of classification algorithms are used for EEG signals, such as support vector machines (SVMs), neural networks (NNs), Bayes' classifier, K-nearest neighbor classifier and linear discriminate analysis (LDA), etc. To improve the performance and efficiency of a classifier, one needs to find the optimal parameter for which the algorithm gives the best performance. Optimization methods such as the PSO method, genetic algorithms and others are used for finding these parameters.

BCI systems are generally not used as a real-time applications because of the low accuracy of pattern recognition algorithms. In particular when the channel number is very large, and with the increment of the channel. A poor arrangement of the channels with high computation complexity influences the accuracy of the classifier. To solve these problems different approaches have been proposed.

Wei and Wang [9] observed that when the number of channels is high, it is very tedious to perform analysis on multi-channel EEG signals. It requires the preparation of a suitable recording, and a tedious calculation which will take a lot of computation time. To solve this problem the authors proposed binary multi-objective particle swarm optimization (BMOPSO) for optimizing the number of channels. The principle objective of this work was to improve the performance of a BCI by reducing the number of channels by considering the problem of channel selection as a search problem. For the experiment they used a dataset which consists of the EEG signals of five healthy subjects. They included three different MI tasks: right hand, left hand and foot moment. The method successfully reduced the number of channels without affecting the accuracy. BMOPSO was not only used for minimizing the number of channels, but also used to maximize the mutual information. The authors found that, through reducing the number of channels, preparation of the recording was easier and the classification algorithm took less time to compute. The proposed method also has some disadvantages, such as a slow learning convergent velocity and, due to the adopting capability of the gradient method, converging to local minima cannot be avoided.

Hasan and Gan [10] describe the effective multi-objective PSO technique to resolve the channel selection problem for BCI. The authors describe two methods for channel selection: multi-objective particle swarm optimization and sequential floating forward search. In this paper the authors used dataset1 of BCI Competition IV. The dataset consisted of EEG signals for seven subjects, each subject having three MI tasks (right hand, left hand and foot). From each subject AR features were calculated and applied as input into an LDA. If the dimension of features was more than 20, then principal components analysis was applied for dimension reduction. In the end, the authors concluded that MOPSO selected a smaller number of channels. However, there are some channels that are selected by both methods as the most important channel for classification.

Lv and Liu [11] observed that when a common special pattern algorithm is used for extracting features from multi-channels, and if the number of channels is large, then the common spatial pattern (CSP) algorithm will address an overfitting problem that is inconvenient for clinical operation. On the other hand, most of the channels contain noisy and redundant data that affect the performance of the classifier. To solve that problem the CSP filters' discrimination and channel number were integrated in a single unit. They then used BPSO for selecting the group of channels. In the experiment the authors used the BCI2003 dataset IV and BCI2005 dataset I. The authors found that maximum accuracy was achieved with 9–14 channels.

Kumar *et al* [12] used a combination of PSO and rough set theory to calculate a minimum number of related features from large dimension extracted features. In order to perform classification on multiclass MI they used a neighborhood rough set classifier. The authors followed four phases of preprocessing and classification. In the first phase the BCI-MI dataset was processed, which was gathered from computerized devices. For this work, the MI datasets were taken from BCI Competition IV Dataset IIa. In the second phase, raw EEG signals were processed in order to improve the quality of signals; for this purpose filter and feature extraction methods were used. In the third step, feature selection was used to reduce the dimensionality of features from the set feature extracted from the wavelet transform method. In this step, the combination of PSO rough set theory was used to select the dominating features. The fourth step used k-fold cross-validation, in which the dataset is divided into two parts, one for training other data for testing and then performing the classification algorithm. The result demonstrates that the proposed technique improves precision more than other comparablee classification algorithms. By using this method the authors achieved significant improvements in terms of sensitivity and positive predictive value. Using the proposed method, the MI task classification accuracy reached up to 80.9848%.

Hsu [13] proposed an automatic artifact elimination method where feature selection is performed using a quantum-behaved PSO algorithm from a set of extracted features. Optimal features are selected through QPSO and these selected features are used as input into the SVM. The author proposed automatic artifact elimination to remove the EOG artifacts and improve the classification accuracy. At the same time a number of features such as spectral power, asymmetry ratio, MFFV and other features are calculated and then combined together. After that QPSO is used to select the features which enhance the classification accuracy.

Xu *et al* [14] used PSO based CSP method for feature extraction. Two parameters, frequency band and time interval, affect the performance of CSP. So, using PSO the authors found the optimal values for the parameters which improve the discriminative ability of CSP.

Filters are used to improve the quality of signals by removing the noise and irrelevant data. The performance of a filter depends on parameters such as the frequency band time interval and others. To improve the performance such parameters need to be optimized. For that, several researchers used PSO to optimize the parameters.

Ahirwal *et al* [15] studies an adaptive noise canceler for improvement of EEG signal filtering. Different versions of PSO were used to design the adaptive noise canceler and perform a comparative analysis on different parameters for a varied range of particles and inertia weights.

6.1.2 Motivation and requirements

BCIs provide communication for those people whose motor cells are damaged and are not able to perform any muscular activity. By using a BCI they can communicate with the outside environment and control devices such as wheelchairs, artificial hands and feet, etc. Designing a BCI is not an easy task, as there are several challenges such as the selection of channels for recording, the dimensionality and complexity of signals, and the performance of the classifier which is used for recognition of imagery action or brain pattern. The aims of this chapter are as follows:

1. To reduce the dimensionality of the data by applying the feature extraction method.
2. To design the ANFIS model for classification which is a combination of NN and fuzzy logic and is capable of handling nonlinearity of data.
3. To improve the performance of ANFIS; the parameters of ANFIS are tuned using PSO.

The remaining sections of this chapter are as follows: in section 6.2 the fundamentals of PSO are briefly explained and in section 6.3 the proposed model is described. Some results of experiments are presented in section 6.4 and we provide some remarks and conclusions in the last section.

6.2 Particle swarm optimization

PSO was designed by Kennedy and Eberhard in 1995 [16] inspired by the communal deportment of bird flocking and fish schooling. PSO is a stochastic optimization method that computes the global best solution using a number of search points called particles. In a PSO system, there is a group of particles, and all particles are moving around a multidimensional search space, and each particle shares its location with other particles and they update their position. Through communication, they reach the best position [17]. The direction of each particle is defined on the basis of a set of particles neighboring the particle and previous experience. The group of particles is organized according to some sort of communication structure or topology, such as the star topology, mesh-topology and wheel topology. There are different versions of PSO and each version is a modified version of a previous one with a high convergence property. The PSO was mainly created for real-valued problems, but most problems are based on discrete values, where the domain of a variable is finite. It includes problems such as integer programming, scheduling and routing. In 1997, Kennedy and Eberhart developed a discrete binary version of PSO for solving discrete optimization problems [18], where the location of each particle is represented by a binary number (0 or 1) and velocity is defined as the probability of change in the state of the particle. This algorithm has a better convergence rate compared to other versions of PSO. In PSO, most of the time it is trapped in the local minima. To solve this problem [19] presented a new version of PSO called guaranteed convergence PSO which includes extra particles that search the region around the current global best to find the final global best. In [20] the authors proposed a new algorithm personal best position PSO where velocity is updated with a new equation. The modification has been carried out through vanishing the global best term within the velocity update equation of the standard PSO. Many PSO versions are not able to solve multiple objective problems because of the way in which they exchange information in order to find the global solution. To solve that problem niche PSOs were introduced to solve complex problems with multi-objective functions; one version of PSO is explained in [21].

PSO looks like an evolutionary algorithm, and it has emerged that PSO is practically identical to the genetic algorithm [22, 23]. A considerable amount of exploration has demonstrated the viability of PSO in improving the persistent and discrete advancement issues. There are different applications of PSO in different fields such as NN training, electric power systems, telecommunications, control, data mining, design, combinatorial optimization, signal processing, and many others. However, when the objective function has a large dimension then most often PSO fails to find an optimal solution. A fall in local minima is the main issue in PSO and to handle this issue several modifications were proposed in the essential particle driven equation [24, 25], to keep the particle velocity or accelerate them using an adaptation technique or randomized method. These proposed modifications work well and have the ability to avoid the fall of local minima. However, PSO still does not guarantee finding a global solution with high dimension search space advantages.

6.2.1 The mathematical model of PSO

PSO is computational, effective and easier to implement compared to other mathematical algorithms and evolutionary algorithms [26]. PSO initializes with a population of a random solutions called particles. Each particle has an individual velocity. All the particles are moving around the search space with a velocity that is dynamically adjusted as per previous behavior. Every particle has a tendency to move toward the best solution in the search space. In PSO once problems are defined, particles are placed in the search space and each one evaluates the fitness value for a given function at its current location. After that particles update their current locations and best according to information shared by all other particles. The next iteration takes place when all particles have moved. With each iteration particles reach close to the optimal solution. In PSO the solution is associated with two vectors, position (X_i) and velocity (V_i). In D dimensionsl search space $X_i = [x_{i1}, x_{i2}, x_{i3}, x_{i4}, \ldots, x_{iD}]$ and $V_i = [v_{i1}, v_{i2}, v_{i3}, v_{i4}, \ldots, v_{iD}]$ are the vectors which are associated with each particle i. X_1 show the location of the first particle. It is necessary to define the movement direction and speed to update the position vector. Vector velocity (V_i) can be used for doing the same thing:

$$V_i^{t+1} = w\, V_i^t + c_1 r_1 \left(P_i^t - X_i^t\right) + c_2 r_2 \left(G^t - X_i^t\right) \tag{6.1}$$

$$X_i^{t+1} = X_i^t + V_i^{t+1}, \tag{6.2}$$

where

c_1 and c_2 are positive constraints,
r_1 and r_2 are two randomly generated numbers with a range [0, 1],
w is inertia weight,
P_i^t is the best position of particle i achieved based on its own experience,
G^t is the best particle position among all,
t is the iteration index.

The position vector represents the location of a particle in the search space and the velocity shows in which direction and at what speed (inertia of movement) the particles move. In each iteration, these two vectors are updated with the above two equations. The new velocity is calculated by the multiplication of the current velocity and a variable called inertia w, and has c_1 and c_2 positive constraints for the last two components in the velocity vector. The first term is called the inertia weight as it maintains the current velocity and moment direction. The second component is the cognitive component, it is also known as an individual component because it considers the particle best position and current position. The third component is a social component; here the particle calculates the distance between its current position and the best position for the whole swarm.

The cognitive and social component have a great impact on the moment of a particle and this impact can be changed by tuning the coefficient. The inertia weight tunes the exploration and exploitation and is normally decreased from 0.9 to 0.4 or 0.2. In real-word problem when searching for global optima, variation in the inertia parameter will balance the exploration and exploitation. To demonstrate the impact of the parameter in PSO, let us consider a simple objective function called a sphere (equation (6.3)). Calculate the average objective value for a different range of inertia weights ([0.9 0.2], [0.9 0.3], [0.9 0.5] [0.5 0.3]) with $c_1 = c_2 = 2$. It was observed that the function quickly converges for adaptive inertia weight in the range [0.9 0.2]. Figure 6.2 shows the convergence of the function for various ranges of inertia weight,

$$f(x) = \sum(x^2). \tag{6.3}$$

Let us also demonstrate the effect of c_1 and c_2 on the convergence of functions. We consider $w = [0.9\ 0.2]$ for $c_1 = 0$, $c_2 = 2$ and $c_1 = 2$, $c_2 = 0$. It was observed that the functions quickly converge toward an optimal solution for $c_1 = 0$ and $c_2 = 2$ and when $c_1 = 2$ and $c_2 = 0$ the particle only searches one part of the search space. Because there is no information exchange between the particles, they are unable to find the global best solution. Figure 6.3 shows convergence of a function for various values of c_1 and c_2.

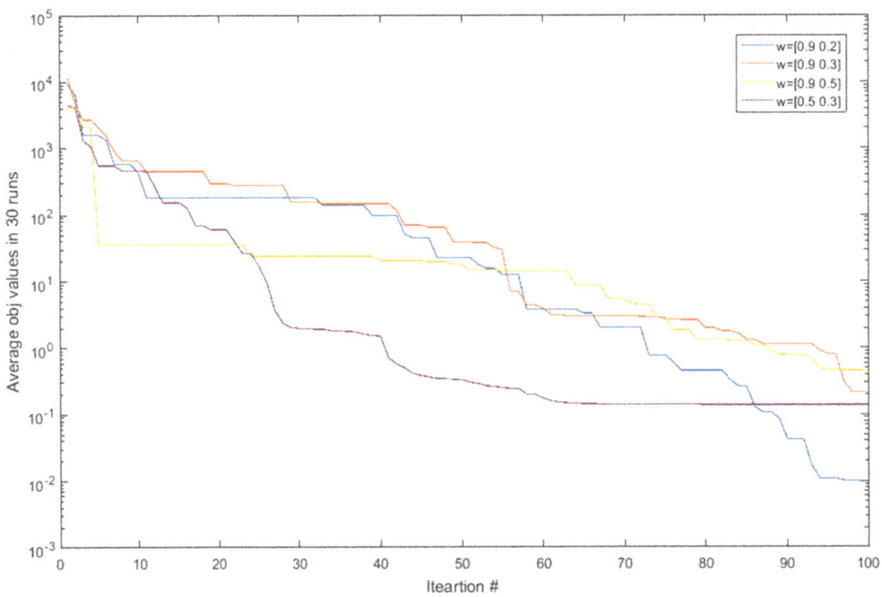

Figure 6.2. Impact of inertia weight in PSO.

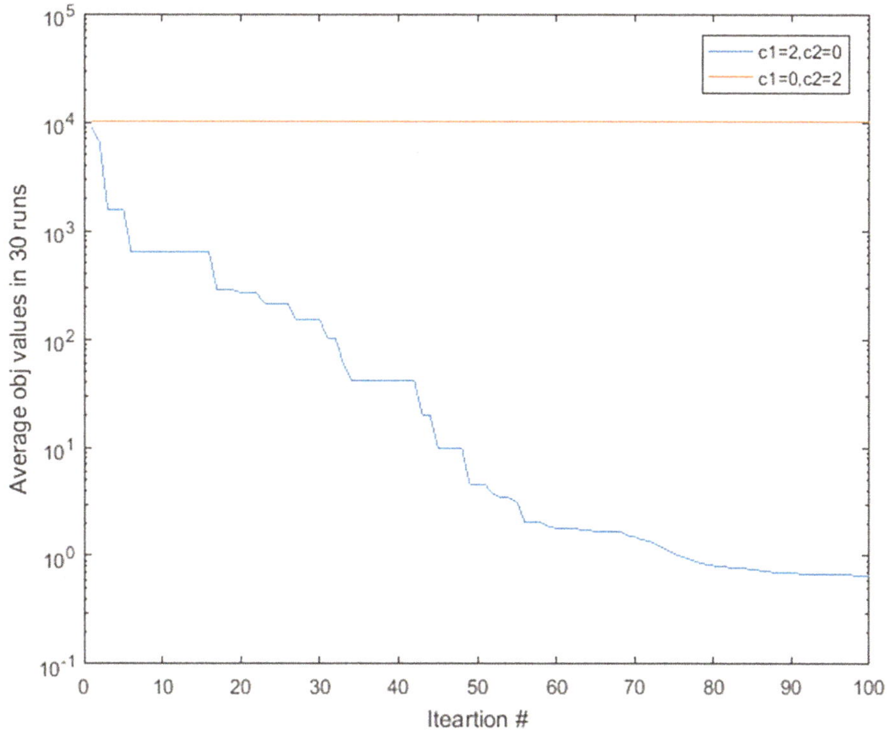

Figure 6.3. Impact of c_1 and c_2 parameter.

6.2.2 Constraint-based optimization

The constraint is a limit applied to the particle, which prevents us from going forever in a particular direction. The constraint applies to multiple variables. Each search space has a boundary and we can use the constraint to simulate this. The set of constraints indicates the upper and lower bound of all parameters. In this way, we can define the boundary for the search space. If we apply the limit to the variable it significantly reduces the size of the search space. The constraint divides the search space into two parts: feasible and infeasible. In figure 6.4, all shaded areas are feasible. An infeasible solution indicates that if we find a solution that violates the constraint then we do not consider it as a desirable solution and it should be avoided. When a large number of constraints divides the space into too many parts, the infeasible region is dominant over the feasible region. This means the solutions for optimization algorithms are more likely to be in a feasible region. To solve this problem the optimization algorithm should find an isolated region and identify promising ones. The constraint optimization problem is represented as the following nonlinear problem.

Minimize

$$f(x_1, x_2, x_3, x_4, \ldots, x_{n-1}, x_n), \tag{6.4}$$

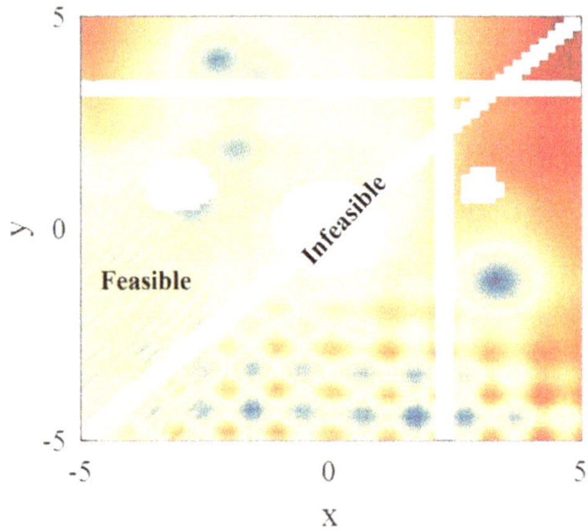

Figure 6.4. Feasible and infeasible regions.

subject to

$$g_i(x_1, x_2, x_3, x_4 \ldots, x_{n-1}, x_n) \geqslant 0\,, \quad i = 1, 2, 3, \ldots, m \tag{6.5}$$

$$h_i(x_i, x_2, x_3, x_4, \ldots, x_{n-1}, x_n) \geqslant 0, \quad i = 1, 2, 3, \ldots, p \tag{6.6}$$

$$lb_i \leqslant x_i \leqslant ub_i, \; i = 1, 2, 3, \ldots, n. \tag{6.7}$$

To handle the constraints, penalty functions are used. These convert the objective to a constraint, meaning that constraints are handled inside the objective function without any modification in the algorithm [27]. The penalty function is represented as follows.

Minimize

$$f(x) + \sigma p(x), \tag{6.8}$$

where

$$p(x) = \begin{Bmatrix} 0 & \text{if } x \in S \\ +\infty & \text{otherwise} \end{Bmatrix}. \tag{6.9}$$

The penalty function returns a 0 if solution x is feasible, moreover in this case there is no penalty. In the case that x is an infeasible solution, the penalty function penalizes it by returning a greater objective value than the actual.

6.3 Proposed method

In the proposed method we perform classification of EEG signal using PSO based ANFIS. Before classification, signals are first acquired from the brain using a 10–20 electrode placement method on the scalp and after that segmentation is performed according to the given trial, and then, using DWT, approximate and detailed coefficients are calculated which are used as features for classification. Figure 6.5 shows the proposed method.

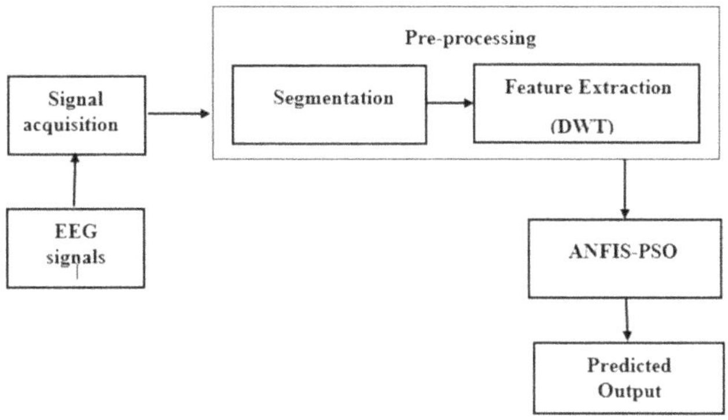

Figure 6.5. Proposed work.

Here 18 features were extracted as detailed and approximate coefficients, these features were applied as input into the ANFIS. The method is briefly explained below.

6.3.1 Materials and methods

In this section we describe the methods and materials used in the proposed work. The complete details are provided below.

6.3.1.1 Dataset

The dataset was acquired from BNCI Horizon-2020 [28]. The dataset related to two-class MI, where signals are recorded for the right hand and foot movement. As per the Graz-BCI training paradigm, a single session is carried out for recording, training and feedback. Sessions have a total of eight runs and of the eight runs five are used for training and three are used for testing. Of a total of 20 trials for a set, per class 50 trials are for training and 30 are for testing. In this recording, process participants had the task of performing sustained (5 s) kinesthetic MI as instructed by a cue. The EEG signal is recorded using an Ag/Agcl electrode, with 512 Hz sampling frequency using a bio-signal amplifier. Electrodes are positioned as per the 10–20 standard. Dataset processing is shown in figure 6.6.

6.3.1.2 Feature extraction

A feature is a distinctive measurement, transform extracted from a set of patterns. Features are used to differentiate each class level with the goal of dimensionality reduction, which reduces the computational complexity. In feature extraction we remove redundant and irrelevant data and improve the quality of signals. EEG signals are random in nature and have a lot of artifacts and noise so, instead of applying these signals as input into a classifier, first it is important to improve the quality of the signals. For this, first signals are filtered by both the time domain as well as spatial domain and then a feature extraction method is deployed. For a better representation of EEG signals, there are various feature extraction methods such as fast Fourier transform (FFT), wavelet transform, principle component analysis

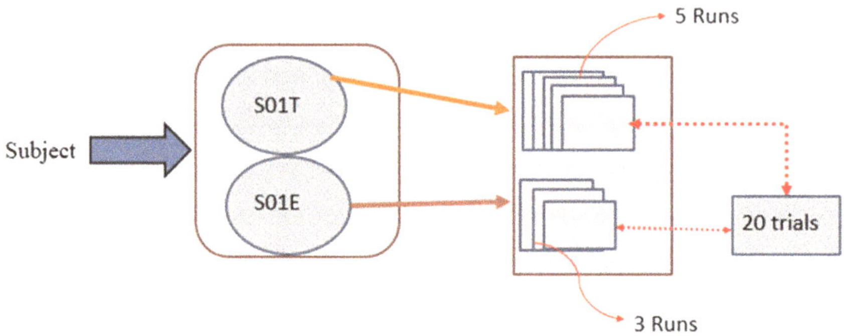

Figure 6.6. Description of dataset processing.

(PCA), independent component analysis (ICA), auto-regression (AR), etc. Among all these methods WT is a widely used feature extraction method because it provides both time and frequency analysis. WT represents EEG signals more accurately. In our work we used a DWT feature extraction method, which is described in detail below.

DWT is a widely used method, with a multiresolution analysis capability in which we perform analysis on both time and frequency [29]. It facilitates the multi-resolution analysis of EEG signals. The signals are split into two parts using a high pass filter and a low pass filter simultaneously. These filters satisfy some condition called the admissibility condition. The approximate and detailed coefficient is computed at each level and the approximate coefficient is again split into two parts, the approximate and detailed coefficient. Repeat this process until a specific level is reached. Wavelet coefficients are an efficient representation of EEG signals [30]. In the context of computation time and complexity of signals, DWT is better compared to other methods. Figure 6.7 represents the nth level of DWT decomposition.

The dilation function $\varnothing_{j,k}(n)$ relies on the low pass filter and the wavelet function $\psi_{j,k}(n)$ relies on the high pass filter which is denoted as

$$\varnothing_{j,k}(n) = 2^{j/2}h(2^j n - k) \tag{6.10}$$

$$\psi_{j,k}(n) = 2^{j/2}g(2^j n - k), \tag{6.11}$$

where $n = 0, 1, 2, 3, \ldots, N - 1, j = 0, 1, 2, 3, \ldots, J - 1, k = 0, 1, 2, 3, \ldots, 2^j - 1$, and $J = \log_2 M$. M represents the length of the EEG signal. Approximate coefficient A_i and detailed coefficient D_i are calculated using the following equations:

$$A_i = \frac{1}{\sqrt{M}} \sum_n x(n) \times \varnothing_{j,k}(n) \tag{6.12}$$

$$D_i = \frac{1}{\sqrt{M}} \sum_n x(n) \times \psi_{j,k}(n). \tag{6.13}$$

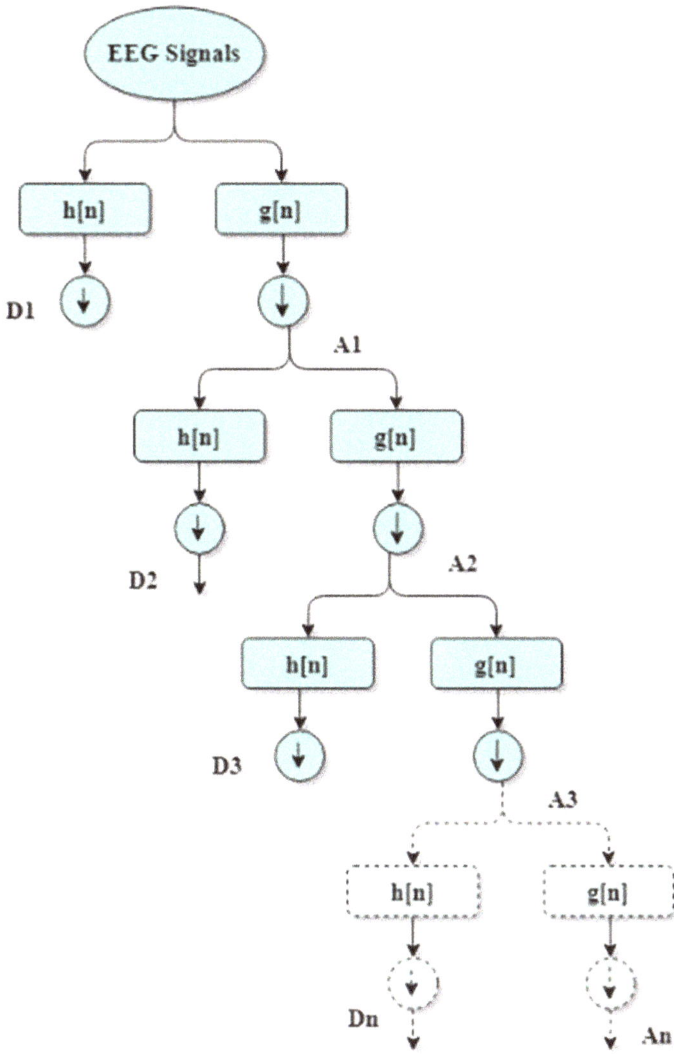

Figure 6.7. DWT decomposition of *n* levels.

6.3.2 Classification

ANFIS is one of the neuro-fuzzy models. Both NN and fuzzy logic are individual systems. When increasing the complexity of the model it is difficult to calculate the membership function and fuzzy rules encourage the development of another approach that has the tendencies of both systems. This approach is called the adaptive neuro-fuzzy system, which gives the advantages of both NN and fuzzy logic. The advantage of fuzzy logic is that fuzzy rules describes real-world problems in a better way. The second thing is interpretability; this means it is easy to explain how every single output value of the fuzzy system is generated. The problem with the fuzzy system is that it needs expert knowledge to create rules and it takes a long time to tune the parameter (membership function). Both problems arise simply because it is not possible to tune the fuzzy system. However, for NN, it trains well but it is remarkably difficult to use earlier learning about the considered system. To overcome the disadvantages of these two systems, some researchers combine both the NN and fuzzy logic systems. The hybrid system ANFIS was proposed by Jang [31].

6.3.2.1 Structure of ANFIS

ANFIS is the standard method for transforming human knowledge or experience into the rule base and database of the FIS [31]. ANFIS extracts features from a dataset and, according to a given error criterion, adjusts the system parameters. For training, a back-propagation gradient descent algorithm in combination with the least squares method is used. FIS is a basic part of ANFIS, which employs fuzzy rules for modeling the qualitative aspects of human knowledge. FIS consists of five functional blocks (figure 6.8).

ANFIS first used the Sugeno model for creating FIS, and for better understanding we have to consider the following if–then rules:

$$\text{Rule 1: if } (x \text{ is } A_1) \text{ and } (y \text{ is } B_1) \text{ then } (f_1 = p_1x + q_1y + r_1)$$

$$\text{Rule 2: if } (x \text{ is } A_2) \text{ and } (y \textit{ is } B_2) \text{ then } (f_2 = p_2x + q_2y + r_2),$$

where x, y and A_i, B_i represent the given input and antecedent, respectively. The f_i includes f_1 and f_2 in terms of p_i, q_i and r_i whose values are found through the training

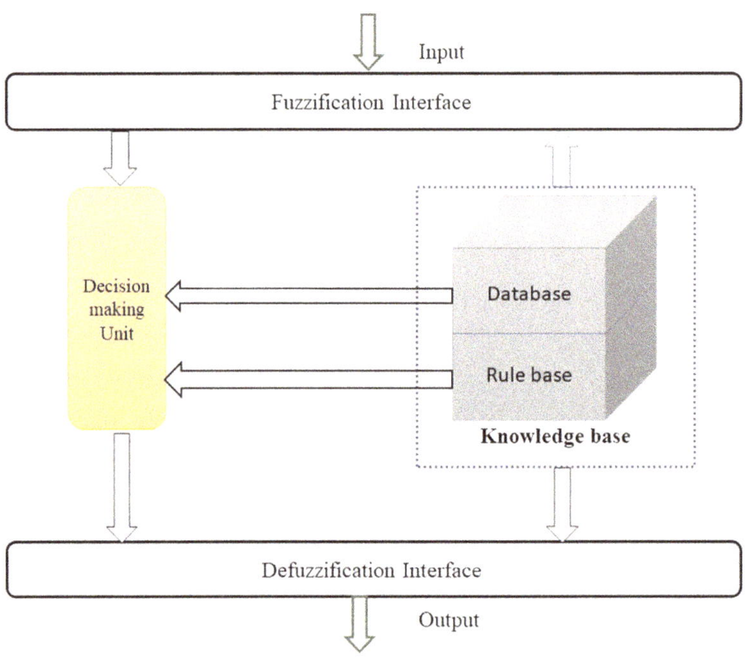

Figure 6.8. Functional block of FIS [31].

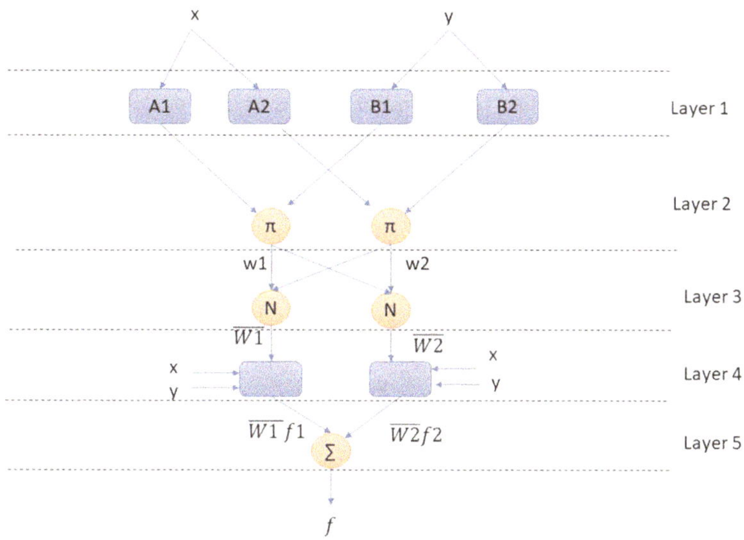

Figure 6.9. Structure of ANFIS [31].

process. The structure for two rules is shown in figure 6.9. Circles represent fixed nodes while squares represent adaptive nodes. Five layers of ANFIS are described.

In the first layer, it receives incoming crisp input and determines rules for each crisp input. All the node are adaptive nodes. This layer is also called the fuzzification layer. The output of this layer is given as

$$O_i^1 = \mu_{A_i}(x) \qquad i = 1, 2, \tag{6.14}$$

where $\mu_{A_i}(x)$ is any membership function that accepts input x and provides a membership value, $\mu_{A_i}(x)$ for the Gaussian membership, the function is given by

$$\mu_{A_i}(x) = \exp\left\{-\left(\frac{x - c_i}{2a_i}\right)^2\right\}, \tag{6.15}$$

where a_i and c_i represent the parameters of the membership function.

Nodes are fixed in the second layer which multiplies the incoming signals and sends the product out, given as

$$w_i = \mu_{A_i}(x) \cdot \mu_{B_i}(y) \qquad i = 1, 2. \tag{6.16}$$

This output is known as the firing strength of the rule.

Nodes in the third layer are fixed as well. Input coming from the second layer is normalized in this layer. We call the output of this layer the normalized firing strength and it is computed as

$$O_i^3 = \overline{w_i} = \frac{w_i}{w_1 + w_2} \qquad i = 1, 2. \tag{6.17}$$

The nodes are adaptive in the fourth layer and their output is the product of the input from the third layer and first-order polynomial as

$$O_i^4 = \overline{w_i}(p_i x + q_i y + r_i) \quad i = 1, 2. \tag{6.18}$$

The fifth layer consists of a single fixed node which aggregates all incoming signals. Hence, the overall output is

$$O_i^5 = \sum_{i=1}^{2} w_i f_i \frac{(\sum_{i=1}^{2} w_i f_i)}{w_1 + w_2} \qquad i = 1, 2. \tag{6.19}$$

6.3.2.2 Learning method

The process of learning in ANFIS is very similar to a NN, where the back-propagation algorithm is used to learn; learning means adjusting the weight of synapses between neurons according to the error criteria. In ANFIS there are two types of parameters: antecedent and consequence. These parameters play an important role in the derivation of rules. On the basis of rules, inputs are mapped into the output. In the rules the 'IF' part consists of the antecedent parameter and the 'THEN' part contains the consequence parameter. The ANFIS learning algorithm is used in order to tune the set of parameters.

In ANFIS, a blend of back-propagation and least squares estimation is used for learning. The algorithm is used to learn both the separate parameters. Back-propagation is used to learn the antecedent (input) parameter and least squares is used for the consequence parameter. The learning procedure consist of two passes: the forward pass and backward pass. In the forward pass the least squares method optimizes the consequent parameter with the predetermined antecedent parameter. After finding the optimal consequent parameter, the backward pass begins immediately. In the backward pass the gradient decent method is used to tune the antecedent parameter. The final output is evaluated by using the consequent parameter. The output is changed according to the antecedent parameter by means of the standard back-propagation algorithm. The output is calculated in terms of MSE and root MSE. Error is calculated in the following way:

$$\text{MSE} = \frac{1}{n}\sum_{k=1}^{n}(y_k - o_k)^2 \tag{6.20}$$

$$\text{RMSE} = \frac{1}{n}\sqrt{\sum_{k=1}^{n}(y_k - o_k)^2}. \tag{6.21}$$

6.3.2.3 ANFIS-PSO

Neuro-FIS is optimized by adapting the membership function parameter and consequent parameter, so objective functions are minimized. The back-propagation algorithm is a widely used adaptation method of FIS that is used to recursively solve the optimization problem. The difficulty with this algorithm is trapping in local minima. To overcome this problem, evolutionary algorithms such as the genetic algorithm and PSO have been used [32, 33]. In this proposed method we perform two-class MI classification based on ANFIS-PSO. Here PSO is used to tune the antecedent and consequent parameters of ANFIS and examine the performance accuracy. The total number of the antecedent parameter is the sum of all the parameters in each membership function. The antecedent parameter (a_i, c_i) is related to the Gaussian membership function shown in equation (6.15). The consequent parameters are output parameters (p_i, q_i and r_i) that are shown in equation (6.18). The mathematical model of ANFIS is quite similar to the NN where the weights are updated according to the error criteria. In ANFIS the antecedent parameters are calculated which are associated with the membership function. In the rules shown in equation (6.1) the 'IF' part contains the membership function parameters and the 'THEN' part contains the linear output variable. These parameters are evaluated as weights; the performance of ANFIS depends on the structural parameter and training related parameter. In this chapter, an ANFIS system is trained by evolutionary algorithms (EAs) such as PSO in order to optimize the classification error. The whole process of this method is shown in figure 6.10. As is evident, the optimization of ANFIS is based on four steps:

1. Initialize the ANFIS system parameters.
2. Evaluate the output of the initial ANFIS and calculate the MSE by comparing the actual output value and target value.

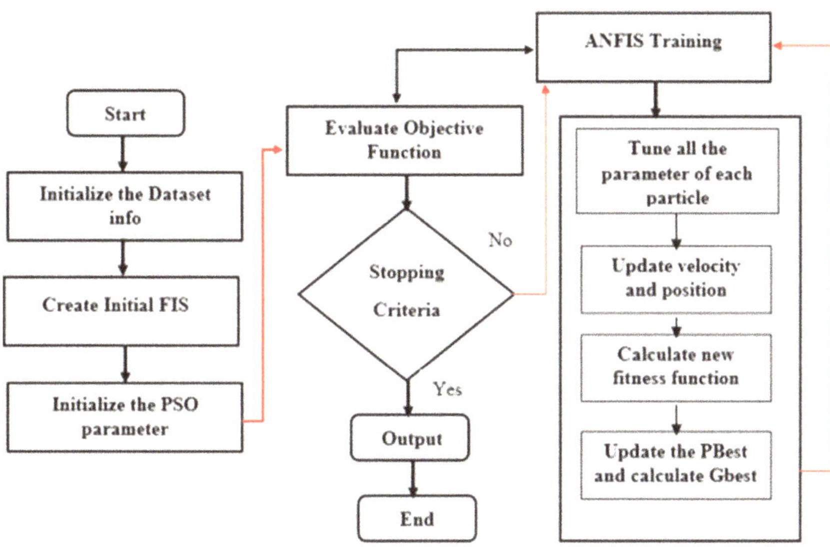

Figure 6.10. Flow diagram of ANFIS-PSO.

3. The output is not too good so in order to minimize the MSE a learning process should be performed.
4. Using PSO, membership functions are optimized in order to minimize the error.

Steps 2 and 3 are repeated until a satisfactory result is obtained.

Moreover, if the number of parameters is large, then the computation times for these algorithms are also high. In this chapter, we used PSO algorithms to tune the membership function. The procedure is shown in figure 6.10.

Algorithm 1. Adopted research methodology

---

1. *Data loading*: Load the EEG data acquired from BNCI Horizon-2020.
2. *Generating basic FIS*: Generate the initial fuzzy inference system.
3. *Training ANFIS*: Tune the parameters of the Gaussian membership function and optimize the MSE value using PSO.
4. *Classification*: Apply ANFIS-PSO on the training and testing data.
5. *Performance evaluation*: Calculate the MSE value for classification on the training and testing data.

END.

6.2.3.4 Optimization problem

ANFIS has two main parameters: antecedent and consequent. To define the problem, we first need to identify these parameters. As mentioned above, the antecedent parameters (a_i, c_i) and consequent parameters (p_i, q_i and r_i) play an important role in the optimization of ANFIS. For each membership function, all the values of parameters are calculated and saved as vectors and applied for training for optimized values. In a vector the number of the variable represents the dimension of the search agent.

In the next step we define the objective function. Objective functions are the fitness of a problem; if we obtain an acceptable value of the objective function training should stop, otherwise we repeat the process. In ANFIS the performance is measured terms of the MSE value. For every sample, MSE values are calculated. To evaluate the overall performance, calculate the average of MSE for all training samples. The average MSE value is calculated according to the equation

$$\text{MSE} = \frac{1}{n}\sum_{k=1}^{n}(y_k - o_k)^2,$$

where y_k is the targeted output and o_k is the actual output predicted by the ANFIS. The length of the training sample is denoted by n and k denotes the kth training sample. So the optimization problem can be defined as

$$\text{Minimize}: f(x) = \text{MSE}. \tag{6.22}$$

So the above function is considered as an objective function and we minimize its value using PSO.

Table 6.1. Average of the approximate and detailed coefficients for channel C3 of subject 1.

Dataset	D1	D2	D3	D4	D5	A5
S01	4.0387	4.2069	3.2889	4.9283	7.3535	76.1838
S02	0.5686	1.0407	2.1799	4.8142	6.9992	84.3975
S03	2.3672	3.1341	4.4636	9.0005	13.4167	67.6180

Table 6.2. Description of the parameters used in PSO.

Description	Considered value/size
Population size	25
Inertia weight	1
Personal learning coefficient	1
Global learning coefficient	2

Table 6.3. Performance of ANFIS.

	Performance of ANFIS			
	Training		Testing	
Subject	MSE	RMSE	MSE	RMSE
S01	1.4724e−08	0.000 121 34	8.5311	2.9208
S02	8.3516e−07	0.000 913 87	2.9647	1.7278
S03	3.4101e−09	5.8396e−05	4.1703	2.0421

6.4 Results

The acquired EEG signals are segmented according to a given trial in the dataset. We use DWT for the decomposition of EEG into five level wavelet coefficients. The average approximate and the detailed coefficients for channel C3 subject 1 are shown in table 6.1.

The extracted features are used as input for ANFIS and ANFIS-PSO. For ANFIS-PSO different parameters are used during the training. The description of all the parameters is provided in table 6.2. In order to perform the classification, the data are divided into two parts, one for training and the other for testing; here 80% of the data are used for training and 20% of the data are used for testing.

The predicted RMSE and MSE values for three subjects are shown in tables 6.3 and 6.4, figures 6.11 and 6.12 shown the training and testing performance of subject 1 for ANFIS and figures 6.13 and 6.14 shown the training and testing performance of subject 1 for ANFIS-PSO.

Table 6.4. Performance of ANFIS-PSO.

	Performance of ANFIS-PSO			
	Training		Testing	
Subject	MSE	RMSE	MSE	RMSE
S01	0.222 39	0.471 58	0.349 15	0.590 89
S02	0.205 23	0.453 02	0.317 52	0.563 02
S03	0.225 36	0.474 72	0.283 06	0.532 03

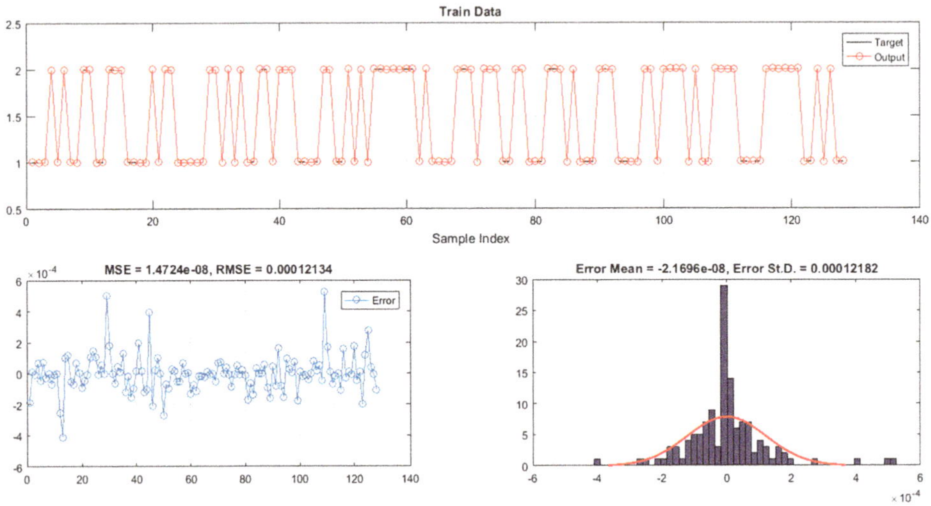

Figure 6.11. Training performance of ANFIS.

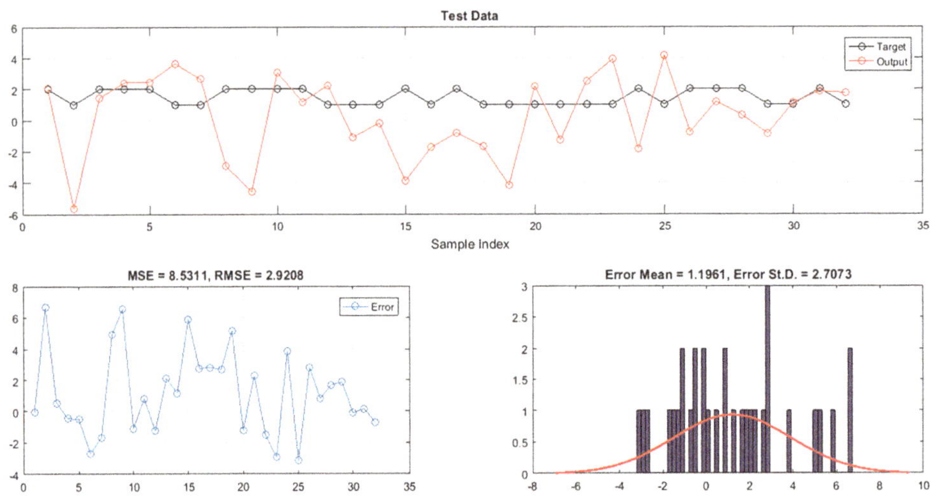

Figure 6.12. Testing performance of ANFIS.

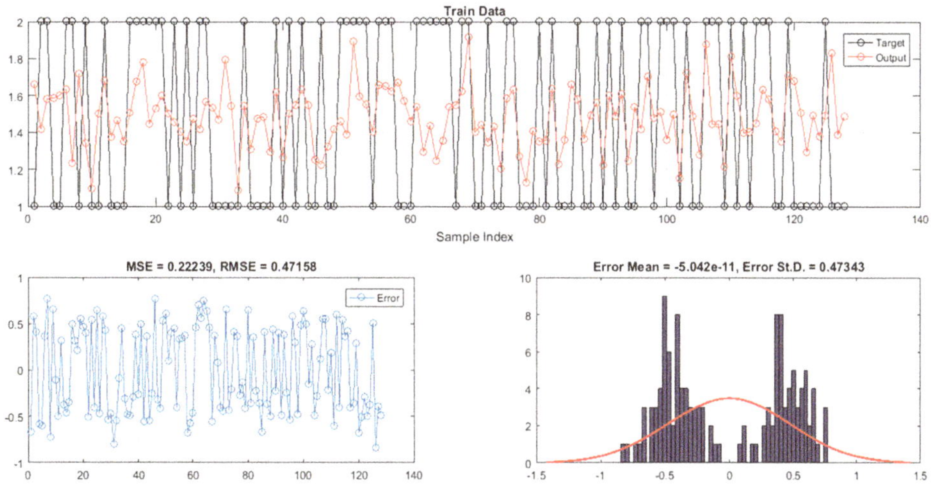

Figure 6.13. Training performance of ANFIS-PSO.

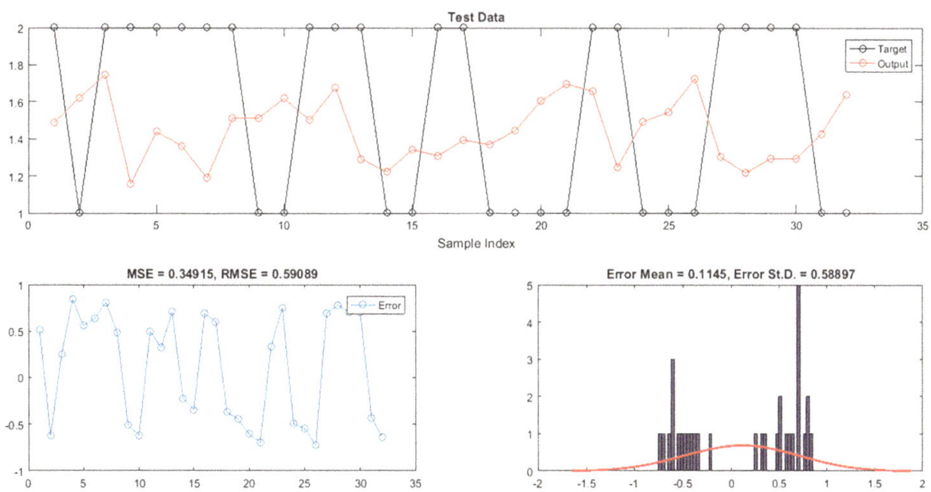

Figure 6.14. Testing performance of ANFIS-PSO.

6.5 Conclusion

In this chapter classification of two-class MI action of the right hand and foot moment is performed using an ANFIS and ANFIS-PSO. Before classification, some preprocessing methods such as segmentation and feature extraction are also experimented. To extract features the DWT method was used, so in this way we calculate six features for each channel (C3, C*z*, C4) and a total of 18 features for each subject. These feature are applied as input for the classifier. It is found that PSO performed well for tuning the parameters of ANFIS with increased accuracy.

References

[1] Mulder T 2007 Motor imagery and action observation: cognitive tools for rehabilitation *J. Neural Transm.* **114** 1265–78

[2] Kübler A, Kotchoubey B, Kaiser J, Wolpaw J R and Birbaumer N 2001 Brain–computer communication: unlocking the locked *Psychol. Bull.* **127** 358

[3] Lotte F, Congedo M, Lécuyer A, Lamarche F and Arnaldi B 2007 A review of classification algorithms for EEG-based brain–computer interfaces *J. Neural Eng.* **4** R1

[4] Pfurtscheller G, Neuper C, Flotzinger D and Pregenzer M 1997 EEG-based discrimination between imagination of right and left hand movement *Electroencephalogr. Clin. Neurophysiol.* **103** 642–51

[5] Chiappa S and Bengio S 2004 HMM and IOHMM modeling of EEG rhythms for asynchronous BCI systems *European Symposium on Artificial Neural Networks ESANN*

[6] Millan J R and Mouriño J 2003 Asynchronous BCI and local neural classifiers: an overview of the adaptive brain interface project *IEEE Trans. Neural Syst. Rehabil. Eng.* **11** 159–61

[7] Penny W D, Roberts S J, Curran E A and Stokes M J 2000 EEG-based communication: a pattern recognition approach *IEEE Trans. Rehabil. Eng.* **8** 214–5

[8] Qin L, Ding L and He B 2004 Motor imagery classification by means of source analysis for brain–computer interface applications *J. Neural Eng.* **1** 135

[9] Wei Q and Wang Y 2011 Binary multi-objective particle swarm optimization for channel selection in motor imagery based brain–computer interfaces *2011 4th International Conference on Biomedical Engineering and Informatics (BMEI)* vol 2 (Piscataway, NJ: IEEE), pp 667–70

[10] Hasan B A S, Jan J Q and Zhang Q 2010 Multi-objective evolutionary methods for channel selection in brain computer interface: some preliminary experimental results *IEEE Congress on Evolutionary Computation (CEC)* pp 1–6

[11] Lv J and Liu M 2008 Common spatial pattern and particle swarm optimization for channel selection in BCI *3rd International Conference on Innovative Computing Information and Control* (Piscataway, NJ: IEEE), p 457

[12] Kumar S U and Hannah Inbarani H 2017 PSO-based feature selection and neighborhood rough set-based classification for BCI multiclass motor imagery task *Neural Comput. Appl.* **28** 239–58

[13] Hsu W-Y 2013 Application of quantum-behaved particle swarm optimization to motor imagery EEG classification *Int. J. Neural Syst.* **23** 1350026

[14] Xu P, Liu T, Zhang R, Zhang Y and Yao D 2014 Using particle swarm to select frequency band and time interval for feature extraction of EEG based BCI *Biomed. Sig. Process. Control* **10** 289–95

[15] Ahirwal M K, Kumar A and Singh G K 2012 Analysis and testing of PSO variants through application in EEG/ERP adaptive filtering approach *Biomed. Eng. Lett.* **2** 186–97

[16] Eberhart R and Kennedy J 1995 A new optimizer using particle swarm theory *MHS'95. Proceedings of the Sixth International Symposium on Micro Machine and Human Science* (Piscataway, NJ: IEEE), pp 39–43

[17] Omrana M, Engelbrechta A and Salmanb A 2005 Particle swarm optimization method for image clustering *Int. J. Pattern Recogn. Artif. Intell.* **19** 297–322

[18] Kennedy J and Eberhart R C 1997 A discrete binary version of the particle swarm algorithm *1997 IEEE International Conference on Systems, Man, and Cybernetics. Computational Cybernetics and Simulation* vol 5 (Piscataway, NJ: IEEE), pp 4104–8

[19] Bergh F V D and Engelbrecht A P 2006 A study of particle swarm optimization particle trajectories *Info. Sci.* **176** 937–71

[20] Singh N and Singh S B 2012 Personal best position particle swarm optimization *J. Appl. Comput. Sci. Math.* **12** 69–76

[21] Kumar A, Singh B K and Patro B 2016 Particle swarm optimization: a study of variants and their applications *Int. J. Comput. Appl.* **135** 24–30

[22] Poli R, Kennedy J and Blackwell T 2007 Particle swarm optimization *Swarm Intell.* 1 33–57

[23] Angline P 1998 Evolutionary optimization versus particle swarm optimization: philosophy and performance difference *Int. Conf. Evolution. Program.* **1447** 601–10

[24] Parsopoulos K and Vrahatis M 2001 Particle swarm optimizer in noisy and continuously changing environments *Artificial Intelligence and Soft Computing* ed M H Hamza (Anaheim, CA: IASTED/ACTA Press) pp 289–94

[25] Hendtlass T 2005 WoSP: a multi-optima particle swarm algorithm *2005 IEEE Congress on Evolutionary Computation* **vol 1** (Piscataway, NJ: IEEE), pp 727–34

[26] Silva A, Neves A and Costa E 2002 Chasing the swarm: a predator–prey approach to function optimisation *Proc. of the MENDEL2002—8th Int. Conf. on Soft Computing Brno, Czech Republic*

[27] Parsopoulos K E and Vrahatis M N 2002 Particle swarm optimization method for constrained optimization problems *Front. Artif. Intell. Appl.* **76** 214–20

[28] Steyrl D, Scherer R, Förstner O and Müller-Putz G R 2014 Motor imagery brain–computer interfaces: random forests vs regularized LDA-non-linear beats linear *Proc. of the 6th Int. Brain-Computer Interface Conf.* pp 241–4

[29] Ren W, Han M, Wang J, Wang D and Li T 2016 Efficient feature extraction framework for EEG signals classification *2016 Seventh International Conference on Intelligent Control and Information Processing (ICICIP)* (Piscataway, NJ: IEEE), pp 167–72

[30] Jahankhani P, Kodogiannis V and Revett K 2006 EEG signal classification using wavelet feature extraction and neural networks *IEEE John Vincent Atanasoff 2006 International Symposium on Modern Computing (JVA'06)* (Piscataway, NJ: IEEE), pp 120–4

[31] Jang J-S R 1993 ANFIS: adaptive-network-based fuzzy inference system *IEEE Trans. Syst. Man Cybernet.* **23** 665–85

[32] Catalão J P d S, Pousinho H M I and Mendes V M F 2011 Hybrid wavelet-PSO-ANFIS approach for short-term electricity prices forecasting *IEEE Trans. Power Syst.* **26** 137–44

[33] Ghomsheh V S, Shoorehdeli M A and Teshnehlab M 2007 Training ANFIS structure with modified PSO algorithm *Mediterranean Conference on Control and Automation* (Piscataway, NJ: IEEE), pp 1–6

Chapter 7

Multi-criterion and topology optimization using Lie symmetries for differential equations

Sailesh Kumar Gupta

Lie addressed the unification problem of all the apparently different integration methods for differential equations and investigated some elegant and simple algebraic structures, commonly known as continuous transformation groups, which hold the key to the problem. This connection with differential equations naturally led to the study of the topological manifolds associated with group structure. Interestingly, the manifolds needed for solving different aspects related to differential equation systems can be thought to be locally some open subsets defined in an n-dimensional Euclidean space where one can choose the local coordinates freely. The key concept turns out to be that of symmetry or the infinitesimal generator associated with the group. Once the generators associated with a system of equations become available, many applications become immediate. For ordinary differential equations (ODEs), the symmetries and their generators always lead to the possibility of going a step forward in the integration of the equation, and if one has a sufficiently large number of symmetries, the complete integration of the differential equation is guaranteed in most cases by the method of quadrature alone. For most partial differential equations (PDEs), one cannot write down the general solution but has to rely on various ansatz such as the similarity solutions, traveling waves, separable solutions and so on. These ansatz methods usually lead to the problem of solving some ODE generated in the process. These methods involve nothing more than looking for solutions that are invariant under a particular group of symmetries associated with the equations. Further, it is well documented in the associated literature that any linear combination of the infinitesimal generators associated with a given system of equations also leads to some group invariant solutions for the same system of equations. Thus, it is easy to understand that we are left with a situation for dealing with infinite group invariant solutions for the system. Hence, we have to

doi:10.1088/978-0-7503-2404-5ch7

devise a mechanism to classify all the inequivalent solutions only and the corresponding problem leads to the optimization problem for the group invariant solutions.

7.1 Introduction

Accurate formulations of different physical problems can be achieved with the help of nonlinear ordinary or partial differential equations (ODEs and PDEs). A vast literature exists on all of the solution methods, such as analytical, numerical or approximate, but the concept of Lie symmetry groups provides a complete analytical understanding of all these apparently different solution methods. Lie [1] began the unification of all these different methods and founded the concept of continuous transformation groups or Lie groups [2–4]. He found that the mathematically challenging nonlinear conditions generated under the action of a continuous transformation group, to preserve the invariance of a differential equation system, could be systematically replaced by some simple linear conditions associated with the generators of the group. The connection of Lie groups with the differential equations naturally led to the study of the topological manifolds [5–7] associated with the group structure. These manifolds can be thought, at least locally, to be some open subsets of an n-dimensional Euclidean space in which the coordinates can be chosen freely, at least locally. The idea of the infinitesimal generator of the group plays an important role in the process. The infinitesimal generators are like a vector field of a given manifold and their flows coincide with the one-parameter groups that they generate. Thus, the starting point of the study of a differential equation using the group techniques basically requires familiarity with two aspects. First, the idea of flow in a vector field and, second, the infinitesimal criterion of invariance for the system under the action of group transformations. The prolongation formula of vector fields becomes the main tool in calculating the symmetry groups of a system [2, 3], and requires the introduction of spaces which also includes the derivatives of different dependent variables along with the independent and dependent variables as coordinates of the manifold. The space of all these variables is called the jet space. Once the symmetry groups are calculated for a system, many applications become immediate. For the ODEs, the existence of Lie symmetries always leads to the possibility of going a step forward in the integration of the equation, and if one has a sufficiently large number of symmetries, then the complete integration of the differential equation is guaranteed in most cases by the method of quadrature alone. Thus, it becomes very important to ascertain the possibility of maximum reduction in order of the equation in this case. When it comes to the PDEs, usually one obtains several arbitrary functions which are not very useful in practical applications. For most PDEs, one cannot write down the general solution but has to rely on various ansatz such as the similarity solutions, traveling waves, separable solutions and so on. These ansatz methods usually lead to solving some ODEs generated in the process. These methods involve nothing more than looking for solutions that are invariant under a particular symmetry group of transformations for the system. Further, it is a well known fact that for a given system of differential equations possessing symmetries, one can have a system of infinite group invariant solutions. This is due to the fact that any linear combination of the symmetry generators of the system again leads to a symmetry associated with the same system. Hence, we have to devise a mechanism to classify all the in-equivalent solutions only, and the problem leads to the optimization of the associated group invariant solutions for the system. The classified solutions give the optimal group invariant solutions [2, 3].

Thus, in this chapter we will try to address the different issues discussed so far with examples. We start with a brief introduction of the fundamentals of topological manifolds and then establish the connection between the groups and differential equations [3]. We discuss the methods to calculate the group invariant solutions for the differential equations and classify them to establish the corresponding optimized solutions. Finally, concluding remarks are given with further reading suggestions.

7.2 Fundamentals of topological manifolds

A Lie group should have the properties of a topological group, that is, one must be able to assign a topology under which the group action and its inverses are both continuous functions. A methodological discussion to explore the analytical structures associated with the Lie groups requires familiarity with the concept of an analytic manifold. Lie's works lead to the concept of defining a Lie group in terms of analytic manifolds whose objects are the differential equations themselves.

7.2.1 Analytic manifolds

The theoretical studies of connected analytic manifolds can be systematically carried out by treating them as some smooth surfaces in a Euclidean space $\mathfrak{R}^n$. Each point in the space is assigned coordinates with the property of possessing continuous mapping with its inverse, which is also continuous. Analytically, if $X \subset \mathfrak{R}^n$ is some open subset of manifold Z and Ψ: $X \rightarrow Y$ is any diffeomorphism such that $Y \subset \mathfrak{R}^n$, then Ψ is a C^∞ map with C^∞ inverses. Thus, for different elements on X, some equivalent objects on Y will always exist. The coordinate charts Θ_ρ: $X_\rho \rightarrow Y_\rho$ assign the structure of a topology to the manifold Z, such that each open subset $W \subset Y_\rho \subset \mathfrak{R}^n$, $\Theta_\rho^{-1}(W)$ is an open subset of Z. Further, if $\alpha \neq \tilde{\alpha}$ are points in Z, then one can have open sets X containing α and $\tilde{X}$, containing $\tilde{\alpha}$ such that $X \cap \tilde{X} = \Phi$ (null set) and further, if the overlap function $\Theta_\sigma \circ \Theta_\rho^{-1}$: $\Theta_\rho(X_\rho \cap X_\sigma) \rightarrow \Theta_\sigma(X_\rho \cap X_\sigma)$ is smooth, then Z is an analytic manifold.

7.2.1.1 Coordinate transformations and maps

If coordinate charts are used to formulate different elements on a manifold, then it must be ensured that the definition is coordinate independent. The problem reduces to that of studying objects under a coordinate transformation. In actual situations, it is found that the studies can be carried out more easily in local coordinates. If one is able to find some special coordinate charts in which the objects under study take some simple form then many complicated calculations can be considerably simplified. A local coordinate map Θ_ρ: $X_\rho \rightarrow Y_\rho$ with a diffeomorphism

$\Psi: Y_\rho \rightarrow \tilde{Y}_\rho$, of $\mathfrak{R}^n$ is called a coordinate transformation. For the manifolds Z and M with the property of smoothness, then this property is retained by the functional map $L: Z \rightarrow M$ if, for $\Theta_\rho: X_\rho \rightarrow Y_\rho \subset \mathfrak{R}^m$ on Z and $\tilde{\Theta}_\sigma: \tilde{X}_\sigma \rightarrow \tilde{Y}_\sigma \subset \mathfrak{R}^n$ on M, then the composite map defined by $\tilde{\Theta}_\sigma \circ L \circ \Theta_\rho^{-1}: \mathfrak{R}^m \rightarrow \mathfrak{R}^n$ is also smooth. If a manifold possess a subset inheriting all its properties, then the subset is a submanifold.

7.2.2 Lie groups and vector fields

7.2.2.1 Group

A group can be defined as a set S endowed with the following properties:

- *Group composition law:* For any two elements a, $b \in S$ taken in a definite sequence we obtain a product element $c \in S$ such that

$$a \cdot b = c \in S.$$

- *Associativity:* The product of any definite sequence of three elements is unambiguous, i.e. if a, b, $c \in S$, then the following holds:

$$a \cdot (b \cdot c) = (a \cdot b) \cdot c.$$

- *Existence of identity:* There is a special element $e \in S$, such that

$$e \cdot k = k = k \cdot e, \quad k \in S.$$

- *Inverses*: For each $k \in S$, there exists a unique inverse k^{-1}, such that

$$k \cdot k^{-1} = k^{-1} \cdot k = e.$$

7.2.2.2 Lie group

An s-parameter Lie group can be defined as a topological group S such that S is also a smooth manifold endowed with the following maps:

- $\Theta: S \times S \rightarrow S, \ \Theta(a, b) = a \cdot b, \ a, b \in S.$
- $i: S \rightarrow S, \ \ i(a) = a^{-1}, \ \ a \in S.$

7.2.2.3 Local transformation group

Each of the transformations of the manifold Z is associated with a Lie group S, if for each $k \in S$ we can find an associated map from Z to itself. Analytically, if S is a group for the local transformations on Z endowed with $\Upsilon \subset S$, such that $\{e\} \times Z \subset \Upsilon \subset S \times Z$, it is associated with a smooth map $\Pi: \Upsilon \rightarrow Z$ such that:

- If a, $b \in S$, and $z \in Z$, then

$$a \cdot (b \cdot z) = (a \cdot b) \cdot z. \tag{7.1}$$

- For all the elements $z \in Z$,

$$e \cdot z = z. \tag{7.2}$$

- For each $a \in S$ and $z \in Z$, if the product $a \cdot z$ is defined, then the following hold

$$a^{-1} \cdot (a \cdot z) = z.$$

7.2.2.4 Vector fields

In a given m-dimensional manifold Z we can define a tangent space to all the points $z \in Z$. The objects defining the space will be the collection of all the tangent vectors at z belonging to different curves passing through the point. It is denoted by the m-dimensional vector space $TZ|_z$ and the elements defining the basis are $\left\{\frac{\partial}{\partial z^1}, \dots, \frac{\partial}{\partial z^m}\right\}$. The corresponding tangent bundle of manifold Z is given by

$$TZ = \underbrace{\bigcup}_{z \in Z} TZ|_z.$$

If $A \in Z$, then $A|_z \in TZ|_z$ defines the corresponding tangent vector at points when $z \in Z$ and $A|_z$ is also assumed to vary smoothly. If written in terms of $(z^1, \dots, z^m)$, i.e. the local coordinates, then we can write the vector field expression

$$A|_z = \xi^1(z)\frac{\partial}{\partial z^1} + \dots + \xi^m(z)\frac{\partial}{\partial z^m} = \sum \xi^i(z)\frac{\partial}{\partial z^i}.$$

The coefficient functions $\xi(z)$ have smoothly varying properties. The hydrodynamic steady fluid flow can be taken as a good example. In this case, the points $z \in Z$ can be associated with the vector $\mathbf{A}|_z$, which physically gives the velocity of the fluid through the point z. For the vector field $\mathbf{A}$, the curve $z = \phi(\varepsilon)$ gives a smooth parametrized integral curve. The related parametrized tangent vector and the corresponding vector field $\mathbf{A}$ coincide at a given point and are given by

$$\frac{d\phi}{d\varepsilon} = \mathbf{A}|_{\phi(\varepsilon)}.$$

The points $z = \phi(\epsilon) = (\phi^1(\epsilon), \dots, \phi^m(\epsilon))$ in local coordinates, satisfy the ODEs

$$\frac{dz^k}{d\varepsilon} = \xi^k(z), \quad k = 1, \dots, m. \tag{7.3}$$

Now it is guaranteed from the theory of differential equations that for smoothly varying $\xi^k(z)$ we can have a unique solution to the initial value problem of equation (7.3) with the initial condition

$$\phi(0) = z_0. \tag{7.4}$$

Again, if we are given smoothly varying functions $h(z)$ with domain $z \in Z$, then $h \cdot A$ becomes a smooth vector field, with $(h \cdot \mathbf{A})|_z = h(z)\mathbf{A}|_z$ and $h \cdot \mathbf{A} = \sum h(z)\xi^i(z)\frac{\partial}{\partial z^i}$.

7.2.2.5 Flow of vector fields and infinitesimal generators

Given a smoothly varying curve C belonging to the manifold Z, parametrized by $\phi(\epsilon)$: $P \to Z$, where P is a sub-interval in $\mathfrak{R}$, we obtain m smooth functions

$\phi(\epsilon) = (\phi^1(\epsilon), \ldots, \phi^m(\epsilon))$. The corresponding tangent vector to the manifold Z at each point $z = \phi(\epsilon)$ of C is given by the derivative $\frac{d\phi}{d\varepsilon} = (\frac{d\phi^1}{d\varepsilon}, \ldots, \frac{d\phi^m}{d\varepsilon})$. Now, for the field $\mathbf{A} \in Z$, the curve C through the point $z \in Z$ is given by $\alpha(\epsilon, z)$. We call it the flow of $\mathbf{A}$ and it has the properties

$$\alpha(\beta, \alpha(\epsilon, z)) = \alpha(\beta + \epsilon, z), \quad z \in Z, \tag{7.5}$$

for all $\beta, \epsilon \in \mathfrak{R}$ with

$$\alpha(0, z) = z \tag{7.6}$$

and

$$\frac{d}{d\epsilon}\alpha(\epsilon, z) = \mathbf{A}|_{\alpha(\varepsilon, z)} \tag{7.7}$$

for all ϵ where defined. Comparing equations (7.5) and (7.6) with equations (7.1) and (7.2), we find the similarity between the group action of the Lie group $\mathfrak{R}$ on Z and the respective flow of $\mathbf{A}$. The transformations defined above are popularly called one-parameter transformation groups and $\mathbf{A}$ is is called the infinitesimal generator. Now, making use of Taylor's theorem, we can write

$$\alpha(\epsilon, z) = z + \epsilon\xi(z) + \mathrm{O}(\epsilon^2),$$

where $(\xi^1(z), \ldots, \xi^m(z))$ are the coefficients of $\mathbf{A}$.

The curves C for the vector field $\mathbf{A}$ are the orbits of the one-parameter group action. Conversely, if a one-parameter transformation group, $\alpha(\varepsilon, z)$, acts on Z then the corresponding generator is obtained by calculating (7.7) at $\varepsilon = 0$ and is given by

$$\mathbf{A}|_z = \left.\frac{d}{d\varepsilon}\right|_{\varepsilon=0} \alpha(\varepsilon, z).$$

In usual practice the flow of $\mathbf{A}$ is referred to as exponentiation and can be expressed as

$$\exp(\epsilon A)z \equiv \alpha(\epsilon, z).$$

This notation allows us to rewrite the properties of the flow of $\mathbf{A}$ listed above for all $z \in Z$ as

$$\exp[(\beta + \epsilon)\mathbf{A}]z = \exp(\beta\mathbf{A})\exp(\epsilon\mathbf{A})z, \qquad \exp(0\mathbf{A})z = z, \qquad \frac{d}{d\epsilon}[\exp(\epsilon\mathbf{A})z] = \mathbf{A}|_{\exp(\varepsilon\mathbf{A})z}.$$

7.2.2.6 Action of vector flows on functions

Let $\mathbf{A} \in Z$ be a vector field and h: $Z \to \mathfrak{R}$ be a mapping. Now the flow generated by $\mathbf{A}$ changes h and the infinitesimal change in the local coordinates can be calculated using Taylor's theorem as

$$h(\exp(\epsilon\mathbf{A})z) = h(z) + \epsilon\mathbf{A}(h)(z) + O(\epsilon^2).$$

The process of differentiation can be continued and if substituted into the Taylor series assuming convergence, we obtain the Lie series for the action of the flow on h and this can be written as

$$h(\exp(\epsilon \mathbf{A})z) = \sum_{k=0}^{\infty} \frac{\epsilon^k}{k!} \mathbf{A}^k(h)(z).$$

7.3. Differential equations, groups and the jet space

We consider a differential equations system Λ with the set of equations given by

$$E_\nu(z, v^{(n)}) = 0, \quad \nu = 1, \ldots, l$$

involving p independent variables $z = (z^1, \ldots, z^p)$, and q dependent variables $v = (v^1, \ldots, v^q)$ with $v^{(n)}$ signifying the nth order derivatives of v with respect to z. Let the respective solutions of the system be given by $v = h(z)$. Let $Z = (z^1, \ldots, z^p) = \mathfrak{R}^p$ be the space representing the independent variables, and $V = (v^1, \ldots, v^q) = \mathfrak{R}^q$ represent the space of the dependent variables. A symmetry group is a local group of transformations S acting on $X \subset Z \times V$, which is some open subset, such that S transforms solutions of E to its own solutions. If the graph of the function $v = h(z)$ is given by

$$\Gamma_h = \{(z, h(z)) \colon z \in \Xi\} \subset Z \times V,$$

then the transformation of the graph Γ_h by g can be written as

$$g.\, \Gamma_h = \{(\tilde{z}, \tilde{v}) = g.\, (z, v) \colon (z, v) \in \Gamma_h\}.$$

Now, for a proper analytical treatment we need to reduce the domain of Ξ and h, so that the transform $g.\, \Gamma_h = \Gamma_{\tilde{h}}$ gives the transformed function $\tilde{v} = \tilde{h}(\tilde{z})$ for all elements g near the identity. Thus, we can say that the function $\tilde{h}$ is the transform of h by g and can be written as $\tilde{h} = g.\, h$.

Now the main task for studying systems like E_ν is to establish a workable criterion to find out the symmetry group for a given system of differential equations with the help of some group transformations. This criterion will be the infinitesimal transformations requiring an appropriate geometrical setting. This idea leads to the visualization of a prolonged space of the space $Z \times V$, so that the final space includes the various partial derivatives occurring in the system in addition to the independent and dependent variables. Now, let the κth order derivative of $h(z)$ be represented by

$$\partial_J h(z) = \frac{\partial^\kappa h(z)}{\partial z^{j_1} \partial z^{j_2} \ldots \partial z^{j_\kappa}},$$

where $J = (j_1, j_2, \ldots, j_\kappa)$ is the set of integers with $1 \leqslant j_\kappa \leqslant p$ indicating the derivatives that are being considered, not necessarily in a given order. Further, let the space $V^{(n)} = V \times V_1 \times \cdots \times V_n$ include the derivatives of the functions $v = h(z)$. The maximum order of the derivative will be n. The resulting total space $Z \times V^{(n)}$ is called the nth jet space. Generally, we are interested only in the system Λ defined on

$X \subset Z \times V$, where the nth jet space has the definition, $X^{(n)} \equiv X \times V_1 \times \cdots \times V_n$. Now, it is straightforward to generalize that if $v = h(z)$ is the graph in X, then the nth prolongation $pr^{(n)}h(z)$ has to be the graph in the nth jet space $X^{(n)}$. Now, to make matters clear we see an example. If, $p = 2$, $q = 1$ with $v = h(i, j)$, we obtain the second prolongation

$$v^{(2)} = pr^{(2)}h(i, j) = (v; v_i, v_j; v_{ii}, v_{ij}, v_{jj}) = \left(h; \frac{\partial h}{\partial i}, \frac{\partial h}{\partial j}; \frac{\partial^2 h}{\partial i^2}, \frac{\partial^2 h}{\partial i \partial j}, \frac{\partial^2 h}{\partial j^2}\right)$$

all evaluated at (i, j). Thus, treating E as the map E: $Z \times V^{(n)} \to \mathfrak{R}^l$, such that the mapping leads to the l-dimensional Euclidean space from the space $Z \times V^{(n)}$, then we can write

$$\Lambda_E = \{(z, v^{(n)}): E(z, v^{(n)}) = 0\} \subset Z \times V^{(n)}.$$

Hence we can say $v = h(z)$ is a solution if

$$\Gamma_h^{(n)} \equiv \{(z, pr^{(n)}h(z))\} \subset \Lambda_E = \{E(z, v^{(n)}) = 0\}.$$

7.3.1 Prolongation of group action and vector fields

The action of S on the nth jet space $X^{(n)}$ leads to $pr^{(S)}$, which is the nth prolongation of S. Thus, its action on the function $v = h(z)$ should lead to the transformed function $\tilde{v} = \tilde{h}(\tilde{z})$. This symmetry condition needs to be connected to the geometric condition relating to that of the corresponding sub-variety Λ_E, so that it becomes invariant of $pr^{(n)}S$. Now keeping in line with the definition of a group transformation, the prolongation of the corresponding infinitesimal generators can be written as

$$pr^{(n)}\mathbf{A}|_{(z,v^{(n)})} = \frac{d}{d\varepsilon}\bigg|_{\varepsilon=0} pr^{(n)}[\exp(\epsilon\mathbf{A})(z, v^{(n)})]$$

for any $(z, v^{(n)}) \in X^{(n)}$. Now, any vector field in X is given by

$$\mathbf{A} = \sum_{i=1}^{p} \xi^i(z, v)\frac{\partial}{\partial z^i} + \sum_{\alpha=1}^{q} \phi_\alpha(z, v)\frac{\partial}{\partial v^\alpha}.$$

Hence we can say that $\xi^{(i)}(z, v)$ and $\phi_\alpha(z, v)$ determines the coefficient of the nth prolongation of $\mathbf{A}$ and can be written as

$$pr^{(n)}\mathbf{A} = \sum_{i=1}^{p} \xi^i(z, v)\frac{\partial}{\partial z^i} + \sum_{\alpha=1}^{q}\sum_{J} \phi_\alpha^J(z, v)\frac{\partial}{\partial v_J^\alpha}.$$

7.3.2 Total derivatives of vector fields and general prolongation formula

It is clear from the discussions so far that we need explicit expressions for the prolonged vector field, $pr^{(n)}\mathbf{A}$. Once this is achieved then we can make a connection between the infinitesimal criterion of invariance of system Λ and the corresponding symmetry groups S attached to it. Now, the group transformations $g_\varepsilon = \exp(\epsilon \mathbf{A})$ for the vector fields $\mathbf{A}$ can be written as

$$(\tilde{z}, \tilde{v}) = g_\varepsilon(z, v) = (\Theta_\varepsilon(z), v)$$

with the components $\Theta^i(z)$ satisfying the equation

$$\left.\frac{d\Theta^i_\varepsilon(z)}{dz}\right|_{\varepsilon=0} = \xi^i(z).$$

Now differentiating the expression with respect to the variable ϵ at $\epsilon = 0$, we obtain the infinitesimal generator $pr^{(1)}g_\varepsilon$. For the vector field $\mathbf{A}$ we obtain the first prolongation

$$pr^{(1)}\mathbf{A} = \sum_{i=1}^{p}\xi^i(z)\frac{\partial}{\partial z^i} + \sum_{j=1}^{p}\phi^j(z, v^{(1)})\frac{\partial}{\partial v_j}, \tag{7.8}$$

where

$$\phi^j(z, v^{(1)}) = \left.\frac{d}{d\varepsilon}\right|_{\varepsilon=0} \sum_{\kappa=1}^{p}\frac{\partial\Theta^\kappa_{-\varepsilon}}{\partial\tilde{z}^j}(\Theta_\varepsilon(z))v_\kappa,$$

with the assumption $g_\varepsilon^{-1} = g_{-\varepsilon}$ in the domain of definition. As the functions are assumed to be smooth, the order of the differentiation can be interchanged and, simplifying, we can write

$$\phi^j(z, v, v_z) \equiv \phi^j(z, v^{(1)}) = -\sum_{\kappa=1}^{p}\frac{\partial\Theta^\kappa}{\partial z^j}v_\kappa,$$

which provides the expression for $pr^{(1)}\mathbf{A}$ in (7.8). Now the above expression can be written in terms of the total derivative of ϕ with respect to z as

$$\phi^j(z, v^{(1)}) = D_j\phi(z, v) = \frac{\partial\phi}{\partial z^j} + v_j\frac{\partial\phi}{\partial v}.$$

7.3.2.1 Definition

Given $H(z, v^{(n)})$, the expression

$$D_iH = \frac{\partial H}{\partial z^i} + \sum_{\alpha=1}^{q}\sum_{J}v^\alpha_{J/i}\frac{\partial H}{\partial v^\alpha_J}, \tag{7.9}$$

where for $J = (j_1, \ldots, j_\kappa)$ the expression $v^\alpha_{J/i} = \frac{\partial v^q_J}{\partial z^i} = \frac{\partial^{k+1}v^\alpha}{\partial z^i \partial z^j \ldots \partial z^{j_\kappa}}$ represents the ith total derivative of H. In (7.9) the summation domain of J goes up to order n.

7.3.2.2 Theorem 7.1

Let

$$\mathbf{A} = \sum_{i=1}^{p}\xi^i(z, v)\frac{\partial}{\partial z^i} + \sum_{\alpha=1}^{q}\phi_\alpha(z, v)\frac{\partial}{\partial v^\alpha}$$

be a vector field in Z, so that the corresponding nth prolongation

$$pr^{(n)}\mathbf{A} = \mathbf{A} + \sum_{\alpha=1}^{q}\sum_{J=1}\phi_{\alpha}^{J}(z, v)\frac{\partial}{\partial v_{J}^{\alpha}}$$

becomes a vector field in the jet space $Z^{(n)}$, with $1 \leqslant j_{\kappa} \leqslant p$, $1 \leqslant \kappa \leqslant n$, then the coefficient functions ϕ_{α}^{J} of $pr^{(n)}\mathbf{A}$ are given by

$$\phi_{\alpha}^{J}(z, v^{(n)}) = D_J\left(\phi_{\alpha} - \sum_{i=1}^{p}\xi^{i}v_{i}^{\alpha}\right) + \sum_{i=1}^{p}\xi^{i}v_{J/i}^{\alpha}, \tag{7.10}$$

with $v_i^{\alpha} = \partial v^{\alpha}/\partial z^i$ etc.

Proof. The theorem will be proved in general by the very simple and effective method of the induction principle. We note that the second jet space $X^{(1+1)}$ belongs to a subspace of the first jet space $(X^{(1)})^{(1)}$. To be precise, we note that the first-order derivative of v_J^{α} gives the second order derivative of the same. Similarly, the space $X^{(n+1)}$ can be treated as a subspace of the nth jet space because its first jet space is given by $(X^{(n)})^{(1)}$. Thus, we can say that the vector fields on the manifold $X^{(n-1)}$ are given by $pr^{(n-1)}\mathbf{A}$, which can then be prolonged to $(X^{(n-1)})^{(1)}$ by the use of first-order prolongation formula. The resulting vector field is now restricted to $Z^{(n)}$ and its subspaces which in turn will determine the expression for $pr^{(n)}\mathbf{A}$. In the process the new nth order coordinates in $(X^{(n-1)})^{(1)}$ are given by $v_{J/\kappa}^{\alpha} = \partial v_J^{\alpha}/\partial z^{\kappa}$ where $J = (j_1, \ldots, j_{n-1})$, $1 \leqslant \kappa \leqslant p$ and $1 \leqslant \alpha \leqslant q$. At this point we use the definition of the total derivative so that the coefficient of $\partial/\partial v_{J/\kappa}^{\alpha}$, in $(pr^{(n-1)}\mathbf{A})^{(1)}$, becomes

$$\phi_{\alpha}^{J/\kappa} = D_{\kappa}\phi_{\alpha}^{J} - \sum_{i=1}^{p}D_{\kappa}\xi^{i}v_{J/i}^{\alpha}. \tag{7.11}$$

Now if we are able to prove that (7.10) solves (7.11) in closed form then the proof is complete. Using induction, we find

$$\begin{aligned}\phi_{\alpha}^{J/\kappa} &= D_{\kappa}\left\{D_J\left(\phi_{\alpha} - \sum_{i=1}^{p}\xi^{i}v_{i}^{\alpha}\right) + \sum_{i=1}^{p}\xi^{i}v_{J/i}^{\alpha} - \sum_{i=1}^{p}D_{\kappa}\xi_{J/i}^{\alpha}\right\}\\ &= D_{k}D_J\left(\phi_{\alpha} - \sum_{i=1}^{p}\xi^{i}v_{i}^{\alpha}\right) + \sum_{i=1}^{p}(D_{\kappa}\xi^{i}v_{J/i}^{\alpha} + \xi^{i}v_{J/i\kappa}^{\alpha}) - \sum_{i=1}^{p}D_{\kappa}\xi^{i}v_{J/i}^{\alpha}\\ &= D_{\kappa}D_J\left(\phi_{\alpha} - \sum_{i=1}^{p}\xi^{i}v_{i}^{\alpha}\right) + \sum_{i=1}^{p}\xi^{i}v_{J/i\kappa}^{\alpha},\end{aligned}$$

with $v_{J/i\kappa}^{\alpha} = \frac{\partial^2 v_J^{\alpha}}{\partial z^i \partial z^{\kappa}}$. Thus, the form of $\phi_{\alpha}^{J/\kappa}$ is like (7.10), which completes the proof. □

7.3.3 Criterion of maximal rank and infinitesimal invariance for differential equations

The given system Λ of equations E_ν will possess the Jacobian matrix $J_E(z, v^{(n)}) = (\partial E_\nu/\partial z^i, \partial E_\nu/\partial v_J^\alpha)$ and if $E(z, v^{(n)}) = 0$, then the maximal rank conditions state that the matrix is of maximum rank with rank l.

7.3.3.1 Theorem 7.2

Let every infinitesimal generator $\mathbf{A} \in S$ on manifold X satisfy the equation

$$pr^{(n)}\mathbf{A}[E_\nu(z, v^{(n)})] = 0, \quad \nu = 1, \ldots, l \quad \text{whenever} \quad E(z, v^{(n)}) = 0, \tag{7.12}$$

then S will also be the symmetry group for the system of equations Λ defined in X, if the system Λ of equations E_ν is of maximal rank.

Proof. Using theorem 7.1 and the condition of maximal rank for the associated Jacobian matrix, the proof is immediate [3].

7.3.4 Differential equations and symmetry groups

Theorem 7.2, along with formula (7.12) gives an effective analytical tool to calculate the symmetry group S of any system Λ. The key idea is to start with an arbitrary infinitesimal generator $\mathbf{A}$. The coefficients ξ and ϕ have a functional dependence on z and v. The corresponding prolonged vector field $pr^{(n)}\mathbf{A}$ will have the transformed coefficients $\phi_\alpha^J(z, v)$. These transformed coefficients involve the partial derivatives of ξ^i and ϕ_α. Hence, one can say that the infinitesimal criterion of invariance (7.12) involves z, v, the z derivative of v, ξ^i, ϕ_α and their higher order partial derivatives with respect to z and v. The natural constraints of the system allow us to eliminate any dependencies among the derivatives of v. The remaining unconstrained partial derivatives of v can then be equated to zero. The resulting PDE system is a set involving the functions ξ^i and ϕ_α. These equations form the symmetry determining equations for the system Λ. The corresponding general solution generates the infinitesimal symmetries for the system Λ. Once the symmetries are available we can use their algebraic properties for different applications. Using exponentiation of the given vector fields, the general explicit expressions for the symmetry groups can be calculated. We will illustrate the procedure with the following example.

7.3.4.1 The Korteweg–de Vries equation (KdV)

This equation [3] has two ($p = 2$) independent variables $z \equiv (z, t)$ and one ($q = 1$) dependent variable v and is given by

$$E(z, v^{(3)}) \equiv v_t + vv_z + v_{zzz} = 0, \tag{7.13}$$

where $v_t = \partial v/\partial z$, $v_{zzz} = \partial^3 v/\partial z^3$ etc. Using theorem 7.2, we can write any vector field as

$$\mathbf{A} = \xi(z, t, v)\frac{\partial}{\partial z} + \eta(z, t, v)\frac{\partial}{\partial t} + \phi(z, t, v)\frac{\partial}{\partial v}.$$

This generates a one-parameter infinitesimal group of (7.13) if

$$pr^{(3)}\mathbf{A}[E(z, v^{(3)})] = 0 \qquad \text{when, } E = 0.$$

Simple calculations give

$$\phi^{t} + \phi^{zzz} + v\phi^{z} + v_z\phi = 0 \tag{7.14}$$

whenever v is a solution of (7.13). In (7.14) the different expressions for ϕ represent the coefficients of prolongations of $\mathbf{A}$ with ϕ^{zzz} representing the coefficients of $\partial/\partial v_{zzz}$ in $pr^{(3)}\mathbf{A}$, etc. Now the coefficient ϕ^{zzz} is given by

$$\phi^{zzz} = D_z^3\phi - v_z D_z^3\xi - v_t D_t^3\eta - 3v_{zz}D_z^2\xi - 3v_{zt}D_z^2\eta - 3v_{zzz}D_z\xi - 3v_{zzt}D_x\eta.$$

Similar expressions can be obtained for other coefficients as well. Once this is done one needs to put all these in (7.14) while making sure to replace v_t by $-(vv_z + v_{zzz})$ in all the expressions before simplifying. The working rule to analyze these equations is to start first from the solution of the coefficients of highest order derivatives equated to zero. Proceeding in this manner we see the coefficient of v_{zzt} is $D_z\eta$. Thus, the solution of $D_z\eta = 0$ gives the solution that η depends on t only. Next the coefficient of v_{zz}^2 gives $\xi_v = 0$, which gives the result $\eta_t = 3\xi_z$ from the coefficient of v_{zzz}. Thus, we obtain the result $\xi = \frac{1}{3}\eta_t z + \sigma(t)$. Next the coefficient of v_{zz} is zero thus making ϕ linear in v and v a function of t only. Next v_z gives the equation

$$-\xi_t - v(\phi_v - \eta_t) + v(\phi_v - \xi_z) + \phi = 0,$$

and the terms without any derivatives of u give

$$\phi_t + \phi_{zzz} + v\phi_z = 0.$$

The final general solution gives

$$\begin{aligned} \xi &= c_1 + c_3 t + c_4 z, \\ \eta &= c_2 + 3c_4 t, \\ \phi &= c_3 - 2c_4 v, \end{aligned}$$

where c are arbitrary constants and hence we obtain the four-dimensional vector fields for the KdV given by

$$\begin{aligned} A_1 &= \frac{\partial}{\partial z} \\ A_2 &= \frac{\partial}{\partial t} \\ A_3 &= t\frac{\partial}{\partial z} + \frac{\partial}{\partial v} \\ A_4 &= 3t\frac{\partial}{\partial t} + z\frac{\partial}{\partial z} - 2v\frac{\partial}{\partial v}. \end{aligned}$$

7.3.5 Differential invariants and the group invariant solutions

7.3.5.1 Definition

For the action of S on $X \subset Z \times V$, the nth order differential invariant of S is given by a smooth function $\Delta\colon X^{(n)} \to \mathfrak{R}$. If $pr^{(n)}g \cdot (z, v^{(n)})$ is defined then the expression for Δ for all $g \in S$ is given by:

$$\Delta(pr^{(n)}g \cdot (z, v^{(n)})) = \Delta(z, v^{(n)}), \ \ (z, v^{(n)}) \in Z^{(n)}.$$

7.3.5.2 Group invariant solutions and steps for calculations

A solution $v = h(z)$ of the system Λ is also the S-invariant solution, if the invariance is maintained for all the elements in S. Precisely, for each $g \in S$, the functions $h(z)$ and $g.\ h$ are the same on their common domains and a solution $v = h(z)$ has a locally S-invariant subset given by the graph $T_h \equiv \{(z, h(z))\} \subset X$. Now we give the algorithmic steps for finding such a solution with s-dimensional orbits of the group.

- Using the basic prolongation formula and the criterion of invariance, find all the vector fields $\mathbf{A}$ for the given system.
- The system Λ is reduced to a transformed system depending on the new $p - s$ independent variables and choosing $s = p - 1$, find all the s-dimensional subgroups of S.
- For a chosen symmetry group S, the set of all functionally independent invariants are constructed and are divided into two categories of new independent $y^i = y^i(z, v)$ and new dependent variables $u^j = v^j(z, v)$, respectively.
- Next we solve for the $p - s$ of the original independent variables z and denote the solution by $\tilde{z}$. Similarly, all the v are also solved in variables y, u and the remaining variables are sorted out as z_r. Further, the z-derivatives of any S-invariant dependent variable v are transformed in terms of new variables and their derivatives as well as z_r, using the chain rule. All of them are then substituted in the original equation $E(z, v^{(n)}) = 0$ so that we find the reduced system $E/S(y, v^{(n)}) = 0$, which will be independent of z_r. The solutions $u = f(y)$ can be reverted back to the original variables and gives the S-invariant solutions $v = h(z)$ for the original system.
- Finally, we repeat all the steps for all the different elements of the symmetry group S. The result is a complete set of group invariant solutions for the original system Λ.

Now we give an example of group invariant solution for the KdV equation. We consider the group of scaling symmetries $(z,\ t,\ v) \to (\varepsilon z,\ \varepsilon^3 t,\ \varepsilon^{-2} v)$, generated by the generator $\mathbf{A}_4$ of the KdV equation. We can find functions $F(z, t, v)$ for any of its generators $\mathbf{A}_i$, such that they satisfy the equations $\mathbf{A}_j F \equiv 0$. This equation can be solved for each of the generators by the method of characteristics and can be written as

$$\frac{dz}{\xi(z, t, v)} = \frac{dt}{\eta(z, t, v)} = \frac{du}{\phi(z, t, v)}. \tag{7.15}$$

The solutions give the invariants for the system corresponding to the generator under consideration. The invariants are then treated as new variables for equation (7.13). Using the procedure enumerated before the global invariants of $\mathbf{A}_4$ in the upper half space ($t > 0$) for the KdV are given by $y = t^{-\frac{1}{3}}z$ and $u = t^{\frac{2}{3}}v$. Now treating u as the function $u(y)$ we obtain the reduced equation

$$u_{yyy} + uu_y - \frac{1}{3}yu_y - \frac{2}{3}u = 0.$$

This equation gives solutions in terms of the second Painlevé transcendent [3] given by equation

$$w_{yy} = \frac{1}{3}w^3 + \frac{1}{3}yw + \kappa$$

with $u = w_y - \frac{1}{6}w^2$ and κ being a constant. Thus, the final similarity solution can be written in terms of the second Painlevé transcend for the KdV.

7.4 Classification of the group invariant solutions and optimal solutions

For each of the s-parameter subgroups $H \subset S$, of a differential equation system in $p > s$ independent variables, we will find group invariant solutions. It is a well established fact [3] that there are always an infinite subgroups associated with a given differential equation system and hence they have infinite group invariant solutions. Thus, it becomes very important to classify the solutions so that only the inequivalent solutions are sorted out for the system, thus giving the optimal group invariant solutions. The process is fairly algorithmic and will be made clear using the cylindrical Korteweg–de Vries (cKdV) equation as an example.

7.4.1 Adjoint representation for the cKdV and optimization of the group generators

The cylindrical Korteweg–de Vries (cKdV) equation originates in different physical situations in non-planar cylindrical geometry and is given by

$$v_t + \frac{v}{2t} + vv_z + v_{zzz} = 0. \tag{7.16}$$

The cKdV has the four-dimensional Lie point symmetry generators [8, 9] given by

$$A_1 = \frac{\partial}{\partial z}$$
$$A_2 = 3t\frac{\partial}{\partial t} + z\frac{\partial}{\partial z} - 2v\frac{\partial}{\partial v}$$
$$A_3 = 2t^{\frac{1}{2}}\frac{\partial}{\partial z} + t^{-\frac{1}{2}}\frac{\partial}{\partial v}$$
$$A_4 = 2zt^{\frac{1}{2}}\frac{\partial}{\partial z} + 4t^{\frac{3}{2}}\frac{\partial}{\partial t} + \left(zt^{-\frac{1}{2}} - 4vt^{\frac{1}{2}}\right)\frac{\partial}{\partial v}.$$

The first step to obtaining optimal group invariant solutions of equation (7.16) following the procedure given in [3] is to find the adjoint representation of the generators. Now for any n-dimensional symmetry algebra S generated by the vector fields $\{\mathbf{A}_1, \mathbf{A}_2, \ldots, \mathbf{A}_n\}$, we can have equivalent elements

$$\left\{\mathbf{A} = \sum_{i=1}^{n} a_i \mathbf{A}_i, \qquad \mathbf{w} = \sum_{i=1}^{n} b_i \mathbf{A}_i\right\} \in S,$$

if any one of the following conditions is satisfied:

- For $g \in G$ we have a transformation $Ad_g(\mathbf{w}) = \mathbf{A}$, where Ad_g is the adjoint of g and is written as $Ad_g(\mathbf{w}) = g^{-1}\mathbf{w}g$.
- There is a constant ϑ, such that $\mathbf{A} = \vartheta\mathbf{w}$.

The use of the Lie series provides an effective tool to find the adjoint system of any symmetry group [3] and can be constructed using the relation

$$Ad_g(\exp(\epsilon\mathbf{A}))\mathbf{w} = \mathbf{w} - \epsilon[\mathbf{A}, \mathbf{w}] + \frac{\epsilon^2}{2!}[\mathbf{A}, [\mathbf{A}, \mathbf{w}]] - \frac{\epsilon^3}{3!}[\mathbf{A}, [\mathbf{A}, [\cdots +]]]\mathbf{w}, \mathbf{A},$$

where $[\mathbf{A}, \mathbf{w}] = \mathbf{A}\mathbf{w} - \mathbf{w}\mathbf{A}$, is the commutator between the two vectors. The tables 7.1 and 7.2 give the commutator table and the adjoint system for the cKdV, respectively. The (i, j)th entry in table 7.1 gives $[\mathbf{A}_i, \mathbf{A}_j]$ and in the table 7.2 the (i, j)th entry indicates $Ad_g(\exp(\epsilon\mathbf{A}_i))\mathbf{A}_j$.

7.4.1.1 Theorem 7.3

Let S be the full vector field group for the cKdV equation with the basis $\{\mathbf{A}_1, \mathbf{A}_2, \ldots, \mathbf{A}_4\}$ and let $\mathbf{A} \in S$ be a subgroup. If $v = h(z)$ is an $\mathbf{A}$-invariant solution to the cKdV, then the transformation $\tilde{h}(z) = gh(z)$ gives a $\tilde{\mathbf{A}}$-invariant solution, where $\mathbf{A} \equiv Ad_g(\mathbf{A}) = g\mathbf{A}g^{-1}$ for all $g \in S$.

Proof. We take any vector $\mathbf{A}_i \in S$. Let $\psi = \{z, t, v\}$ denote the set of all variables for the cKdV equation and $T(\epsilon)$ denote a symmetry mapping generated by vector $\mathbf{A}_i$ such that

$$T(\varepsilon)\colon \psi \mapsto \exp(\varepsilon\mathbf{A}_i)\psi = \tilde{\psi}.$$

We further define the action of $T \equiv T(\varepsilon)$ on any smooth function $F(\psi)$ by

$$TF(\psi) = F(T\psi) = F(\tilde{\psi}). \tag{7.17}$$

Now using (7.17) and assuming the existence of T^{-1} we can write

$$\tilde{\mathbf{A}}F(\psi) = T\mathbf{A}F(\psi) = T\mathbf{A}T^{-1}F(\tilde{\psi}). \tag{7.18}$$

Using the inverse transform theorem we can write $F(\tilde{\psi}) = h(\psi)$ and $F(\psi) = h(\tilde{\psi})$ and this is always true for group invariant solutions generated by Lie symmetry groups [2, 3] and hence we can write, from (7.18),

Table 7.1. Commutator table for cKdV.

	A_1	A_2	A_3	A_4
$\mathbf{A}_1$	0	$\mathbf{A}_1$	0	$\mathbf{A}_3$
$\mathbf{A}_2$	$-\mathbf{A}_1$	0	$\frac{A_3}{2}$	$\frac{3}{2}\mathbf{A}_4$
$\mathbf{A}_3$	0	$-\frac{A_3}{2}$	0	0
$\mathbf{A}_4$	$-\mathbf{A}_3$	$-\frac{3}{2}\mathbf{A}_4$	0	0

Table 7.2. Adjoint action for cKdV.

Ad_g	A_1	A_2	A_3	A_4
$\mathbf{A}_1$	$\mathbf{A}_1$	$\mathbf{A}_2 - \epsilon\mathbf{A}_1$	$\mathbf{A}_3$	$\mathbf{A}_4 - \epsilon\mathbf{A}_3$
$\mathbf{A}_2$	$e^{\varepsilon}\mathbf{A}_1$	$\mathbf{A}_2$	$e^{-\frac{\varepsilon}{2}}\mathbf{A}_3$	$e^{-\frac{3}{2}\varepsilon}\mathbf{A}_4$
$\mathbf{A}_3$	$\mathbf{A}_1$	$\mathbf{A}_2 + \frac{\varepsilon}{2}\mathbf{A}_3$	$\mathbf{A}_3$	$\mathbf{A}_4$
$\mathbf{A}_4$	$\mathbf{A}_1 + \epsilon\mathbf{A}_3$	$\mathbf{A}_2 + \frac{3}{2}\epsilon\mathbf{A}_4$	$\mathbf{A}_3$	$\mathbf{A}_4$

$$\tilde{\mathbf{A}}h(\tilde{\psi}) = T\mathbf{A}T^{-1}h(\psi). \tag{7.19}$$

Now an invariant is a function $h(z, t, v)$ for the cKdV if it satisfies the equation

$$\mathbf{A}_i h \equiv 0. \tag{7.20}$$

Hence using (7.19) and (7.20) we can say that if $f = h(z)$ is an invariant solution for $\mathbf{A}$ then $\tilde{f} = h(\tilde{z})$ will be an invariant solution for $\tilde{\mathbf{A}}$, when

$$\tilde{\mathbf{A}} = T\mathbf{A}T^{-1} = \exp(\epsilon\mathbf{A}_i)\mathbf{A}\exp(-\epsilon\mathbf{A}_i) \equiv Ad_{\mathbf{A}_i}(\mathbf{A}). \qquad \square$$

7.4.1.2 Optimal system of subalgebras for the cKdV

The concept of optimal systems of subalgebras was first used in [2]. Starting from the generators of the cKdV group S, any general generator $\mathbf{A} \in V$, where V is the space of all the generators, can be written as

$$\mathbf{A} = c_1\mathbf{A}_1 + c_2\mathbf{A}_2 + c_3\mathbf{A}_3 + c_4\mathbf{A}_4.$$

The coefficients $c_i (i = 1, 2, 3, 4)$ are arbitrary constants. It is obvious that the transformation has infinite subalgebras corresponding to which there are infinite group invariant solutions. To tackle this problem one looks for subalgebras that are similar, meaning they can be connected by some transformation of the group thereby also connecting the group invariant solutions generated by each member of a similar class. Thus, it will be sufficient to take any one member of the similar class as a representative of the class. The set of all inequivalent classes gives the optimized subalgebra for the system.

Now following the procedure in [3] and using the tables 7.1 and 7.2 we will find the optimal system of subalgebras and their corresponding representatives for equation (7.16). The main idea is to simplify as many coefficients of the vector (c_i), as much as possible, by application of the adjoint map given in table 7.2. For the cKdV, the function $\zeta(\mathbf{A}) = c_2$ is found to remain an invariant under the action of the full adjoint (Ad_g) group and can be expressed as

$$\zeta(Ad_g(\mathbf{A})) = \zeta(\mathbf{A}), \qquad \mathbf{A} \in V, \qquad g \in S.$$

Thus, at first we must start by considering different values of c_2. We suppose $c_2 \neq 0$ and set $c_2 = 1$. The transformed vector is then acted on by $Ad_g(\exp(-\frac{2}{3}c_4\mathbf{A}_4))$ so that the coefficients of $\mathbf{A}_4$ vanish and we obtain a new vector

$$\mathbf{A}' = Ad_g\left(\exp\left(-\frac{2}{3}c_4A_4\right)\right)\mathbf{A} = c_1'\mathbf{A}_1 + c_3'\mathbf{A}_3 + \mathbf{A}_2$$

for some other constants c_1', c_3' depending on the original constants c_1, c_3 and c_4. The new vector is acted by $Ad_g(\exp(-2c_3'\mathbf{A}_3))$ so that it is transformed to $\mathbf{A}'' = c_1''\mathbf{A}_1 + \mathbf{A}_2$, which can again be transformed by $Ad_g(\exp(c_1''\mathbf{A}_1))$ so that the coefficients of $\mathbf{A}_1$ vanish and the new vector A''' is equivalent to $\mathbf{A}_2$. Thus, we see that under the adjoint action the vector $\mathbf{A}$ with $c_2 \neq 0$ generates subalgebras that are equivalent to the subalgebra spanned by $\mathbf{A}_2$. The rest have $c_2 = 0$ and are of the form

$$\mathbf{A} = c_1\mathbf{A}_1 + c_3\mathbf{A}_3 + c_4\mathbf{A}_4.$$

Next we suppose $c_4 \neq 0$ and put $c_4 = 1$. The vector is then transformed by $Ad_g(\exp(c_3\mathbf{A}_1))$ so that the coefficients of $\mathbf{A}_3$ vanish and we obtain the new transformed vector

$$A' = c_1'\mathbf{A}_1 + \mathbf{A}_4.$$

No further simplification is possible for this vector and therefore the coefficient c_1' can be chosen only as 0 or ± 1. Hence, any one-dimensional vector with $c_2 = 0$ and $c_4 \neq 0$ is spanned by either

$$\mathbf{A}_4, \qquad \mathbf{A}_4 + \mathbf{A}_1 \qquad \text{or} \quad \mathbf{A}_4 - \mathbf{A}_1.$$

All the remaining cases can be solved similarly [9] and we obtain an optimized system of subalgebras for the cKdV to be spanned by the vectors

$$\mathbf{A}_1, \qquad \mathbf{A}_2, \qquad \mathbf{A}_3, \qquad \mathbf{A}_4, \qquad \mathbf{A}_4 + \mathbf{A}_1, \qquad \mathbf{A}_4 - \mathbf{A}_1.$$

Again we note that $(x, t, v) \to (-x, t, -v)$ gives the discrete symmetry associated with the cKdV which maps $\mathbf{A}_4 - \mathbf{A}_1$ to $\mathbf{A}_4 + \mathbf{A}_1$, and thus reduces the number of inequivalent subalgebras for the cKdV equation to five, i.e.

$$\mathbf{A}_1, \qquad \mathbf{A}_2, \qquad \mathbf{A}_3, \qquad \mathbf{A}_4, \qquad \mathbf{A}_4 + \mathbf{A}_1, \tag{7.21}$$

which gives the optimal system of subalgebras for the cylindrical Korteweg–de Vries (cKdV) equation.

7.4.2 Calculation of the optimal group invariant solutions for the cKdV

Having found the optimized system of one-dimensional subalgebras for the cKdV, we are now in a position to calculate the corresponding optimized system of group invariant solutions in this case. However, it must be noted that each solution will have a singularity at $t = 0$. Now we proceed with each of the generators in the optimal subalgebra (7.21):

i. $\mathbf{A}_1 = \frac{\partial}{\partial z}$

The invariants for the generator are given by $y = t$, $u = v$ and the corresponding reduced equation is

$$\frac{du}{dy} + \frac{u}{2y} = 0.$$

Simple integration gives the solution $u(y) = a/\sqrt{y}$, where a is a constant. Thus, in terms of original variables the solution becomes

$$v = at^{-\frac{1}{2}}. \tag{7.22}$$

ii. $\mathbf{A}_2 = 3t\frac{\partial}{\partial t} + z\frac{\partial}{\partial z} - 2v\frac{\partial}{\partial v}$.

For this generator $\xi(z, t, v) = z$, $\eta(z, t, v) = 3t$, $\phi(z, t, v) = -2v$. The invariants for the generator are given by $y = zt^{-\frac{1}{3}}$ and $u = t^{\frac{2}{3}}v$. The reduced equation for (7.16) is given by

$$\frac{d^3u}{dy^3} + \frac{du}{dy}\left(u - \frac{y}{3}\right) - \frac{u}{6} = 0. \tag{7.23}$$

The solutions of equation (7.23) transform to the group invariant solution

$$v = t^{-\frac{2}{3}}u\left(zt^{-\frac{1}{3}}\right). \tag{7.24}$$

iii. $\mathbf{A}_3 = 2t^{\frac{1}{2}}\frac{\partial}{\partial z} + t^{-\frac{1}{2}}\frac{\partial}{\partial v}$.

The invariants in this case are $y = t$ and $u = z/2t - v$. The corresponding reduced equation is given by

$$\frac{du}{dy} + \frac{u}{y} = 0$$

with solutions $u = b/y$, where b is an arbitrary constant. The final solution in terms of v is

$$v = \frac{z}{2t} - \frac{b}{t}. \tag{7.25}$$

iv. For $\mathbf{v}_4$, the invariants are $y = zt^{-\frac{1}{2}}$ and $u = vt - z/2$. The reduced equation after two integrations gives

$$\left(\frac{du}{dy}\right)^2 = -\frac{u^3}{3} + mu + n,$$

where m and n are arbitrary constants. It has solution $u = -12\wp(y)$, where the Weierstrass elliptic functions ($\wp$) [4] satisfy the equation

$$\left(\frac{d\wp}{dy}\right)^2 = 4\wp^3 - g\wp - g_1, \tag{7.26}$$

where g and g_1 are the invariants for $\wp(y)$. Now, if a, b and c are the roots of equation (7.26) then we can have the following situations:

- When $a < b < c$ then the solution for $t > 0$ gives the cnoidal wave solution [4]

$$v = \frac{z + 2\delta}{2t} + \frac{2\delta}{s^2 t} dn^2\left(\sqrt{\frac{\delta}{6s^2}}\, zt^{-\frac{1}{2}}, s\right), \tag{7.27}$$

  where $dn(y, s)$ is the Jacobian elliptic function with amplitude $\beta = (c - a)/2$ and the modulus $s = [(c - b)/(c - a)]^{\frac{1}{2}}$.

- If $a = b < c$, then for the boundary conditions $u(y) \to a$, i.e. $\frac{du}{dy}$, $\frac{d^2u}{dy^2} \to 0$ as $y \to \infty$ we obtain the famous soliton-like solution [4],

$$v = \frac{z + 2a}{2t} + \frac{1}{t}(c - a)\mathrm{sech}^2\left(\sqrt{\frac{c - a}{12}}\, zt^{-\frac{1}{2}}\right). \tag{7.28}$$

- If $a = b = c$, then we have the solution with some constant ν,

$$v = \frac{z}{2t} - \frac{12}{t(zt^{-\frac{1}{2}} - \nu)^2}. \tag{7.29}$$

v. For $\mathbf{A}_4 + \mathbf{A}_1$, the invariants are $y = zt^{-\frac{1}{2}} + 1/4t$ and $u = vt - yt^{\frac{1}{2}}/2 - t^{-\frac{1}{2}}/8$. The resulting reduced equation can be integrated to give the first Painlevé transcendent,

$$\frac{d^2u}{dy^2} + \frac{u^2}{2} - \frac{y}{4} = \sigma,$$

where σ is a constant. The corresponding solution in terms of v are given by,

$$v = \frac{zt^{-\frac{1}{2}}}{2} + \frac{1}{8t} + \frac{t^{-\frac{3}{2}}}{8} + \frac{1}{t}u\left(zt^{-\frac{1}{2}} + \frac{1}{4t}\right). \tag{7.30}$$

The solutions for v between equations (7.22)–(7.30) are the required optimized system of solutions. By definition, under suitable group action, any of the group invariant solutions lead to some other group invariant solutions for the same equation and an example for the cKdV can be found in [9].

7.5 Concluding remarks

Historically, the transformational properties of some group actions on a given space were found to have the properties of Lie groups. Cartan [5] was the first to use coordinate charts in building a concrete foundation of the concept of a manifold. The general analytical methods using the concepts of manifolds appeared in the work of Palais [10], in which group invariant solutions are also discussed. Subsequently, the concepts of quotient manifolds were also introduced by Olver [3]. Although Lie had knowledge of the adjoint representation of Lie groups, Ovsiannikov [2] used it to classify the group invariant solutions and gave the concept of optimized solutions for differential equations. The technique was used further by Patera *et al* [11], who applied it to some very important symmetry groups in physics. Further generalization techniques were proposed by Ovsiannikov [2] and the duo Bluman and Cole [12], and also by Ames [13], and are popularly known as the nonclassical methods. A direct method for PDEs was introduced by Clarkson and Kruskal [14]. The analytical connections of all these methods were proved by Levi and Winternitz [15], while Nucci and Clarkson [16] showed that it is not quite as general as the nonclassical approach. Galaktionov [17] has provided a further promising generalization called nonlinear separation.

References

[1] Lie S 1888 *Theorie der Transformationsgruppen* (Leipzig: Teubner)

[2] Ovsiannikov L V 1982 *Group Analysis of Differential Equations* (New York: Academic)

[3] Olver P J 1993 *Applications of Lie Groups to Differential Equations* (New York: Springer)

[4] Ibragimov N H 1994 *Lie Group Analysis of Differential Equations* vol 1 (Boca Raton, FL: CRC Press)

[5] Cartan E 1930 *La Théorie des Groupes Finis et Continus et l, Analysis Situs Mém. Sci. Math.* 42 (Paris: Gauthier-Villars)

[6] Chevalley C 1946 *Theory of Lie groups I* (Princeton, NJ: Princeton University Press)

[7] Warner F W 1971 *Foundations of Differentiable Manifolds and Lie Groups* (Berlin: Springer)

[8] Zakharov N S and Korobeinikov V P 1980 Group analysis of the generalised Korteweg–de Vries–Burger's equation *J. Appl. Math. Mech.* **44** 668

[9] Gupta S K and Ghosh S K 2017 Classification of optimal group-invariant solutions: cylindrical Korteweg–de Vries equation *J. Optim. Theory Appl.* **173** 763–9

[10] Palais R S 1957 A global formulation of the Lie theory of transformation groups *Memoirs of the American Mathematical Society Number 22* (Providence, RI: American Mathematical Society)

[11] Patera J, Winternitz P and Zassenhaus H 1975 Continuous subgroups of the fundamental groups of physics. I. General method and the Poincaré group *J. Math. Phys.* **16** 1597–614

[12] Bluman G W and Cole P D 1969 The general similarity solution of the heat equation *J. Math. Mech.* **18** 1025–42

[13] Ames W F 1965 *Nonlinear Partial Differential Equations in Engineering* (New York: Academic)

[14] Clarkson P and Kruskal Z 1989 New similarity reductions of the Boussinesq equation *J. Math. Phys.* **30** 2201–13

[15] Levi D and Wintemitz P 1989 Nonclassical symmetry reduction: example of the Boussinesq equation *J. Phys.* A **22** 2915–24

[16] Nucci Z C and Clarkson P 1992 The nonclassical method is more general than the direct method for symmetry reductions. An example of the Fitzhugh–Nagumo equation *Phys. Lett.* A **164** 49–56

[17] Galaktionov V A 1990 On new exact blow-up solutions for nonlinear heat conduction equations with source and applications *Diff. Int. Eq.* **3** 863–74

Chapter 8

Learning classifier system

Kapil Kumar Nagwanshi

The learning classifier system (LCS) is a technique which utilizes the power of genetic algorithms and machine learning to make decisions based on specific rules. The general components of LCSs include rule based and decision systems, learning algorithms and rule discovery systems. The current chapter aims to discuss different learning classifier systems for optimization. To understand each of these classifier systems some datasets are required. MATLAB® has been utilized to simulate the behavior and track the use of each of these classifiers. Examples of such learning systems include tree, K-nearest-neighbor, support vector machine discriminant, Bayes' and ensemble classifiers. The degree of each of these classifiers has also been chosen to describe the performance. The principal component analysis may used to reduce the learning dimension to obtain a faster result. Multicore CPU and GPU play a significant role in speeding up the system's performance. Note, classifiers can be used with every classification problem. This chapter also helps readers to choose the best classifier for their work. Some other tools are also discussed at the end of the chapter. The performance of the system depends on the dataset used. Case studies suggest the type of result, such as a scatter plot, confusion matrix, parallel plot and ROC curves. Later in this chapter, Python is introduced for developing the learning classification. This chapter also aims to describe cloud-based products such as BigML® and Microsoft® AzureML® for the optimization of classification problems. The chapter ends with concluding remarks including benchmarking.

8.1 Introduction

The Internet is full of data. If you want to know some details about any particular topic, it will be able to suggest anything from a few items to a very long list. From a downloaded search, concluding a result can be a straightforward task in a few cases, but for all other cases it will be a complex problem. There are a significant

doi:10.1088/978-0-7503-2404-5ch8

number of patterns available to determine a result. These patterns can be obtained from a different dataset either created by an individual person or available on the Internet for free or a fee. Learning classifier systems (LCSs) are effective in recognizing such patterns to solve a variety of problems from difficult to elementary questions.

Holland [1] introduced the concept of the LCS. Later, in 1988, a guest editorial was written by Goldberg and Holland [2] which described the utilization of genetic algorithms (GAs) and machine learning. In the same editorial, they also explained the GAs and classifier systems with a significant set of references. According to Urbanowicz and Moore [3], 'A learning classifier system (LCS) is defined as a model of rule based machine learning technique based on evolutionary computing that consolidates a discovery component, for example, genetic algorithm (GA) with a learning component such as supervised, unsupervised or reinforcement learning'. Similarly, the classifiers are defined by mapping the input stimuli to the output action through a population of condition–action–payoff rules. LCSs are suitable for classification as well as optimization of any problem domain that is solved or approximated by a distributed set of local approximations [4]. This chapter starts with the background of LCS, followed by a MATLAB demonstration of a different set of classifiers using a graduate admissions dataset under a Creative Common licence [5].

8.2 Background

In the year 2000 Holland *et al* [6] presented the answer to what a learning classifier system was, according to the best researchers of the time. Holland divided the classifier algorithm into three parts: the first part addresses parallelism and coordination, the second part gives the credit assignment and the third part describes the rule discovery. Then he described how classifier systems deal with these issues, followed by rule implementation and the description of future directions. Booker described the importance of classification mechanisms in parallelism as well as standardization, that allows construction of a block technique for data processing through the use of differentiation in conflict management.

Riolo [6] found that the main features of such a classifier system should be as follows: (i) A communications message board. (ii) A representation of understanding based on rules. (iii) A contest to activate guidelines, biased by inputs, past performance and anticipated outcome projections. (iv) Parallel firing of rules, with endogenously emerging consistency and coordination of operation as a rapidly growing state created and preserved by the bid handling dynamics. Only at the effector interface is explicit conflict resolution exclusively implemented. (v) Temporal difference techniques of some sort are used to allocate credit, e.g. the traditional algorithms of the bucket-brigade, some-sharing system, Q-learning algorithms, etc. Note that various types of lines of credit may be assigned simultaneously, e.g. traditional power is anticipated to payoff depending on past performance, some rule of payoff accuracy or consistency [7], or a certain measure of the capacity to forecast a subsequent state [8, 9]. (vi) Heuristics suitable for the modeling scheme create the finding of the rule. Examples include activated bonding to detect the surprise-triggered forecast of asynchronous causal links [10–12], or traditional GAs with mutation and recombination. Holmes *et al* [13] adapted LCS for the design, implementation and evaluation of EpiCS, for knowledge discovery in epidemiological monitoring of a national child automobile passenger protection program. The paper was concluded with significant detail of regression analytics.

Urbanowicz and Moore [3] explain that if the problem is of a high degree of complexity, then LCS can solve the problem efficiently. The LCS has be applied in biology, computer science, artificial intelligence, machine learning, evolutionary biology, evolutionary computation, reinforcement learning, supervised learning, evolutionary algorithms and GAs to build the solution domain as a learning classifier system. Zang *et al* [14] utilized XCS with memory conditions (absent in standard XCS) to solve 'maze environments and the perceptual aliasing problem' with robust output.

8.3 Classification learner tools

Several tools are available to describe the learning classifier system. In this chapter, the author will describe the MATLAB® Classification Learner App, BigML®, and Microsoft® AzureML®. In this text, the author has used these tools wherever it is applicable to affect the output of the classifier system.

The readers need to have a licensed or demo version of MATLAB to run their experiments. It is recommended that the latest version of MATLAB is installed. Currently, readers can try the version MATLAB 2019a. If the reader's organization has already purchased MATLAB and other cloud-based tools, it is straightforward to start working on it. The BigML and AzureML tools have been used in order to understand the solution graphically. BigML and AzureML are both cloud-based services one can use by merely creating a trial account on the latest websites.

8.3.1 MATLAB®: classification learner app

The Classification Learner app of MATLAB® trains models to classify data through a diverse set of classifiers to examine supervised machine learning. It will start with data exploration, then selection of features, followed by specific validation schemes if any, and then the models are trained to evaluate outcomes. Automated training could be undertaken to search for the most elegant form of classification, including decision trees, discriminating assessment, support vector machines (SVMs), logistic regression, nearest neighbors, naive Bayes and classification of an ensemble. Learning can be accomplished by providing a piece of input information with a number of observations and associated responses, such as labels or categories. Training is used to train a template to produce fresh information responses. The trained model can also be used in workspace programming with the export of produced MATLAB code [15–17].

8.3.2 BigML®

BigML® is an advanced tool for machine learning, which contributes a collection of robustly engineered algorithms established to elucidate real-life problems. The algorithms based on supervised learning have been utilized to solve classification, regression and time series forecasting while unsupervised learning has been used to provide cluster analysis, anomaly detection, topic modeling and principal components analysis by implementing a single, standardized framework. In this chapter, the author has chosen BigML for its applicability for understanding the characteristic features of different classifiers. BigML provides a broad set of modeling scenarios to analyze and compare, which makes analysis easy. One can directly open an evaluation account to understand BigML through the webpage https://bigml.com [18–20].

8.3.3 Microsoft® AzureML®

The Microsoft Azure Machine Learning Studio is a collaborative, drag-and-drop tool you can use to build, test and deploy predictive analytics solutions on data of interest. It provides a facility to publish models as web services that can efficiently be utilized by custom apps or BI tools such as Excel. In most solutions, one needs to provide the dataset as an input. Next, one is expected to split the dataset into the training dataset and testing dataset, followed by choosing one or more algorithm to train using the train model block. Then the output of the train model block will be validated by the score model block. Finally, the production of the score model is evaluated by the evaluate model block to obtain the output in the form of performance parameters such as ROC, confusion matrix, etc. Readers can obtain a subscription through the AzureML website https://studio.azureml.net [21].

8.4 Sample dataset

Let us describe each classifier type using the graduate admissions dataset as shown in table 8.1 [5]. In this chapter the author is not going to justify the predictions of different models. This chapter focuses on the learning process and its steps to obtain the output. The dataset consists of 500 rows and nine columns, namely, Serial No., GRE Score, TOEFL Score, University Rating, Sum of Product (SOP), LOR, CGPA, Research and Chance of Admit. Of these nine columns, Serial Number has no role in prediction, so this column is deleted from the dataset. The column with the heading Chance of Admit will act as the response class and the remaining columns act as predictors.

Figure 8.1(a) shows the graph of 61 response classes of the original dataset and figure 8.1(b) shows eight response classes after rounding-off the values in the Chance of Admit column from the original dataset. This alteration of data has been done in order to understand the classification learner problem.

Table 8.1. Sample data from the graduate admission dataset.

Serial No.	GRE score	TOEFL score	University rating	SOP	LOR	CGPA	Research	Chance of admit
1	337	118	4	4.5	4.5	9.65	1	0.92
2	324	107	4	4	4.5	8.87	1	0.76
3	316	104	3	3	3.5	8	1	0.72
4	322	110	3	3.5	2.5	8.67	1	0.8
5	314	103	2	2	3	8.21	0	0.65
6	330	115	5	4.5	3	9.34	1	0.9
7	321	109	3	3	4	8.2	1	0.75
8	308	101	2	3	4	7.9	0	0.68
9	302	102	1	2	1.5	8	0	0.5
10	323	108	3	3.5	3	8.6	0	0.45
11	325	106	3	3.5	4	8.4	1	0.52
12	327	111	4	4	4.5	9	1	0.84
13	328	112	4	4	4.5	9.1	1	0.78
14	307	109	3	4	3	8	1	0.62
15	311	104	3	3.5	2	8.2	1	0.61

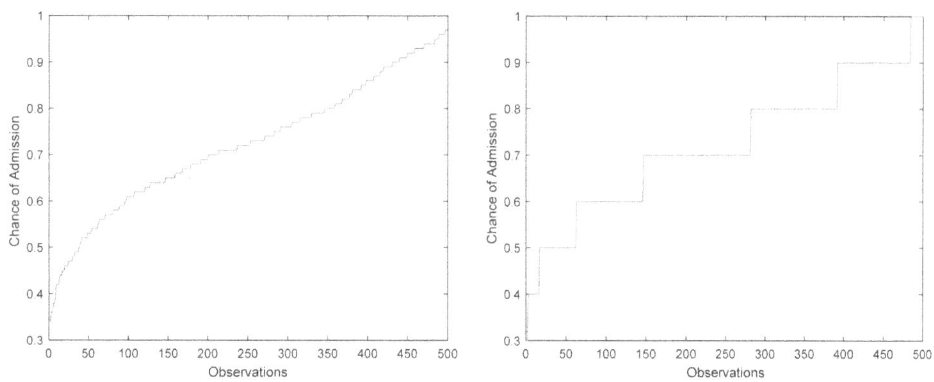

(a) Original Dataset. (b) Simplified Dataset.

Figure 8.1. Graduate admission dataset with (a) 61 classes and (b) 8 classes.

8.4.1 Splitting the dataset

The admission dataset can be split manually or one can run the following MATLAB code to split the dataset for training into the Training dataset, and for testing into the Testing dataset to evaluate the algorithms:

```
admissiontable = readtable('Admission_Predict.csv');
save('admissiontable1');
load admissiontable;
[m,n] = size(admissiontable); P = 0.70;
idx = randperm(m);
Training = admissiontable(idx(1:round(P*m)),:);
 save('Training');
Testing = admissiontable(idx(round(P*m)+1:end),:);
save('Testing');
```

After running the above code, the admissions table is split into the Training table with 350 rows and the Testing table with 150 rows. On the other hand, AzureML provides a Dataset Split module which by default splits the dataset into a 50:50 ratio of training:testing. The subscriber can change this value by selecting this module.

8.5 Learning classifier algorithms

This section deals with learning classifier algorithms in detail with the help of different tools. The following algorithms are discussed here:

- Logistic regression classifiers.
- Decision tree classifiers.
- Discriminant analysis classifiers.
- Support vector machines classifiers.
- Nearest neighbor classifiers.
- Ensemble classifiers.

Before going to test each of the above stated models it is required to load the Admission prediction training dataset using the following MATLAB command (see section 8.4.1 on splitting dataset for the sample dataset):

```
» load Training;
```

After running the above command the variable names were modified as `GREScore`, `TOEFLScore`, `UniversityRating`, `SOP`, `LOR`, `CGPA`, `Research` and `ChanceOfAdmit` to make them valid MATLAB identifiers. The original names are saved in the `VariableDescriptions` property. Now the stated table (Dataset) is available for training of the classification learners. Subsequently, we can now start the classification learner app and start the new session from the workspace with the `admissiontable` dataset available in the workspace. The app will automatically select the predictors and response (by default the last column). One can select any number of predictors from the available list. After selecting predictors and responses, it is necessary to validate the dataset by pressing the Start Session button. By default, the cross-validation is five-fold to protect against overfitting of the data.

The Classification Learner starts with the scatter plotting of the data. After this step, it is necessary to select the classifier(s) to see the classification performance. A scatter plot is useful to check the predictability of the response. The plot has the option to choose different x and y parameters. Figure 8.2(a)–(f) show the scatter plot to examine seven variables for predicting the response by choosing different options on the different x and y parameters under Predictors to reflect the distribution of response variable `ChanceOfAdmit` in different colors.

Note: In MATLAB the classification learner app does not support the regression classifier, and for this purpose there is a separate app called the regression learner. The regression learner is not able to give prediction in the form of a confusion matrix and ROC directly, so to see this the output of AzureML has been used.

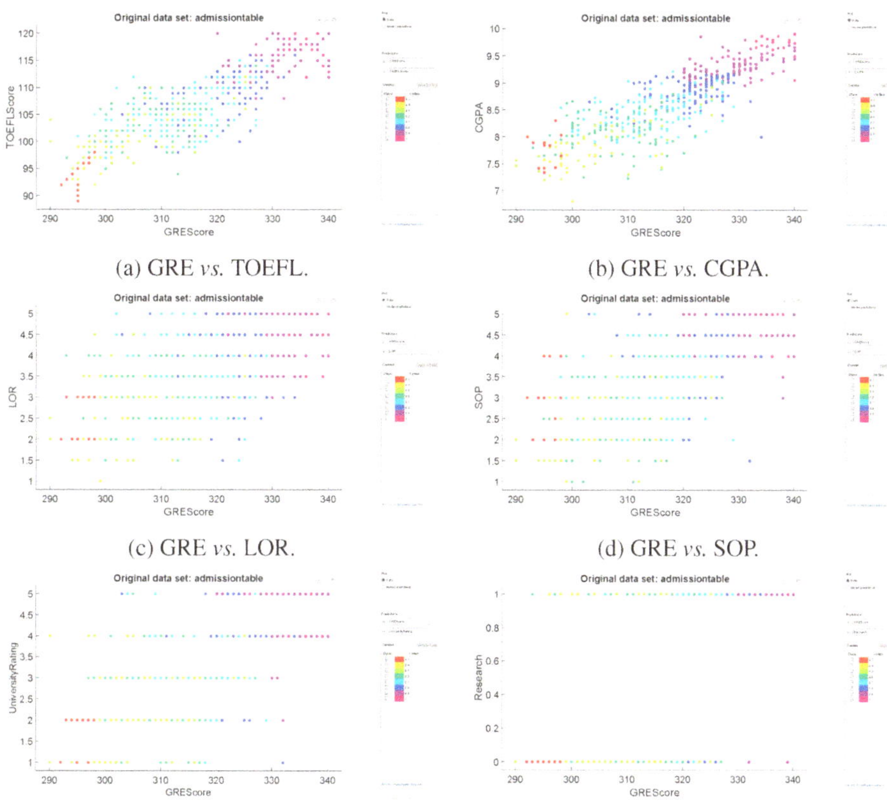

(a) GRE *vs.* TOEFL. (b) GRE *vs.* CGPA.

(c) GRE *vs.* LOR. (d) GRE *vs.* SOP.

(e) GRE *vs.* University Ranking. (f) GRE *vs.* Research.

Figure 8.2. Showing the response in color coded format with respect to GRE versus other predictors.

8.5.1 Logistic regression classifiers

This section will explain the concept of the logistic regression classifier with the help of AzureML as well as MATLAB. The logistic function shown in equation (8.1) is also recognized as the sigmoid function that statisticians use to express the characteristics of population growth in ecology, propagating and maximizing the environmental carrying potential. It is an S-shaped curve that can take any real-valued amount and trace it to a value between 0 and 1, but never at those boundaries exactly:

$$f(t) = \frac{1}{1 + e^{-t}}. \tag{8.1}$$

Figure 8.3 shows the logistic regression flow using AzureML. Here the author has selected a multiclass logistic regression classifier and a two-class logistic regression classifier. The admissions dataset is divided into training and testing datasets. The training model takes two inputs, the left-hand side uses the algorithm, and the right-hand side uses the training dataset. The output of the Train Model block is provided as the left-hand side input of the score model, and the right-hand output of the split data module is taken as the right-hand input of the score model block. The Evaluate model block can take two sources of information for comparison of the algorithm. In this case, because the nature of the algorithm is varied, i.e. one is multiclass and the other is two-class, so the current author took two different Evaluate model blocks.

The output of the Evaluate model for multiclass logistic regression is shown as a confusion matrix (see section 8.6.1 for more details on the confusion matrix) shown in figures 8.1 and 8.7, and for a two-class regression model it is shown in figure 8.4 as

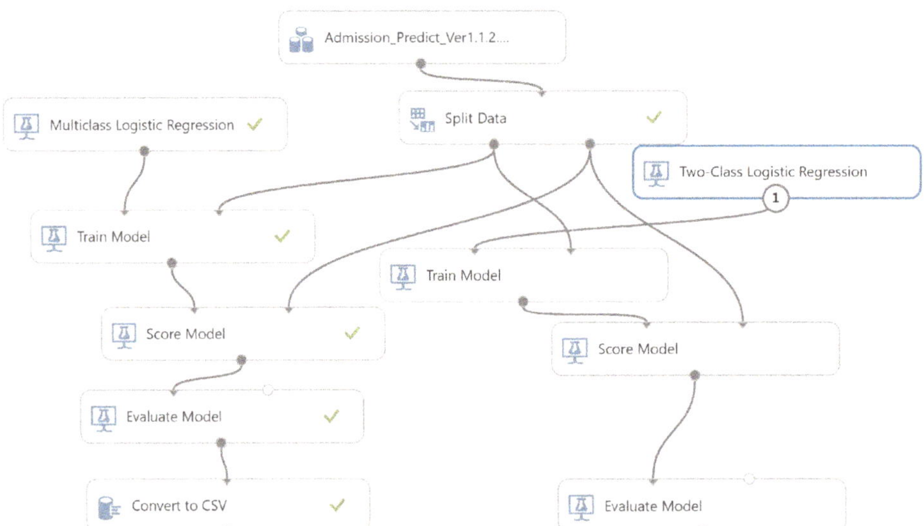

Figure 8.3. Multiclass and two-class logistic regression experimentation in AzureML.

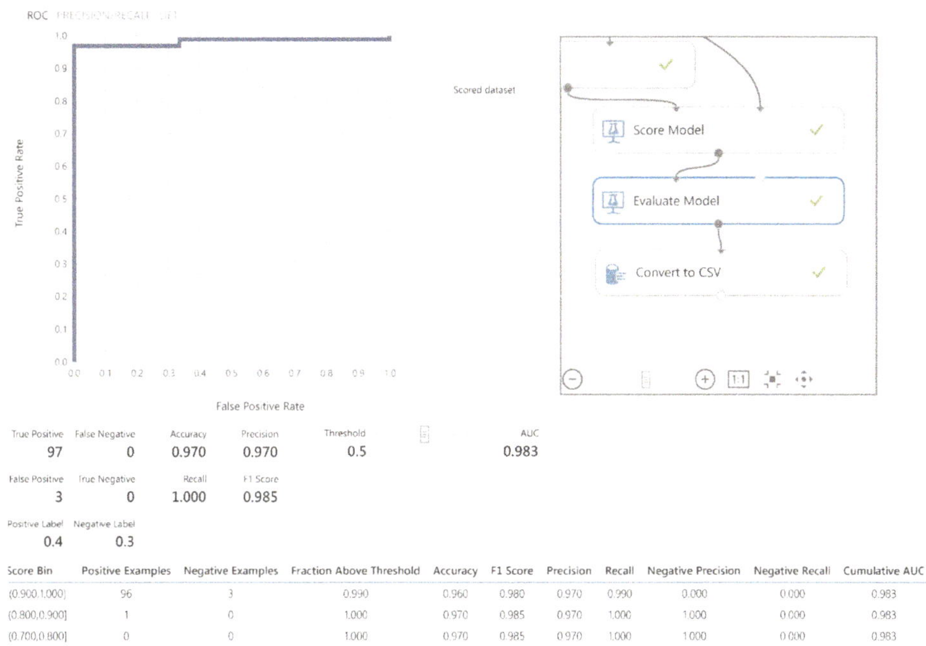

Figure 8.4. ROC two-class logistic regression experimentation in AzureML.

Table 8.2. The confusion matrix and results obtained from AzureML. (ALL: average log loss.)

	Predicted class										
Class	0.3	0.4	0.5	0.6	0.7	0.8	0.9	1	ALL	Precision	Recall
0.3	0	0	0	2	1	0	0	0	2.55	0	0
0.4	0	0	0	0	2	0	0	0	2.87	0	0
0.5	0	0	0	4	4	1	0	0	1.88	0	0
0.6	0	0	0	3	12	0	0	0	1.33	0.33	0.2
0.7	0	0	0	0	21	6	1	0	1.13	0.47	0.75
0.8	0	0	0	0	5	10	3	0	1.14	0.45	0.56
0.9	0	0	0	0	0	5	16	0	0.84	0.67	0.76
1	0	0	0	0	0	0	4	0	2.59	0	0

ROC (see section 8.6.2 on the receiver operating characteristic). Further, the output can be taken as the `.csv` file shown in table 8.2 to download the results of a different set of algorithms.

It is assumed that the reader is aware of the logistic function, therefore, the author does not intend to show the logistic function plot, although readers can go to https://www.geeksforgeeks.org/understanding-logistic-regression/ for more details. One can also assume that the target variable is categorical. Based on the number of classes, logistic regression can be categorized as follows: (i) Binomial logistic regression is used if there are two possible target classes 0 or 1, for example admitted versus not admitted. (ii) Multinomial logistic regression is used when the target variable can have three or more possible classes which are not ordered, i.e. the types have no quantitative significance, for example IPL cricket teams in India can be 'Chennai Super King' versus 'Rajasthan Royals' versus 'Kings Eleven'. (iii) Ordinal logistic regression is used if target classes are ordered, for example a test score can be described as 'very poor', 'poor', 'good' or 'very good'. For this purpose, each category can be given a score, such as 0, 1, 2 and 3.

Algorithm 1. Logistic regression classifier

---

1. Initialize the weights at $t = 0$ to $w(0)$.
2. **repeat** $t = 0 : n$.
3. Compute the gradient: $\nabla E_{in} = -\frac{1}{N}\sum_{n=1}^{N}\frac{y_n x_n}{1 + e^{y_n w^T(t) x_n}}$.
4. Update the weights: $w(t+1) = w(t) - \eta \nabla E_{in}$.
5. **until** Stopping Condition is false.
6. **return** final weights w.

Consider w_i to be the weights or coefficients, x_i to be the input. h represents the hypothesis set that a selected classifier brings and θ is a sigmoid or logistic function. Then the regression function s, shown in figure 8.5, can be given by

$$s = \sum_{i=0}^{d} w_i x_i. \tag{8.2}$$

Now we come to linear classification. For this, the characteristic function for h has binomial values and the shape of the curve is like]. Second, for linear regression the characteristic function for h has binomial values and the shape of the curve is like _/¯. Finally, for logistic regression the characteristic function for h has continuous values and the shape of the curve is like ∫.

Logistic regression is presented as an algorithm. One is a classification technique based on supervised learning; therefore, it helps us to converge those uncertain posteriors with a differentiable decision function. For logistic regression, $w(0)$ can be

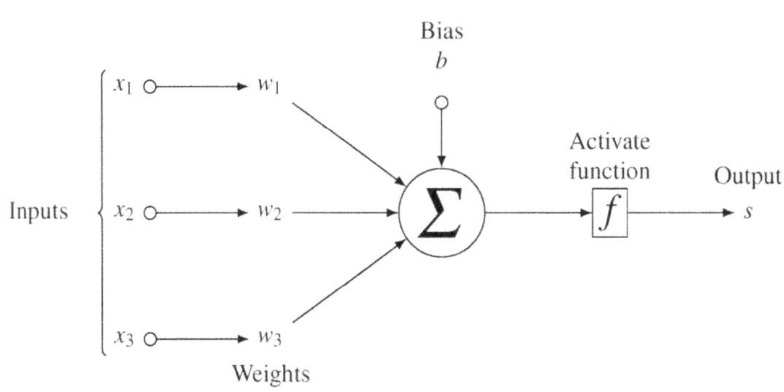

Figure 8.5. Logistic regression function for four inputs with inputs $x_0 = b$ and $w_0 = 1$.

0, it makes sense to start at $\eta = 0.5$ because it is the most uncertain state. Again, maybe it is not a logistical regression issue as other concepts such as neural networks, as the issue of logistical optimization is smoothly convex. Generally speaking, we use a blend of requirements as a stop condition. Let us use the ubiquitous threshold ($\Delta w <$ threshold), for example. The stochastic gradient descent is from time to time a stronger strategy. It is extremely effective and often produces excellent outcomes. The conjugate gradient is the best of all the derivative-based techniques. The concept is to optimize the second order without effectively calculating the Hessian.

In MATLAB based experimentation on the same dataset, figure 8.6 exhibits different graphical results for the LR classifier. The responses have been plotted for predictor versus residual, predicted response versus residual, true response versus predicted response, etc. The confusion matrix obtained from the Evaluate model as a .csv file, shown in table 8.2, also gives output in the form of average log loss, precision and recall. In this way, one can use a logistic regression classifier to obtain and validate the applicability of the algorithms used (figure 8.7).

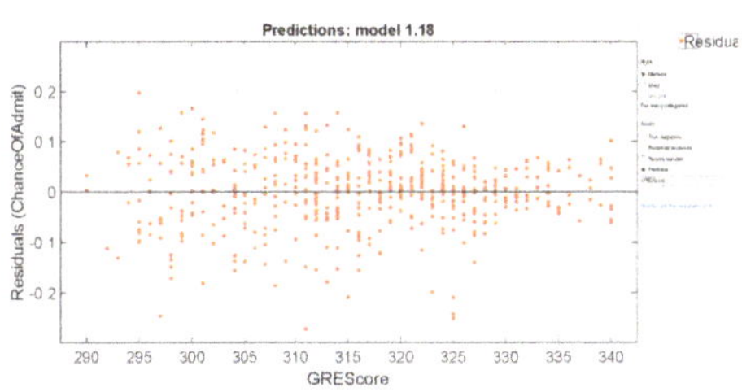

(a) Predictor(GREScore) *vs.* Residual.

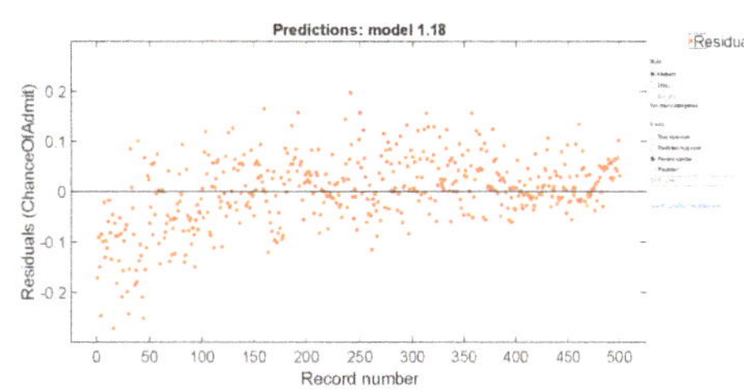

(b) Predictor(Record#) *vs.* Residual.

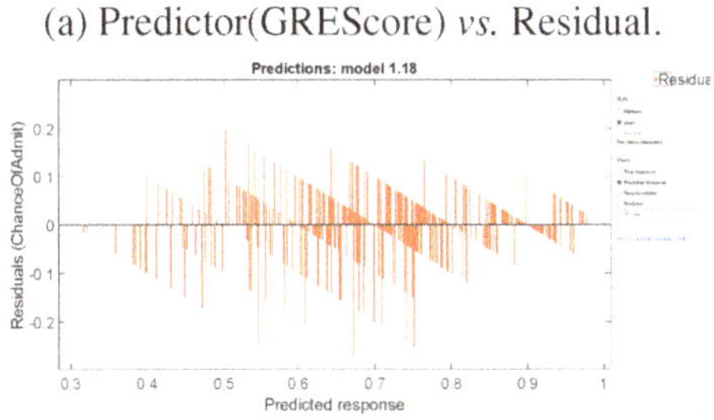

(c) Predicted Response *vs.* Residual.

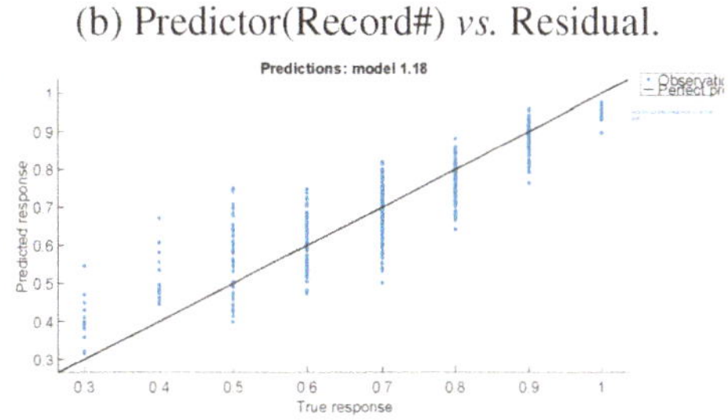

(d) True Response *vs.* Predicted Response.

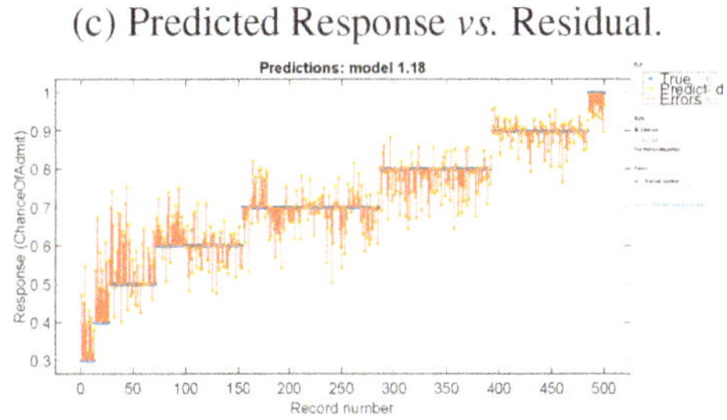

(e) Record# *vs.* Response.

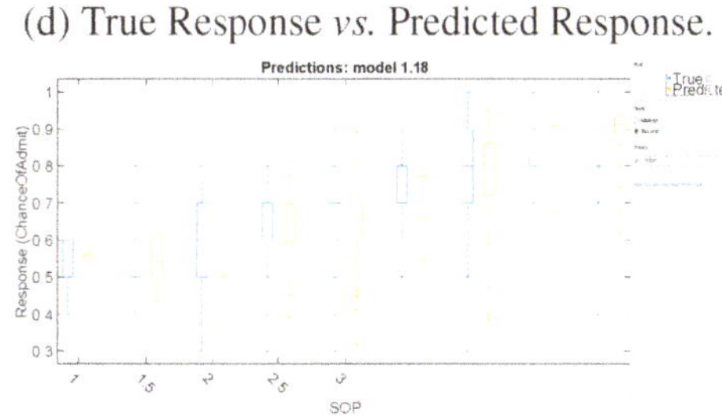

(f) SOP *vs.* Response.

Figure 8.6. Showing the responses in logistic regression classifiers.

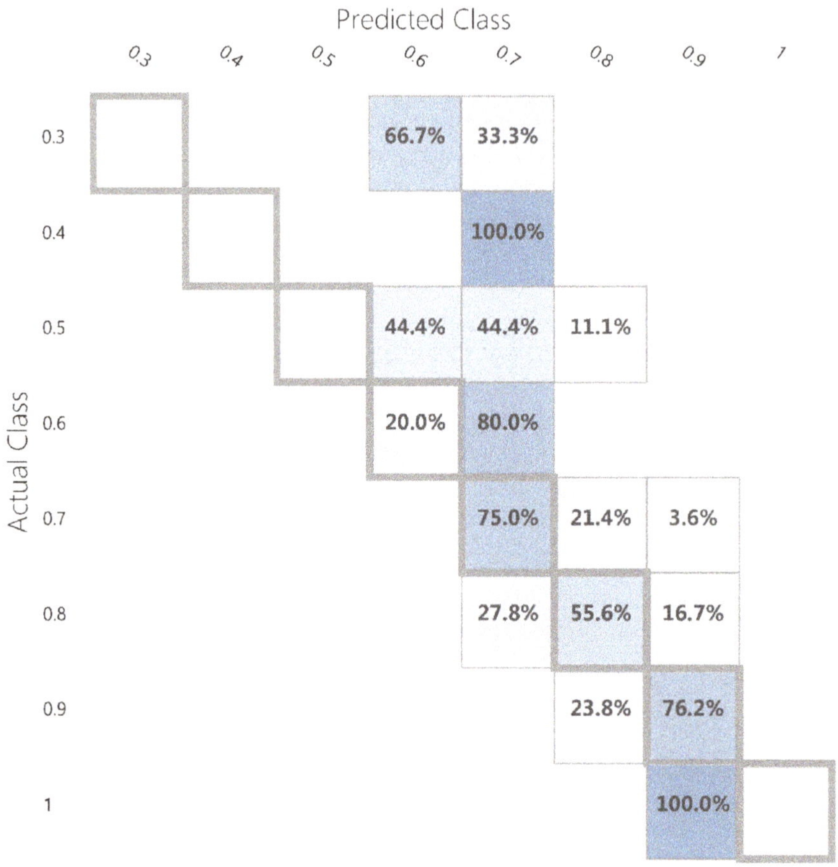

Figure 8.7. Confusion matrix multiclass logistic regression experimentation in AzureML.

8.5.2 Decision tree classifiers

The classification tree algorithm or decision tree algorithm refers to the family of algorithms based on supervised learning that can be used to solve regression and classification issues. With a faster training time, the decision tree belongs to a non-parametric white box class of machine learning methods which does not rely on probability distribution hypotheses to express ongoing decision-making logic based on the SOP or disjunctive normal form prototype, and is not readily accessible in the black box class of algorithms along with the artificial neural networks. The time complexity of decision trees depends on the amount of records and the amount of features in the dataset. Decision trees can accurately handle high-dimensional information. The purpose of this method is to produce a training model for predicting answers by learning the rules of decision making deduced from the training dataset. Compared to imminent classification algorithms, the tree algorithm is straightforward and attempts to determine the solution using representation of the tree. The tree's internal nodes constitute a trait and a class tag for the leaf node. Figure 8.8(a) gives sample code snippet and corresponding tree diagram for the tree classification algorithm in figure 8.8(b)–(c). Algorithm 2 gives the pseudocode for the tree classifier. Initially, it will begin by organizing the most fitting trait of the dataset at the root of the tree, accompanied by splitting the training set into subsets such that each subset comprises data with the same attribute value. This operation operates on each subset indefinitely until leaf nodes are found in all tree branches.

Algorithm 2. Tree classifier

1. Arrange the fittest attribute using attribute selection measures (ASM), such as Gini index, information gain or gain ratio to split the records of the dataset at the root of the tree.
2. Make that attribute a decision node and divide the training set into subsets.
3. Every subset comprises data with the same value for an attribute.
4. Repeat step 1 and step 2 on each subset until it finds leaf nodes in all the branches of the tree.

All the internal nodes in a decision tree act like decision nodes which contain features or attributes, while leaf nodes give the response or outcome through branches which act as decision rules (see figure 8.8(a) and (b)). A topmost decision node is known as the root of the tree.

The measurement of of the attribute in the attribute selection measure choice or splitting rules is a method for selecting the splitting criteria that best compute the partition. It enables us to manage breakpoints on a specified node for tuples. By defining the specified dataset, ASM offers a class for each function and chooses the highest resulting function as a dividing variable. The information gain, gain ratio and Gini index are some common choice criteria [22, 23].

Information gain is a contraction in entropy that measures the distinction between pre-split entropy and the dataset's median post-split entropy based on the function numbers given. Suppose that P_i is the probability that an arbitrary tuple exists in $D \in C_i$. The median quantity of data needed to differentiate a tuple class tag in D is Info(D) given by equation (8.3), in equation (8.4) $\text{Info}_A(D)$ is the anticipated data required to classify a D tuple depending on the characteristic partitioning A, $\frac{|D_j|}{|D|}$ characterizes the weight of the jth partition and the attribute A with the peak information gain, and v is the amount of discrete values in trait A. Equation (8.5) can determine Gain(A). The dividing attribute at node N is selected:

$$\text{Info}_A(D) = -\sum\nolimits_{i=1}^{m} p_i \log_2 p_i \tag{8.3}$$

$$\text{Info}_A(D) = \sum\nolimits_{j=1}^{v} \frac{|D_j|}{|D|} \times \text{Info}(D_j) \tag{8.4}$$

$$\text{Gain}(A) = \text{Info}(D) - \text{Info}_A(D). \tag{8.5}$$

According to Navlani [24], Gain(A) is appropriate for an attribute with many implications. But for a uniquely identified attribute, such as the AADHAR number

```
IF cgpa > 8.73745 AND
    IF cgpa <= 9.0625 AND
        IF gre_score <= 330 AND
            IF sop > 3.75 AND
                IF gre_score > 319 AND
                    IF gre_score > 321 AND
                        IF toefl_score <= 113 AND
                            IF lor <= 4.75 THEN
                                chance_of_admit = 0.8
```

(a)

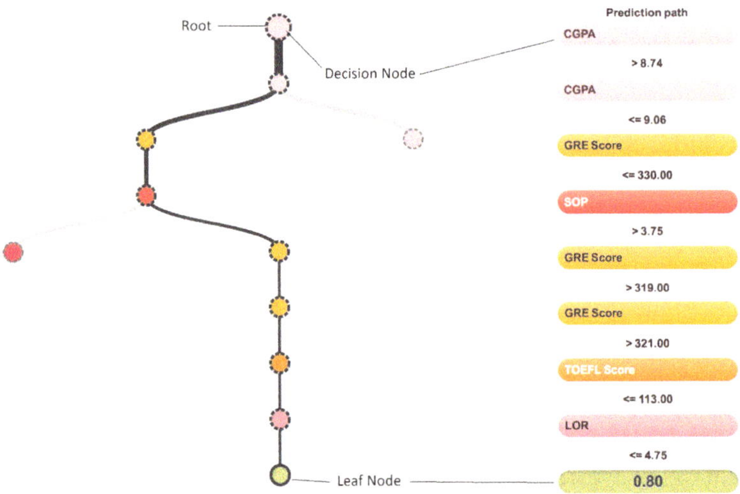

(b)

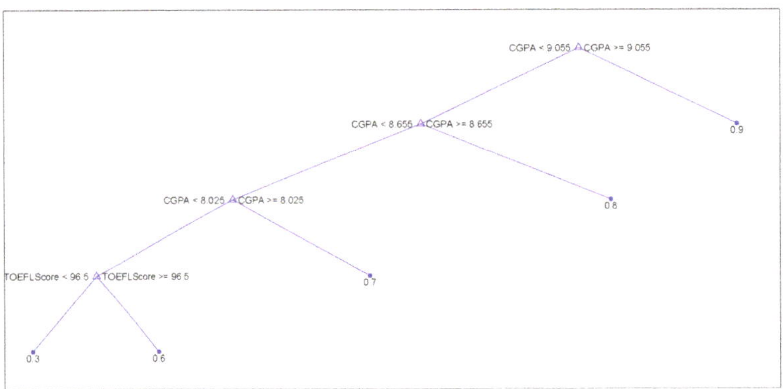

(c)

Figure 8.8. Classification tree algorithm: (a) sample rules; (b) sample tree diagram in BigML; and (c) sample tree diagram in MATLAB (pruning level 9 of 13).

of any student, it has zero Info(D). Due to the tidy partitioning, it maximizes the benefit of data Gain(A) and creates unusable partitioning. The gain ratio GainRatio(A) provided in equation (8.7) can be used to resolve this problem, it handles the bias issue by using divided data to normalize the information gain $\mathrm{SplitInfo}_A(D)$ as shown in equation (8.6). For trait A, the maximum value of the gain ratio GainRatio(A) is chosen as a splitting attribute:

$$\mathrm{SplitInfo}_A(D) = -\sum_{j=1}^{v} \frac{|D_j|}{|D|} \times \log_2 \frac{|D_j|}{|D|} \tag{8.6}$$

$$\mathrm{GainRatio}(A) = \frac{\mathrm{Gain}(A)}{\mathrm{SplitInfo}_A(D)}. \tag{8.7}$$

The Gini index provides another way of splitting the decision tree. For a given partition of data D into D_1 and D_2, the Gini index of D can be estimated by

$$\mathrm{Gini}_A(D) = \frac{|D_1|}{|D|}\mathrm{Gini}(D_1) + \frac{|D_2|}{|D|}\mathrm{Gini}(D_2), \tag{8.8}$$

with

$$\mathrm{Gini}(D) = 1 - \sum_{i=1}^{m} p_i^2. \tag{8.9}$$

If a continuously valued attribute is in the exercise, then select a couple of neighboring values as a prospective split point with a lower Gini index being taken as the split point; on the other side, if it has a discrete valued attribute, a subset giving a minimum Gini index has been chosen as a dividing attribute.

8.5.3 Discriminant analysis classifiers

Discriminant analysis is a classification technique based on Bayes' theorem to estimate the probability estimates that distinct classes produce data based on distinctive Gaussian distributions of predictors in each of the response classes. Linear discriminant analysis aims to map the higher dimensional space characteristics into a smaller dimensional space. If the classes are well separated and if the classes are linearly separable, then other classification techniques such as logistic regression work well, but if the case is complicated with no linearly separated classes, in this case linear discriminant analysis works well. It gives a method to produce classifiers based on labeled training data for a predictive model of group membership to apply on new cases. Unlike some other classifiers which work on a two-dimensional plane, the LDA can work well with higher dimensionality and hyperplanes. The discriminant analysis further produces an equation that can be used to classify new examples [25]. Because of space limitations, the author recommends reading an article written by Tharwat *et al* [26]. It is an excellent tutorial with a graphical representation of the LDA algorithm. The five general steps for performing a linear discriminant analysis are presented in algorithm 3.

Algorithm 3. Discriminant analysis classifier

---

1. Estimate the d-dimensional mean vectors for the distinct classes from the dataset.
2. Compute the scatter matrices (in-between-class and within-class scatter matrix).
3. Compute the eigenvectors (e_1, e_2, ..., e_d) and corresponding eigenvalues (λ_1, λ_2, ..., λ_d) for the scatter matrices.
4. Sort the eigenvectors by decreasing eigenvalues and choose k eigenvectors with the largest eigenvalue to form a $d \times k$-dimensional matrix W (every column represents an eigenvector).
5. Use this $d \times k$ eigenvector matrix to transform the samples onto the new subspace. This can be summarized by the matrix multiplication: $Y[n, k] = X[n, d] \times W[d, k]$ (where X is a $n \times d$-dimensional matrix representing the n samples, and y are the transformed nk-dimensional samples in the new subspace).

Readers can also use the RSS feed by Raschka to go through each of the above steps in detail using Python [27]. LDA is a classification technique that is easy and efficient. Many extensions and variants to the method are available because it is easy and well known. Some popular extensions lead to: (i) regularized discriminant analysis (RDA), which introduces regularization in variance estimation which is effectively covariance, moderating the impact of various factors on LDA; (ii) flexible discriminant analysis (FDA), where nonlinear input components such as splines are used; and (iii) quadratic discriminant analysis (QDA), in which, when there are numerous input factors, each category utilizes its own variance or covariance estimate.

The initial design was termed as discriminant analysis of the linear discriminant or Fisher's discriminant analysis. Multiple discriminant analysis has been linked to the multiclass variant. All of these are now called linear discriminant analysis [28].

8.5.4 Support vector machine classifiers

A support vector machine (SVM) is technically assigned by a separate hyperplane as a discriminative classifier. In other words, considering the labeling of training data (supervised learning), the classifier produces an ideal hyperplane categorizing fresh instances. This hyperplane is a row separating a plane into two sections in two-dimensional spaces in which each category lies on either hand. In principle, the SVM algorithm is applied using a kernel. Hyperplane learning in linear SVM is done by transforming the problem using some linear algebra outside the scope of this introduction to SVM. A strong understanding is that, instead of the results themselves, the linear SVM can be interpreted using the inner product of any two variables. The inner product around two vectors is the sum of each pair of input values multiplying. For example, the inner product of the vectors [3, 4] and [6, 7] is $4 \times 7+3 \times 6$ or 46. The equation for building a prediction for a new input using the dot-product flanked by the input (x) and for each support vector (x_i) can be given by

$$f(x) = B_0 + \sum(a_i \times (x, x_i)). \tag{8.10}$$

Equation (8.10) encompasses determining the inner products of a new input vector (x) with all support vectors in the training data. The coefficients B_0 and a_i must be projected from the training data by the learning algorithm. In linear kernel SVM the dot-product is known as the kernel which is given by

$$K(x, x_i) = \sum(x \times x_i). \tag{8.11}$$

The kernel describes the resemblance or measure of range between fresh information and vectors of assistance. The dot-product is the degree of resemblance used for linear SVM or a linear kernel since the distance is a linear input combination. Other kernels, such as polynomial kernels and radial kernels, can be used to transform the input space into higher dimensions. This is called the trick of the kernel or kernel trick. It is suitable to use more complex kernels as it allows lines to differentiate the classes that are bent or even more complex. This in turn can lead to more accurate classifiers. Further, the SVM classifiers can be divided into the following. (i) Polynomial kernel SVM, where we can use a polynomial kernel instead of the dot-product, for instance $K(x, x_i) = 1 + \sum(x \times x_i)^d$, and where the degree of polynomial to the learning procedure must be defined by the side. When $d = 1$ this will contribute to the linear kernel. The polynomial kernel enables input space for curved rows. (ii) Radial kernel SVM, which is more complicated. For example, one can see the value of $K(x, x_i) = e^{(-\gamma \times \Sigma(xx_i^2))}$ with γ as some learning parameter. A worthy default value for γ is 0.1, and γ lies between $0 < \gamma < 1$. The radial kernel is quite global and can generate complicated areas inside the feature space, such as closed polygons in 2D space.

An optimization method must be used to solve the SVM model. A quantitative optimization method could be used to search for the hyperplane coefficients. This is inefficient and is not the strategy used in SVM applications, such as LIBSVM, that are commonly used. You could use stochastic gradient descent if you implement the algorithm as an workout. Specialized optimization processes exist that re-formulate the issue of optimization to be a issue of quadratic programming. The sequential minimum optimization approach is the most common technique for SVM planning. It divides the issue into sub-problems that can be analytically rather than numerically resolved.

8.5.5 Nearest neighbor classifiers

The nearest neighbor principle continually achieves strong efficiency amongst the many techniques for statistical pattern recognition without any hypothesis regarding the distribution of training examples. It includes a number of favorable as well as adverse instances. A fresh sample is categorized by calculating the distance from the closest situation, which then determines the ranking of the sample. The KNN classifier expands this concept by taking the k-points closest to the center and giving these the symbol K. Tiny and odd links (typically 1, 3 or 5) are frequently selected. No matter how long k measurements help decrease the impacts of loud points in the training dataset, cross-validation is used for the selection of k. By following algorithm 4 m, we can introduce a KNN template.

Algorithm 4. Nearest neighbor classifier

1. Load the data State Initialize the value of k.
2. **for** getting the predicted class, iterate from 1 to the total number of training data points **do**.
3. Calculate the distance between the test data and each row of training data.
   Here we will use Euclidean distance as our distance metric since it is the most popular method. The other metrics that can be used are Chebyshev, cosine, etc.
4. Sort the calculated distances in ascending order based on distance values.
5. Get the top k rows from the sorted array.
6. Get the most frequent class of these rows.
7. **return** the predicted class.

Many methods are available to improve the efficiency and velocity of the next classification. One strategy for this issue is the pre-sorting of the practice sets (such as *kd*-trees or cells of Voronoi). Another alternative is to select a subset of trained information so as to approximate the Bayes' ranking in accordance with the 1-NN law (through the subgroup). As k can now be restricted to one and redundant data points taken from the training set, this can produce substantive velocity gains. As k can now be restricted to 1 and redundant data points are deleted from the practice pack, important increases in velocity can be achieved. These methods of information alteration can also enhance efficiency with the removal of misclassification points.

8.5.6 Ensemble classifiers

Ensemble methods are meta algorithms that mix various machine learning techniques in a single predictor model to reduce (bagging) variability or prejudice (boosting), or to enhance (stacking) projections. Sequential ensembles, for example AdaBoost, are techniques in which primary learners are sequentially produced. Sequential approaches' fundamental motive is to take advantage of the dependency among base learners. By weighing earlier wrong instances with a greater weight, general efficiency can be improved. (ii) Parallel ensemble techniques such as the random forest are produced in linear by the basic learners. The fundamental motive of parallel techniques is to use the autonomy of grassroots students because errors can be dramatically decreased by average.

The majority of ensemble techniques use a single basic learning algorithm to make homogeneous, uniform and basic students. Some techniques also use heterogeneous learners which lead to heterogeneous sets. The basic learners have to be precisely as feasible and as varied as necessary in order to make ensemble processes more precise than any of their individual participants.

8.6 Performance

The dataset has been converted to obtain the visualization of results in a simple form, and this is the reason why the results are not up to the mark. Figure 8.9 shows the prediction response in color coded format with respect to GRE versus TOEFLScore, where the legends mean: • = correct data and × = incorrect data.

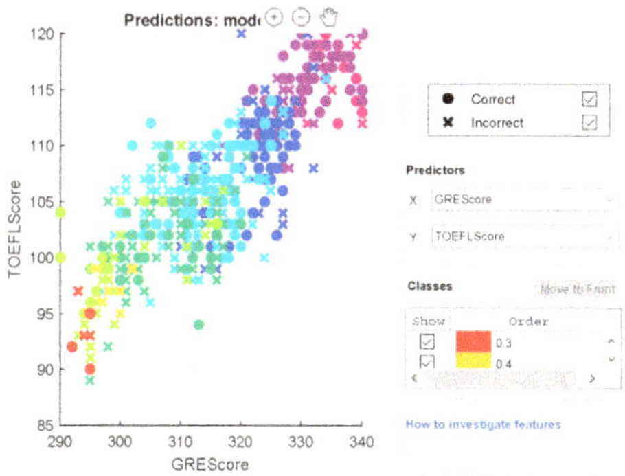

(a) Decision Trees

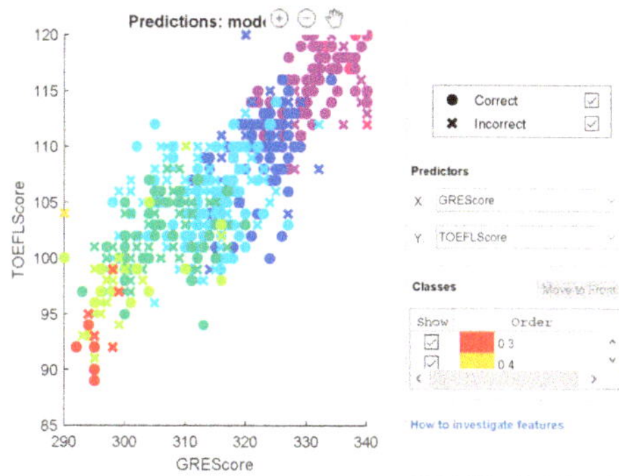

(b) Support Vector Machines.

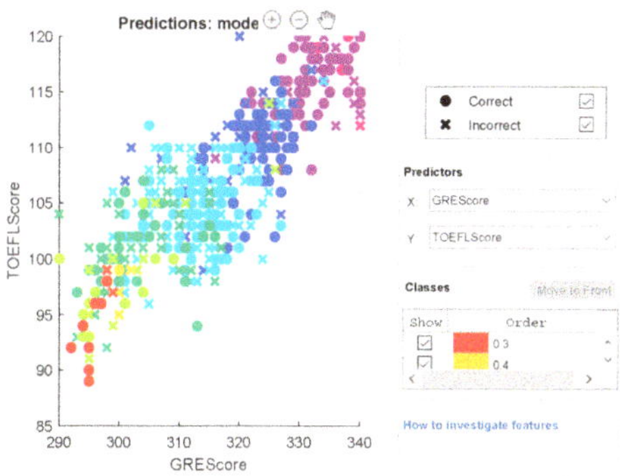

(c) Nearest Neighbor

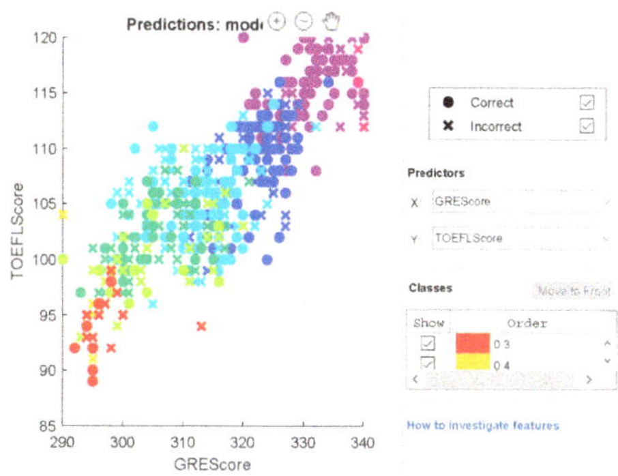

(d) Discriminant Analysis.

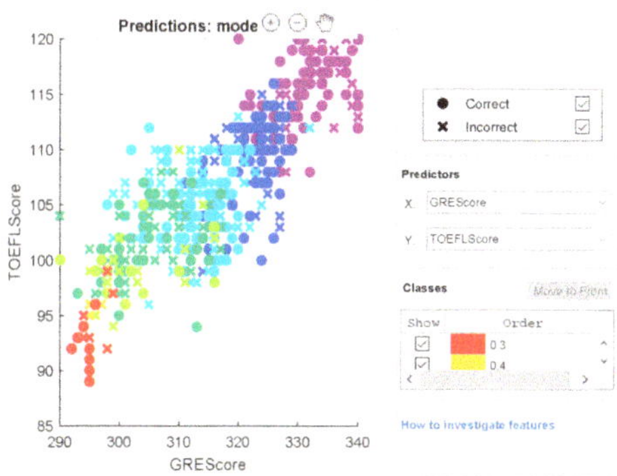

(e) Ensemble.

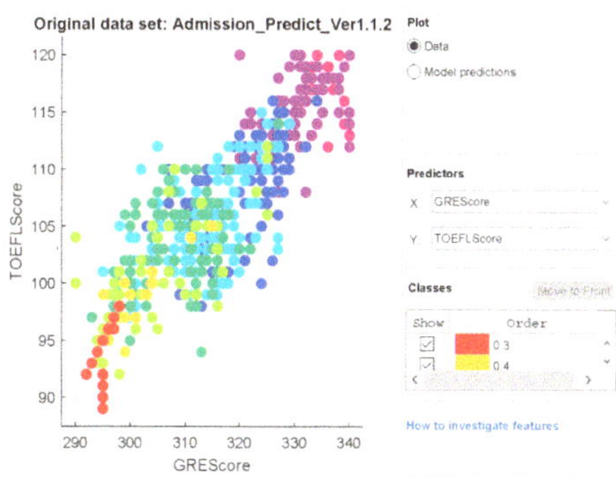

(f) Original data.

Figure 8.9. Showing the prediction response in color coded format with respect to GRE versus TOEFLScore (• = correct data and × = incorrect data).

In this figure, readers can compare the visuals of the decision tree, SVM, NN, discriminant analysis and ensemble with the original data.

Similarly, figure 8.10 shows the prediction response in color coded format of two classes with respect to GRE versus TOEFLScore. This figure exhibits the results of scatter plots for the SVM classifier only. In this set of plots, the author has recorded

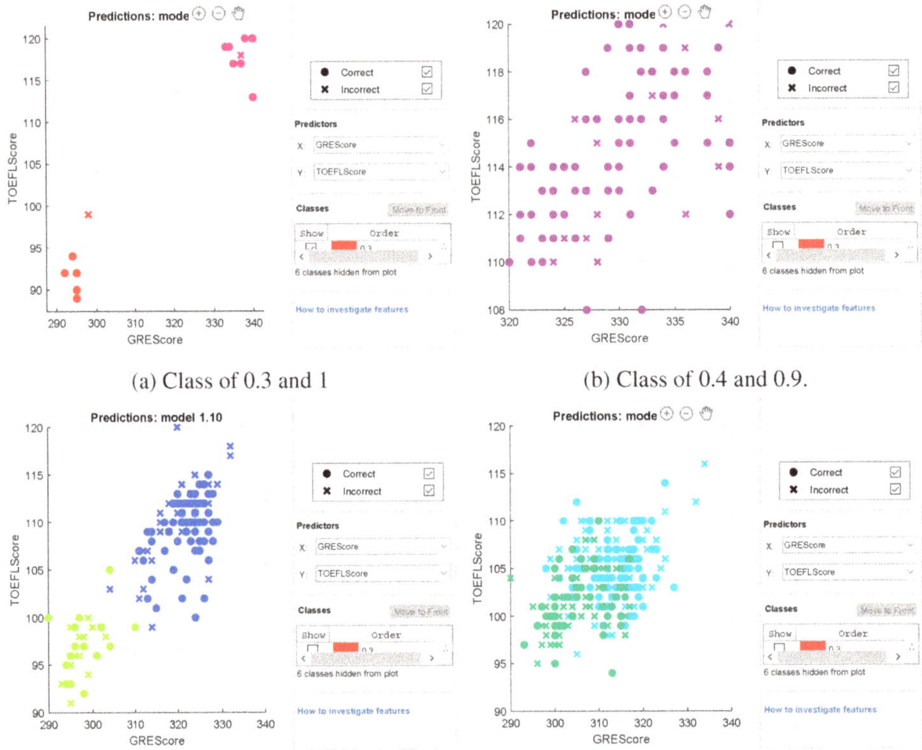

(a) Class of 0.3 and 1

(b) Class of 0.4 and 0.9.

(c) Class of 0.5 and 0.8

(d) Class of 0.6 and 0.7.

Figure 8.10. Showing the prediction response in color coded format of two classes with respect to GRE versus TOEFLScore (• = correct data and × = incorrect data).

predictions for two–two classes (0.3–1, 0.4–0.9, 0.5–0.8 and 0.6–0.7) so that the readers may be able to see the responses and count the correct and incorrect responses as well. The readers can plot the plots for all the available classifiers. All the tree tools discussed support different styles of scatter plots.

8.6.1 Confusion matrix

For measuring the performance of learning classifier systems, several measures are used that depend on four major parameters. A confusion matrix (see table 8.3) consists of the four results from the binary classification and is a two by two table.[1]

1. True positive (TP): true is identified as true (correct identification).
2. True negative (TN): false is identified as false (correct identification).
3. False positive (FP): true is identified as false (wrong identification).
4. False negative (FN): false is identified as true (wrong identification).

[1] The definitions used in this chapter have been taken from the website: https://classeval.wordpress.com/introduction/basic-evaluation-measures/.

Table 8.3. Confusion matrix and performance measurements for different classifiers.

		Predicted footprint match			
		Positive	**Negative**	Total number of incorrect predictions = PP + FN	$\text{EER} = \frac{\text{FP} + \text{FN}}{\text{P} + \text{N}}$
Observed footprint match	**Positive**	True positive (TP)	False positive (FP)	Total number of matches = TP + FP	$\text{PPV} = \frac{\text{TP}}{\text{TP} + \text{FP}}$ $\text{FDR} = \frac{\text{FP}}{\text{TP} + \text{FP}}$ $\text{FDR} = 1 - \text{PPV}$
	Negative	False negative (FN)	True negative (TN)	Total number of non-matches = TN + FN	$\text{NPV} = \frac{\text{TN}}{\text{TN} + \text{FN}}$
		All with footprint match P = TP + FN	All footprint non-match N = FP + TN	Total number of correct predictions = TP + TN	$\text{ACC} = \frac{\text{TP} + \text{TN}}{\text{P} + \text{N}}$ $\text{ACC} = 1 - \text{EER}$
		$TPR = \frac{TP}{P}$ $FNR = \frac{FN}{P}$	$TNR = \frac{TN}{N}$ $FPR = \frac{FP}{N}$ $FPR = 1 - TNR$	Informedness = TPR + TNR − 1	Markedness = PPV + NPV − 1

Therefore, the number of correct identification cases represented by P and the number of false identification cases represented by N are given by

$$P = TP + FN \tag{8.12}$$

$$N = FP + TN. \tag{8.13}$$

Figure 8.11 gives the confusion matrix for (a) decision tree, (b) SVM, (c) NN, (d) discriminant analysis, (e) ensemble and (f) logistic regression. This confusion matrix is plotted for the number of observations or frequency of occurrence. Diagonal responses have been shown in green, i.e. if the true class and predicted classes are the same, and red observations show the true classes that were confused with predicted

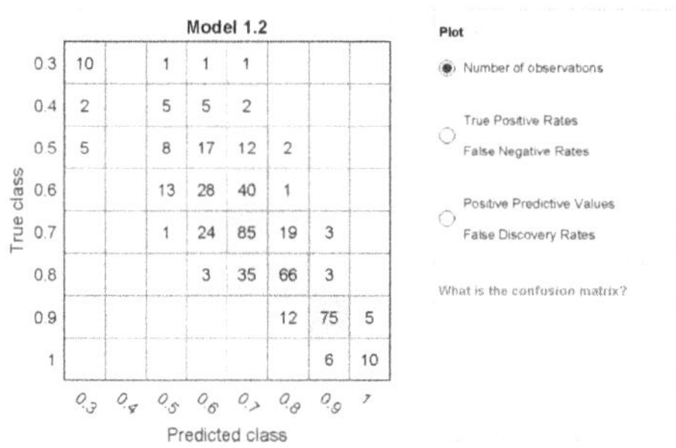

(a) Decision Trees

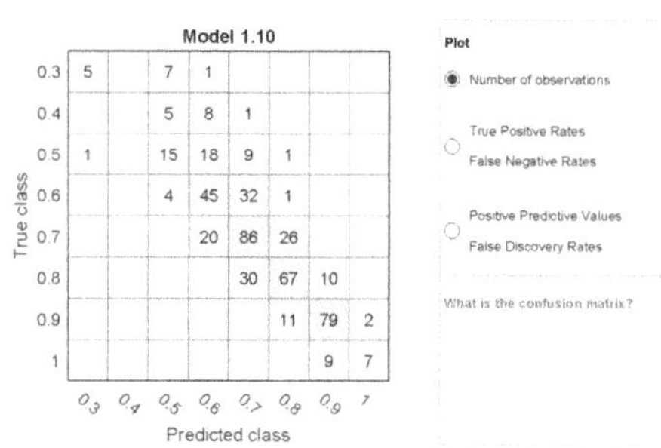

(b) Support Vector Machines.

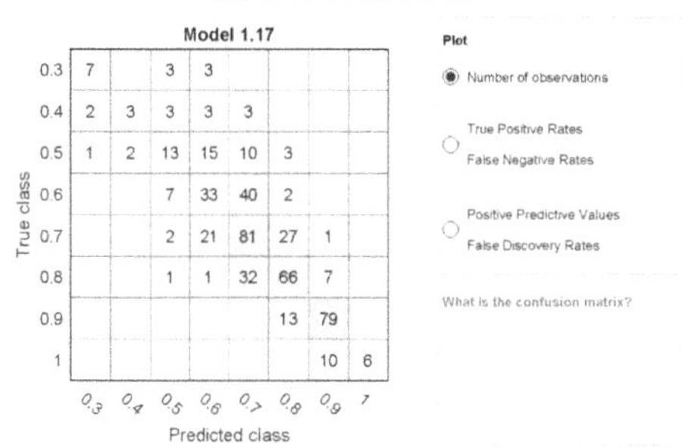

(c) Nearest Neighbor

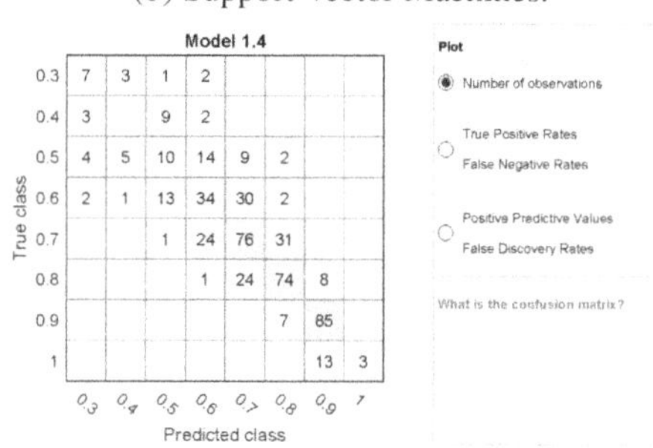

(d) Discriminant Analysis.

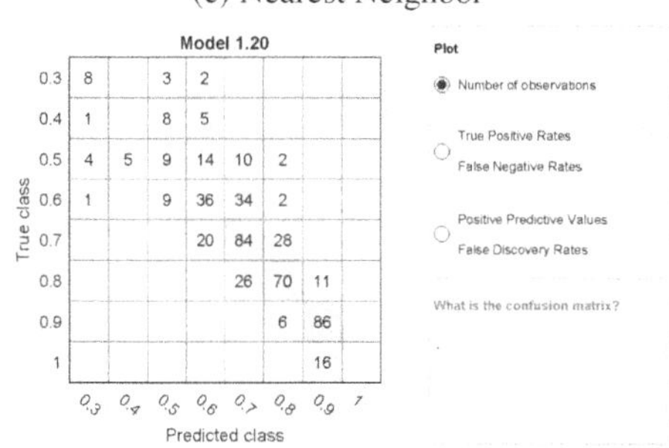

(e) Ensemble.

	Predicted Class										
Class	**0.3**	**0.4**	**0.5**	**0.6**	**0.7**	**0.8**	**0.9**	**1**	**ALL**	**Precision**	**Recall**
0.3	0	0	0	2	1	0	0	0	2.55	0.00	0.00
0.4	0	0	0	0	2	0	0	0	2.87	0.00	0.00
0.5	0	0	0	4	4	1	0	0	1.88	0.00	0.00
0.6	0	0	0	3	12	0	0	0	1.33	0.33	0.20
0.7	0	0	0	0	21	6	1	0	1.13	0.47	0.75
0.8	0	0	0	0	5	10	3	0	1.14	0.45	0.56
0.9	0	0	0	0	0	5	16	0	0.84	0.67	0.76
1	0	0	0	0	0	0	4	0	2.59	0.00	0.00

(f) Logistic Regression(AzureML).

Figure 8.11. Confusion matrix for a number of observations.

classes. The confusion matrix for the number of observations for the logistic regression algorithm has been taken from the AzureML tool.

Definition 1[2]. Sensitivity, recall or true positive rate 'is calculated as the number of correct positive predictions divided by the total number of positives. ... The best sensitivity is 1.0, whereas the worst is 0.0':

$$\text{TPR} = \frac{\text{TP}}{\text{P}} = \frac{\text{TP}}{(\text{TP} + \text{FN})}. \tag{8.14}$$

Definition 2. Specificity or true negative rate (TNR) 'is calculated as the number of correct negative predictions divided by the total number of negatives. ... The best specificity is 1.0, whereas the worst is 0.0':

$$\text{TNR} = \frac{\text{TN}}{\text{N}} = \frac{\text{TN}}{(\text{FP} + \text{TN})}. \tag{8.15}$$

Definition 3. Miss rate or false negative rate (FNR) is the ratio between false negative and false negative + true positive predictions:

$$\text{FNR} = \frac{\text{FN}}{(\text{FN} + \text{TP})}. \tag{8.16}$$

Definition 4. Precision or positive predictive value (PPV) 'is calculated as the number of correct positive predictions divided by the total number of positive predictions. ... The best precision is 1.0, whereas the worst is 0.0':

$$\text{PPV} = \frac{\text{TP}}{(\text{TP} + \text{FP})}. \tag{8.17}$$

Definition 5. Negative predictive value (NPV) is measured as the number of correct negative predictions divided by the total number of of negative predictions. The best NPV is 1.0, the worst is 0.0:

$$\text{NPV} = \frac{\text{TN}}{(\text{TN} + \text{FN})}. \tag{8.18}$$

Definition 6. False discovery rate (FDR) 'is calculated as the number of false positive predictions divided by the total false positive and true positive predictions. The best FDR is 0.0, while the worst is 1.0. It can also be calculated as (1 − PPV)':

[2] Definitions are taken from https://classeval.wordpress.com/introduction/basic-evaluation-measures/.

$$\text{FDR} = \frac{\text{FP}}{(\text{FP} + \text{TP})} = 1 - \text{PPV}. \tag{8.19}$$

Definition 7. Fall-out or false positive rate (FPR) 'is calculated as the number of incorrect positive predictions divided by the total number of negatives. The best false positive rate is 0.0 whereas the worst is 1.0':

$$\text{FPR} = \frac{\text{FP}}{\text{N}} = \frac{\text{FP}}{(\text{FP} + \text{TN})} = 1 - \text{TNR}. \tag{8.20}$$

Definition 8. Informedness is the distance from (i.e. measured perpendicular to) the random line joining coordinate (0,0) and coordinate (1,1) in the ROC. You can then use markedness as the second metric to identify your classification system's general value:

$$\text{Informedness} = \text{Sensitivity} + \text{Specificity} - 1 \tag{8.21}$$

$$\text{Markedness} = \text{Precision} + \text{NPV} - 1. \tag{8.22}$$

Definition 9. Error rate (ERR) 'is calculated as the number of all incorrect predictions divided by the total number of the dataset. The best error rate is 0.0, whereas the worst is 1.0':

$$\text{EER} = \frac{\text{FP} + \text{FN}}{\text{TP} + \text{TN} + \text{FP} + \text{FN}} = \frac{\text{FP} + \text{FN}}{\text{P} + \text{N}}. \tag{8.23}$$

Definition 10. Accuracy (ACC) 'is calculated as the number of all correct predictions divided by the total number of the dataset. The best accuracy is 1.0, whereas the worst is 0.0. It can also be calculated by 1 – ERR':

$$\text{ACC} = \frac{(\text{TP} + \text{TN})}{(\text{P} + \text{N})} = 1 - \text{EER}. \tag{8.24}$$

Figure 8.12 shows the confusion matrix for PPV versus FDR, and figure 8.13 shows the confusion matrix for TPR versus FNR. We only explain the confusion matrix for four algorithms and the remaining two algorithms are left as an exercise for the reader. The matrix in figure 8.12 determines the horizontal characteristics of

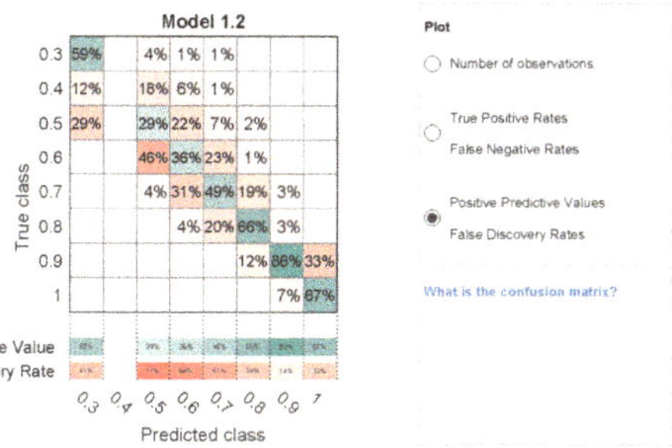

(a) Decision Trees

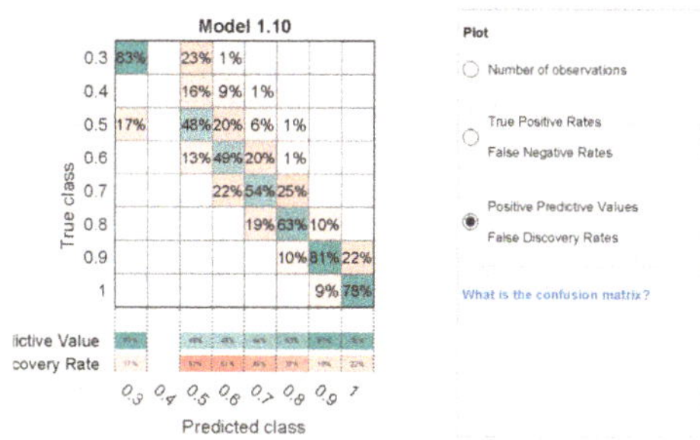

(b) Support Vector Machines.

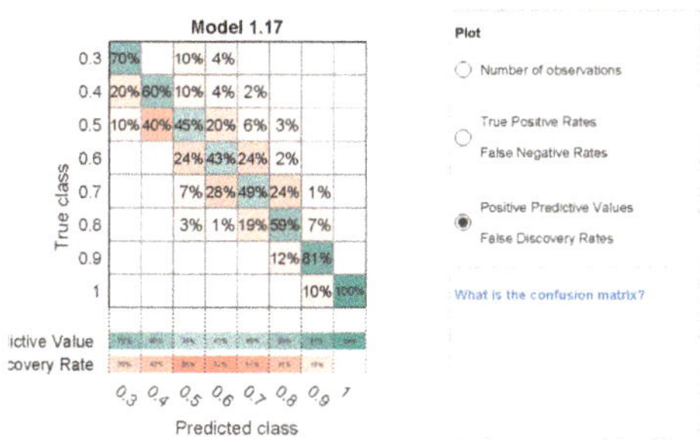

(c) Nearest Neighbor

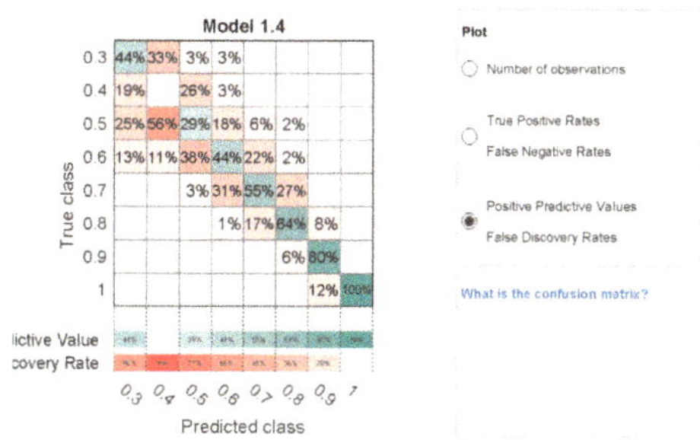

(d) Discriminant Analysis.

Figure 8.12. Confusion matrix for PPV versus FDR.

the predicted classes. The green result is known as the positive predictive value (see definition 5), while the false discovery rate (see definition 6) is represented by the red shaded results. The matrix in figure 8.13 determines the vertical characteristics of true classes. The green shaded result is known as the TPR (see definition 1), while FNRs (see definition 3) are in red.

8.6.2 Receiver operating characteristic

The receiver operating characteristic curves obtained in this experiment provide the trade-off between sensitivity (TPR, see definition 5) and fall-out (FPR, see definition 49) or specificity (TNR, see definition 5) as FPR = 1 − TNR (see equation (8.20)) for all possible cut-offs for a test or its combinations. Also, the area under the ROC curve (AUC) presents an idea about the advantage of employing the test(s), which is a measure of the usefulness of an experiment. The best cut-off has the highest true positive rate together with the lowest false positive rate leading to a greater area.

Figure 8.14 displays the receiver operating characteristics for the decision tree, SVM, NN, and discriminant analysis classifiers for positive class 0.3. One can easily see the different values of ROC and AUC. The right-hand panel is helpful in the creation of the other plots as well.

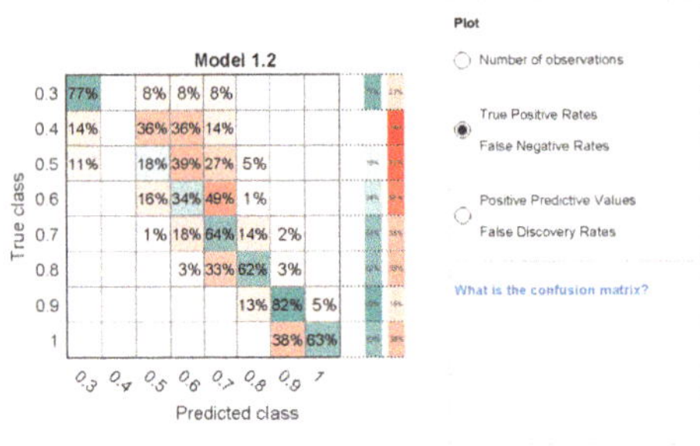

(a) Decision Trees

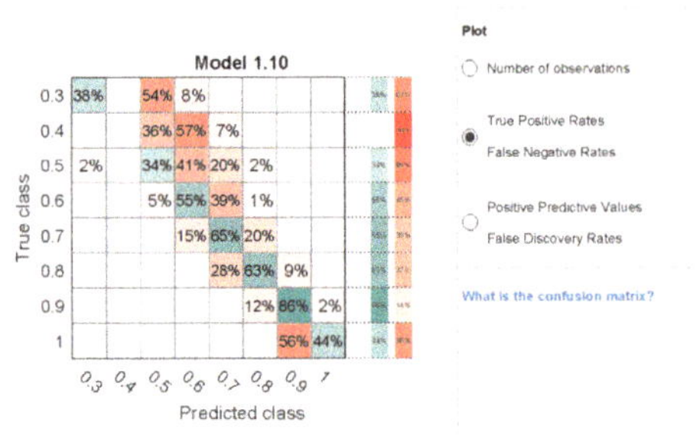

(b) Support Vector Machines.

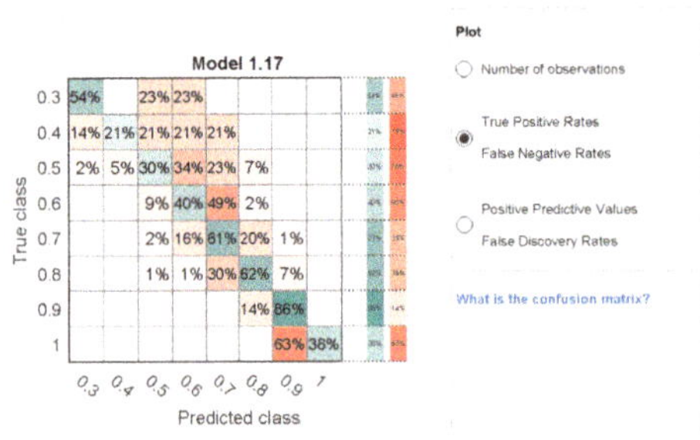

(c) Nearest Neighbor

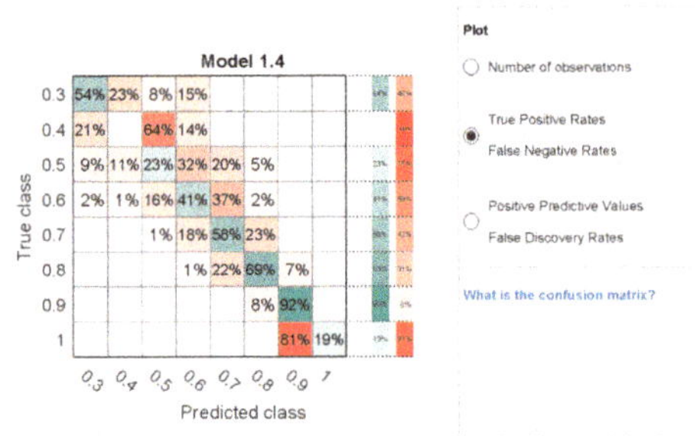

(d) Discriminant Analysis.

Figure 8.13. Confusion matrix for TPR versus FNR.

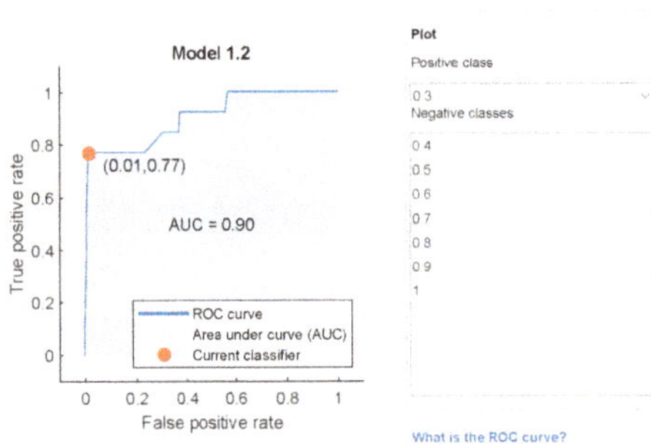

(a) Decision Trees

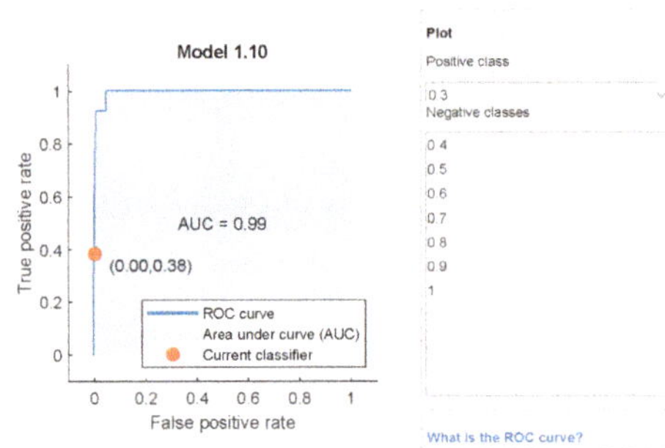

(b) Support Vector Machines.

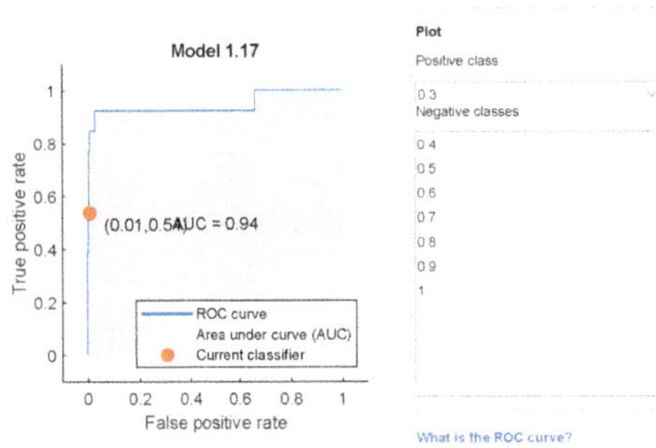

(c) Nearest Neighbor

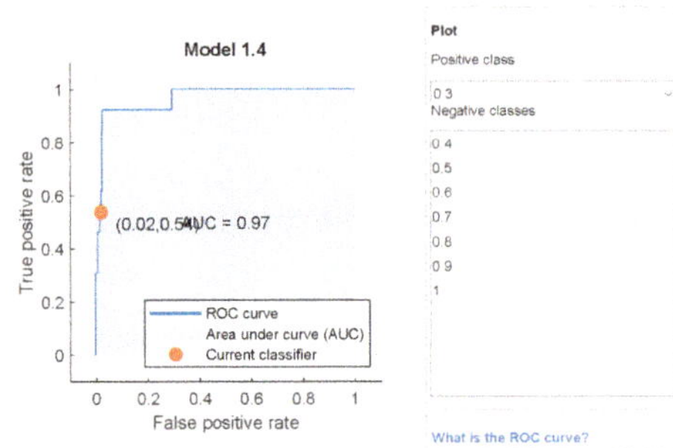

(d) Discriminant Analysis.

Figure 8.14. ROC for positive class 0.3.

8.6.3 Parallel plot

Figure 8.15 demonstrates the parallel characteristics plot for the decision tree, SVM, NN, discriminant analysis classifiers and data only. 'A parallel coordinate plot displays as a line or chart each column in the data table. A line point reflects each of

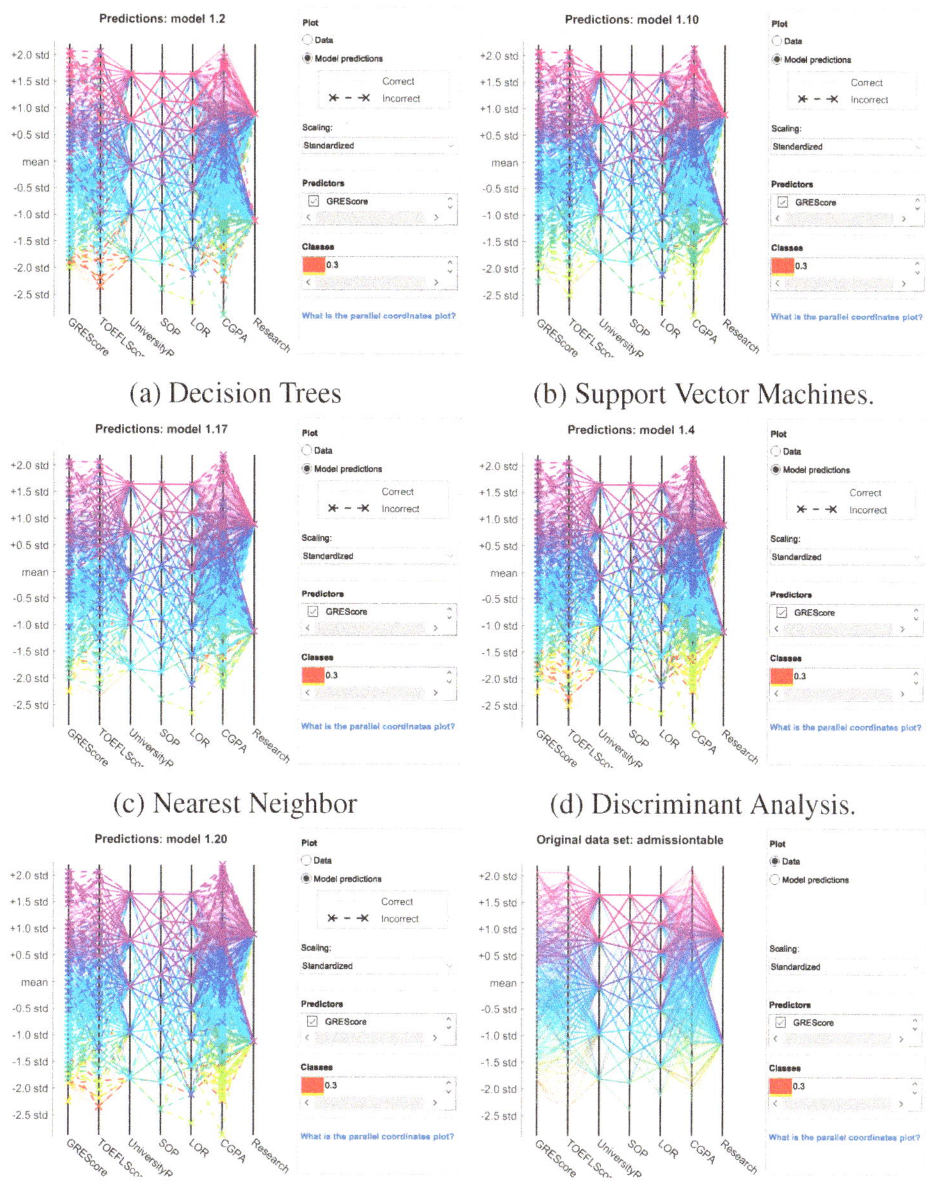

(a) Decision Trees (b) Support Vector Machines. (c) Nearest Neighbor (d) Discriminant Analysis. (e) Ensemble. (f) Original data.

Figure 8.15. Parallel plots.

a row's attributes. This makes parallel coordinate plots comparable in appearance to row graphs, but there is a significantly distinct manner in which information is transformed into a plot. The values are always standardized in a parallel coordinate plot. This implies that the smallest value in the respective column is set at 0 percent for each stage along the x-axis and the largest value is placed at 100 percent along the y-axis in that row. The size of the different columns is completely separate, so comparing the height of the curve in one column with the height of the curve in another column is meaningless' [29, 30].

8.7 Conclusion

This chapter covered an introduction and background to LCS. The focus of this chapter is to understand the LCS practically, therefore, MATLAB, BigML and AzureML have been discussed to understand how the results from learning classifiers can be obtained. The dataset and its processing, such as splitting into training and testing datasets has been covered in detail with MATLAB code as well as AzureML procedure. Six learning classifiers, namely logistic regression classifiers, decision tree classifiers, discriminant analysis classifiers, support vector machines classifiers, nearest neighbor classifiers and ensemble classifiers have been discussed. The logistic regression classifier is included in MATLAB as an independent classifier app, so the author has used AzureML to address various performance matrices and also MATLAB. The remaining five classifiers have been tested using MATLAB's Classification Learner App.

In the latter half of this chapter, different visualizations of performance have been shown, which include scatter plots, prediction plots, tree diagrams, code snippets, confusion matrices, parallel plots and receiver operating characteristics. Table 8.4 at

Table 8.4. Performance comparison of LCS.

Classifier	Accuracy	Prediction speed (ops)	Training time (sec)	Remarks
Decision trees	52.8	46 000	0.89	#Split 20, Gini's diversity index
Discriminant analysis	58.2	29 000	0.69	Linear, full covariance exponential kernel, MSE = 0.0048
Logistic regression	0.069 RMSE	12 000	15.41	MAE = 0.0522, R squared = 0.80, basis function is constant
Support vector machines	60.8	5200	1.93	Linear kernel, box constraint level = 1, multiclass method, one-vs-one
Nearest neighbor	57	17 000	2.64	$N = 10$, Euclidian distance, square inverse distance weight
Ensemble	59	2400	3.96	Subspace ensemble dimension 4; descriminant learning, 30 learners

the end of this chapter has been provided on the basis of experimentation carried out by the author for the readers to justify their choices of algorithms. The LCS system of MATLAB is a potent tool for researchers to obtain results quickly, and hence the comparison of different classifiers is meaningful in the chapter. BigML is a purely automated tool with almost negligible user intervention. AzureML is a robust tool with a high degree of flexibility.

Acknowledgments

The author is grateful to his colleague Dr S P Dubey from Rungta College of Engineering and Technology, Bhilai, India, who provided insight and expertise that greatly assisted in the research, although they may not agree with all of the interpretations/conclusions of this chapter. He is immensely grateful to his mentor Dr G R Sinha, who always motivates him for research and development, and academic excellence.

References

[1] Holland J 1976 Adaptation *Progress in Theoretical Biology* ed R Rosen and F M Snell (New York: Academic)

[2] Goldberg D E and Holland J H 1988 Genetic algorithms and machine learning *Mach. Learn.* **3** 95–9

[3] Urbanowicz R J and Moore J H 2009 Learning classifier systems: a complete introduction, review, and roadmap *J. Artif. Evol. Appl.* **2009** 736398

[4] Butz M V 2015 Learning classifier systems *Springer Handbook of Computational Intelligence* (Berlin: Springer) pp 961–81

[5] Acharya M S, Armaan A and Antony A S 2019 *A Comparison of Regression Models for Prediction of Graduate Admissions ICCIDS 2019: IEEE Int. Conf. on Computational Intelligence in Data Science Chennai, India,*

[6] Holland J H *et al* 2000 What Is a Learning Classifier System? *Learning Classifier Systems* (Berlin: Springer)

[7] Wilson S W 1995 Classifier fitness based on accuracy *Evol. Comput.* **3** 149–75

[8] Riolo R L 1991 Lookahead planning and latent learning in a classifier system *Proceedings of the First International Conference on Simulation of Adaptive Behavior on From Animals to Animats* (Cambridge, MA: MIT Press), pp 316–26

[9] Stolzmann W 1998 Anticipatory classifier systems *Genet. Program.* **98** 58–64

[10] Holland J H 1983 Escaping brittleness *Proc. Second Int. Workshop on Machine Learning*

[11] Robertson G G and Riolo R L 1988 A tale of two classifier systems *Mach. Learn.* **3** 139–59

[12] Holland J H, Holyoak K J, Nisbett R E and Thagard P R 1989 *Induction: Processes of Inference, Learning, and Discovery* (Cambridge, MA: MIT Press)

[13] Holmes J H, Durbin D R and Winston F K 2000 The learning classifier system: an evolutionary computation approach to knowledge discovery in epidemiologic surveillance *Artif. Intell. Med.* **19** 53–74

[14] Zang Z, Li D and Wang J 2015 Learning classifier systems with memory condition to solve non-Markov problems *Soft Comput* **19** 1679–99

[15] Zebin T, Scully P J and Ozanyan K B 2017 Inertial sensor based modelling of human activity classes: feature extraction and multi-sensor data fusion using machine learning algorithms *eHealth 360°* (Cham: Springer), pp 306–14
[16] Noor N Q M, Sjarif N N A, Azmi N H F M, Daud S M and Kamardin K 2017 Hardware Trojan identification using machine learning-based classification *J. Telecom. Electron. Comput. Eng. (JTEC)* **9** 23–7
[17] Maleki M, Manshouri N and Kayikçioğlu T 2017 Application of PLSR with a comparison of MATLAB classification learner app in using BCI *2017 25th Signal Processing and Communications Applications Conf. (SIU)*
[18] Nagwanshi K K and Dubey S 2018 Statistical feature analysis of human footprint for personal identification using BigML and IBM Watson analytics *Arab. J. Sci. Eng.* **43** 2703–12
[19] Kessel M, Ruppel P and Gschwandtner F 2010 BIGML: a location model with individual waypoint graphs for indoor location-based services *PIK-Praxis Informationsverar. Kommunikation* **33** 261–7
[20] Zainudin Z and Shamsuddin S M 2016 Predictive analytics in Malaysian dengue data from 2010 until 2015 using BigML *Int. J. Adv. Soft. Comput. Appl.* **8** 18–30
[21] Chappell D 2015 New White Paper: Introducing Azure Machine Learning *A Guide for Technical Professionals, Microsoft Corporation* http://davidchappellopinari.blogspot.com/2015/08/new-whitepaper-introducing-azure.html
[22] Mántaras R L D 1991 A distance-based attribute selection measure for decision tree induction *Mach. Learn.* **6** 81–92
[23] Liu W Z and White A P 1994 The importance of attribute selection measures in decision tree induction *Mach. Learn.* **15** 25–41
[24] Navlani A 2018 Decision Tree Classification in Python https://www.datacamp.com/community/tutorials/decision-tree-classification-python
[25] Hallinan J S 2012 Data mining for microbiologists *Systems Biology of Bacteria* ed C Harwood and A Wipat (Methods in Microbiology vol 39) (Amsterdam: Elsevier) pp 27–79
[26] Tharwat A, Gaber T, Ibrahim A and Hassanien A E 2017 Linear discriminant analysis: a detailed tutorial *AI Commun.* **30** 169–90
[27] Raschka S 2014 Linear Discriminant Analysis—Bit by Bit https://sebastianraschka.com/Articles/2014_python_lda.html
[28] Sawla S 2018 Linear Discriminant Analysis https://medium.com/@srishtisawla/linear-discriminant-analysis-d38decf48105
[29] Li M, Zhen L and Yao X 2017 How to read many-objective solution sets in parallel coordinates *IEEE Comput. Intell. Mag.* **12** 88–100
[30] Hussain A and Vatrapu R 2014 Social data analytics tool (SoDaTo) *Advancing the Impact of Design Science: Moving from Theory to Practice* ed M C Tremblay, VanderMeer D, Rothenberger M, Gupta A and Yoon V (Cham: Springer)

Chapter 9

A case study on the implementation of six sigma tools for process improvement

Bonya Mukherjee, Rajesh Chamorshikar and Subrahmanian Ramani

A blast furnace (BF) is one of the primary processes for the production of crude steel in an ore based integrated steel plant (ISP). In a BF, iron ore is reduced to pure molten iron by supplying hot air blasts through the tuyers at the bottom of the furnace, just above the hearth. A redox reaction takes place due to heat and mass transfer and, along with the molten iron and slag that are tapped from the bottom of the furnace, BF gas is emitted from the top of the BF as a by-product of the smelting process. The raw BF gas carries a dust load of 25–40 g N^{-1} m^{-3}, which is removed in a gas cleaning plant (GCP) by spraying with water at high pressure through nozzles located at various levels of a gas scrubber. The water removes the dust from the BF gas and exits from the bottom of the scrubber in the form of slurry, is cooled in a cooling tower and goes to radial settling tanks (RSTs) for removal of dust and is then recycled back to the scrubber for cleaning of BF gas. The clean BF gas is then fed into the gas network for consumption as fuel gas in various units of the plant. Since April 2012, the ISP in question was facing the problem of a high concentration of suspended solids in the clean water supply of the GCPs of BFs. The average suspended solids concentration was 277 ppm against the specification of 100 ppm, which was adversely affecting the functioning of the GCPs, leading to insufficient removal of dust from BF gas. There were a number of factors that could have impacted the concentration of total suspended solids (TSS) in the supply water to GCPs. These factors were: insufficient removal of dust from BF gas in the dust catchers of BFs; a high concentration of TSS in the fresh make-up water supply itself; and insufficient and inefficient settling of dust and suspended solids in the RSTs.

To narrow down the probable root cause/causes within the shortest possible time, the authors decided to take up the above problem as a six sigma project. Based on

doi:10.1088/978-0-7503-2404-5ch9

the past year's performance data for the BF dust catchers, GCPs and RSTs, various six sigma tools and techniques were used to find and filter out the critical inputs that influence the output to a large extent. Thus, it was possible to quickly carry out a root cause analysis (RCA) and it was found that the problem was mainly due to certain operational lacunae in the RSTs and necessary remedial actions were taken. The TSS levels in the supply water were effectively brought down to less than 100 ppm, and have been maintained at those levels since April 2013. The financial benefits have been to the tune of Rs. 1 Crore per annum, as validated by the finance department. Other benefits have included the reduced choking of nozzles/throttle assembly and U-seals, as well as gas lines, reduced frequency of unscheduled/ emergency breakdowns/shutdowns in GCPs, and better removal of dust from BF gas, leading to reduced specific fuel consumption and better furnace productivity at the BF gas consumer end. The non-tangible benefits have been a renewed motivation in the workforce, due to the elimination of non-value adding activities such as attending to repetitive unscheduled repairs and sudden breakdowns.

9.1 Introduction

9.1.1 Generation and cleaning of BF gas

BF is one of the primary processes for the production of crude steel in an ore based ISP. In a BF, counter current heat and mass exchange takes place for smelting of iron ore using coal and coke to produce hot metal or molten iron in the liquid state. The solid raw materials such as iron ore lumps, sinter and coke, and flux materials such as limestone are charged from the top and hot air blast is supplied through the tuyers at the bottom of the furnace, just above the hearth. A redox reaction takes place due to heat and mass transfer and along with the molten iron and slag that are tapped from the bottom of the furnace, BF gas is emitted from the top of the BF as a by-product of the smelting process. This gas has a calorific value of 780–800 kcal N^{-1} m^{-3} and is utilized as one of the primary fuel gases in an ISP. The raw BF gas carries a dust load of 25–40 g N^{-1} m^{-3}. This dust content has to be removed or minimized before the raw BF gas can be used as a fuel in the various furnaces and stoves of an ISP.

9.1.1.1 BF gas cleaning process

Each BF is equipped with a dry dust catcher in which more than 50% of the dust load is separated from the gas, and a GCP which is either dry type or wet type.

The functions of the GCPs of BFs are:

a. To reduce the temperature of BF gas.
b. To remove dust content from BF gas using a wet scrubber and atomizer.
c. To maintain the top pressure of BF.

The primary function of the GCPs of BF is to remove the suspended dust particles from the raw BF gas, before it is supplied to the consumer as fuel. The raw BF gas emerging from the BFs has a dust load of around 25–40 g N^{-1} m^{-3}. More than 50% of the dust is removed in the dry dust catcher. The raw BF gas containing the remaining 10–15 g N^{-1} m^{-3} of dust then goes to the scrubber where the gas flows from bottom to top and water is sprayed on it from nozzles at four levels at a rate of 800–1000 m^3 h^{-1}. After the scrubber, the raw BF gas passes through a set of parallel venturi atomizers where again water is sprayed on the gas, to remove the remaining dust. The dust content of clean BF gas is less than 5 mg N^{-1} m^{-3}. The clean BF gas is then fed into the gas network for consumption as fuel gas in various units of the plant.

9.1.1.2 Problems encountered in cleaning of BF gas

The slime water containing the dust removed from the BF gas collects in a common slime channel from where it is distributed to radial settling tanks (RSTs) for settling of the suspended solids and the clarified overflow water is supplied (pumped) back to the scrubbers of the GCPs, after cooling in cooling towers and adding fresh make-up water to compensate for evaporation and other process losses. In this particular case, the slime water was distributed to five RSTs, to RSTs 1 and 2 through one route and to RSTs 3, 4 and 5 through an alternative route from a common slime water channel. As per standard operating practices (SOPs), the TSS levels of supply water to GCP scrubbers, after passing through the RSTs and the addition of fresh make-up water, should be less than 100 ppm, on average. But, for the duration of a year, it was observed that the TSS of the supply water to the GCPs was as high as 280–300 ppm, which was affecting the functioning of the GCPs in the following ways:

a. Choking of the nozzles in the scrubbers
b. Erosion of nozzle, throttle assembly and atomizer shell plate surfaces
c. Insufficient removal of dust content from BF gas, thus resulting in insufficient cleaning of BF gas

The authors decided to use a six sigma (DMAIC) process to bring about process improvement by reducing the concentration of TSS in the supply water, which was adversely affecting the operation and equipment health of the GCPs of the BFs. Various six sigma tools and techniques were used to find and filter out the critical inputs that influenced the output to a large extent.

9.2 Problem overview

Since April 2012, the ISP in question had been facing the problem of a high concentration of suspended solids in the supply water of BF GCPs 1–6. The average suspended solids concentration was 277 ppm against the specification of 100 ppm. (See figure 9.1)

The TSS in supply water started increasing from the month of April 2012 and was drastically high in the months of May 2012 and June 2012. The objective of the six sigma project was to reduce the TSS in the water supplied for gas cleaning from an average 277 ppm to 130 ppm by May 2013, on the basis of past performance data. The project was started in December 2012 and the performance improvement started from April 2013.

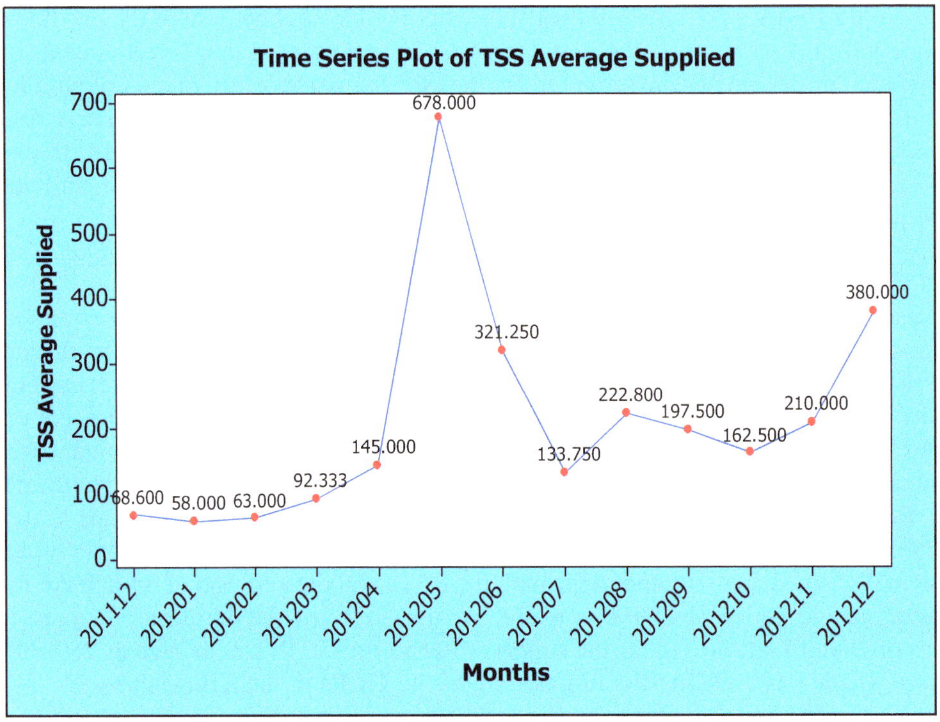

Figure 9.1. Monthly TSS (in ppm) in the supply water of GCPs 1–6.

9.3 Project phase summaries

9.3.1 Definition

The definition phase is the beginning phase of the six sigma improvement process. In this phase, the project team identifies and defines the problem, creates a map of the process and outlines the focus of the project for themselves and the customers of the process. The project team measures both the process baseline and the process entitlement.

Process baseline: The current performance of a process over a timeline before any change is made the input variable(s). This baseline data can also be used as a yardstick for measuring future efficiency in the process.

Process entitlement: The historical best performance of a process that a project team is 'entitled' to achieve. It is also based on design data or technical specifications of the process/equipment. The entitlement analysis also helps to decide whether the project should address process improvement (DMAIC) or process redesign (design for six sigma).

On analyzing one year's data for the TSS of the supply water to GCPs 1–6 from December 2011 to December 2012 in comparison to historical data, the average or baseline TSS was found to be 277 ppm, while the entitlement was 68 ppm, based on standard norms. The *P*-value of 0.021 (see figure 9.2) also shows that the probability distribution was not normal for the average TSS of the supply water from April 2012.

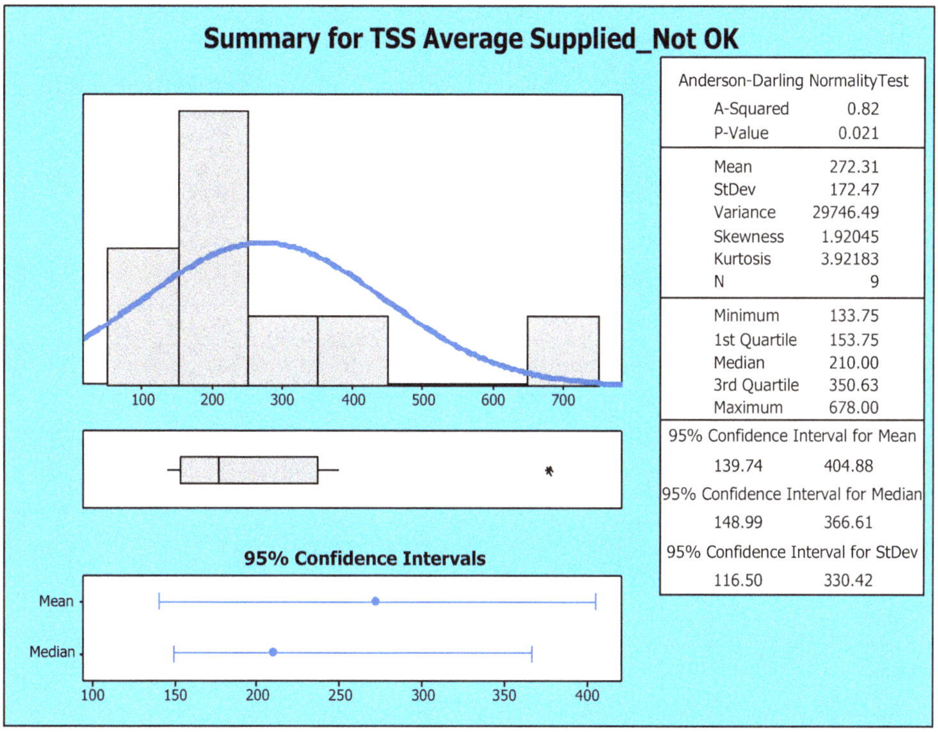

Figure 9.2. Graphical summary for the period April 2012–December 2012 before the start of the project.

However, a graphical summary (see figure 9.3) of the TSS data for the four months before April 2012, i.e. December 2011 to March 2012 shows that the average TSS was not only below 100 ppm, but also the distribution was normal, i.e. the process capability was within the limits.

In a hypothesis test, a *P*-value establishes the significance of the results. Hypothesis tests are used to establish the validity of a claim made about a dataset. This claim is known as the null hypothesis.

The *P*-value is a number between 0 and 1 and can be interpreted as such:

a. A *P*-value of less than 0.05: strong evidence against the null hypothesis and hence the null hypothesis can be rejected.
b. A large *P*-value of more than 0.05: weak evidence against the null hypothesis, so the null hypothesis cannot be rejected outright and can be further investigated statistically for validity.
c. *P*-values very close to 0.05 can go either way.

9.3.2 Measurement

In the measurement phase of a lean six sigma improvement process, the current performance of the process, the magnitude of the problem and the probable factors affecting the problem are measured.

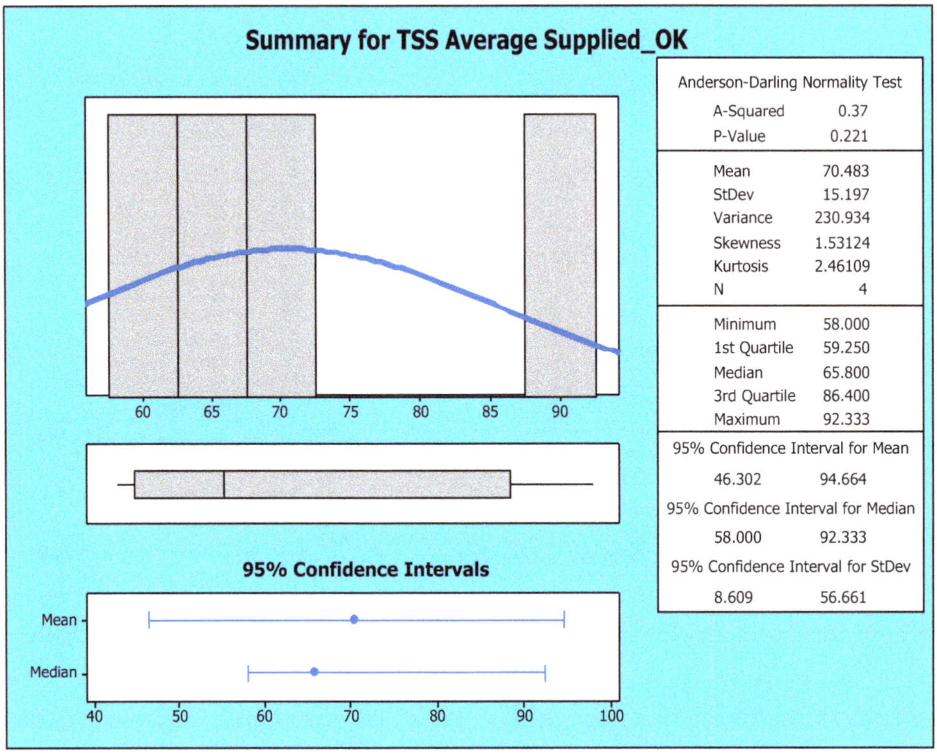

Figure 9.3. Graphical summary for the period December 2011–March 2012.

Various six sigma tools were used in the measurement phase, namely the process flow diagram, time series plot, regression and correlation analysis, I/O worksheet, and C and E matrix, to identify the causes of high TSS in the supply water.

Regression analysis (fitted line plot): Fitted line plots were drawn between the measured TSS of the return water from GCPs 1–6 to RSTs 1 and 2 and the return water to RSTs 3, 4 and 5 to check whether the TSS of the return water had any bearing on the TSS of the supply water. A close correlation was found between the TSS of the return water before RSTs and supply water after RSTs, as is evident from figures 9.4 and 9.5.

A fitted line plot is a statistical technique for regression analysis (linear) to find the best fit line for a set of data points. This is used when experimental data or historical process data are plotted and the data points are scattered across the plot area.

The formula for fitted line plot or regression line is $y = mx + b$ where m is the slope and b is the y-intercept.

R^2 or the coefficient of determination is a statistical measure of how close the data are to the fitted line and how much the variation in dependent variables can be explained by the variation in the independent variable.

An R^2 value of 50% and above usually indicates a better correlation between the dependent and independent variables. A low R^2 value usually fails to indicate any

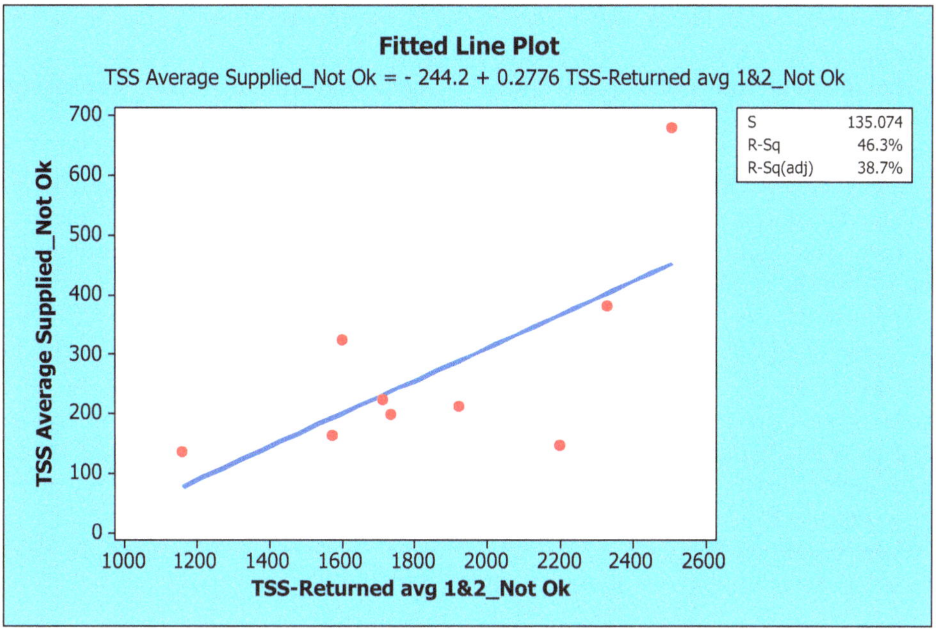

Figure 9.4. Fitted line plot between the average TSS of supply water and the average TSS of return water to RSTs 1 and 2; with an R^2 value of 46.3%.

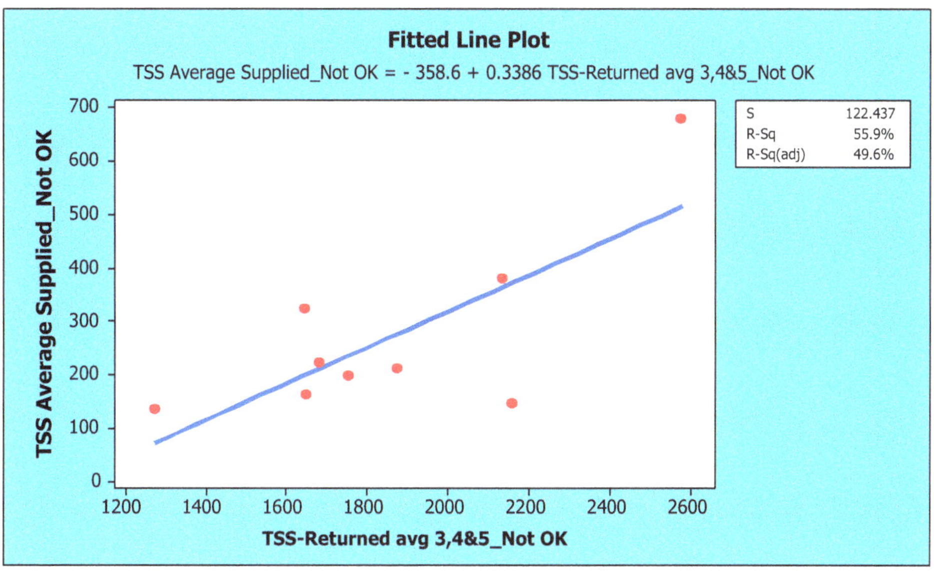

Figure 9.5. Fitted line plot between the average TSS of supply water and the average TSS of return water to RSTs 3, 4 and 5; with an R^2 value of 55.9%.

straightforward correlation between a single dependent variable with an independent variable.

The R^2 value, however, is not always straightforward and, hence, statisticians have to rely on adjusted R^2 values to establish any kind of correlation between dependent and independent variables.

After a close correlation was established between the TSS levels of supply water and the TSS levels of return water from the scrubbers, it was deemed prudent to examine other factors that might be contributing to the increased TSS of the return water from the scrubbers of the GCPs.

The raw BF gas passes through a dry dust catcher, where more than 50% of the dust is removed before it enters the scrubbers. The quantity of dust removed can be measured from the average number of dump cars that are filled up with dust removed from the dust catchers periodically. It was felt that inefficient operation of the dust catchers might cause insufficient removal of dust from the BF gas, which might have in turn resulted in a higher dust load in the scrubbers, thereby adversely impacting the TSS of both return water and supply water.

Therefore, fitted line plots were drawn between the average TSS of supply water and the average number of dump cars (see figure 9.6).

Also, separate fitted line plots were drawn between:

1. The average TSS of return water to RSTs 1 and 2 and the average number of dump cars (see figure 9.7)
2. The average TSS of return water to RSTs 3, 4 and 5 and the average number of dump cars (see figure 9.8)

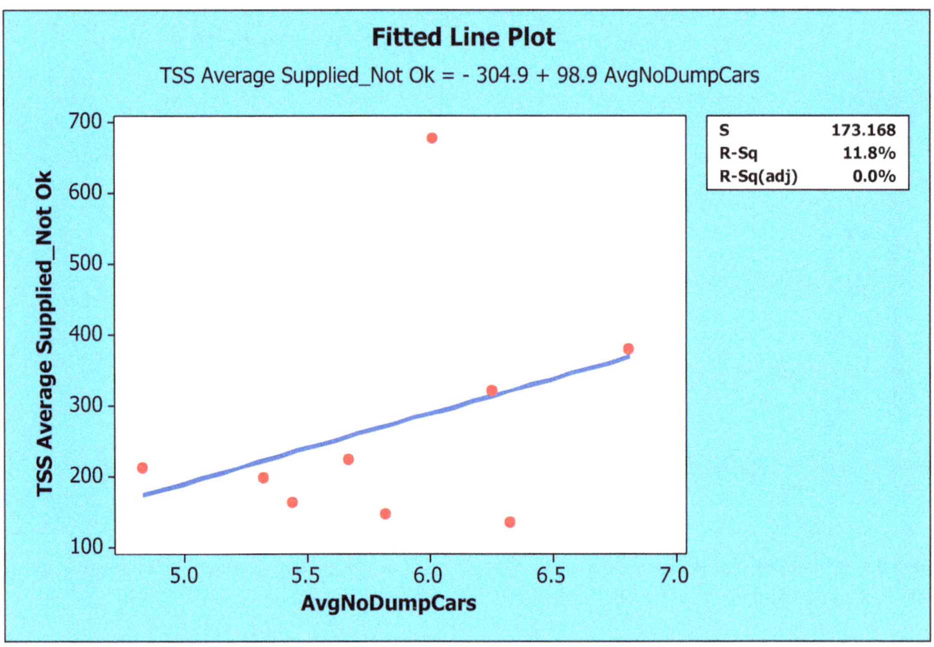

Figure 9.6. Fitted line plot between average TSS of supply water and the average number of dump cars from the dust catchers; with an R^2 value of 11.8%.

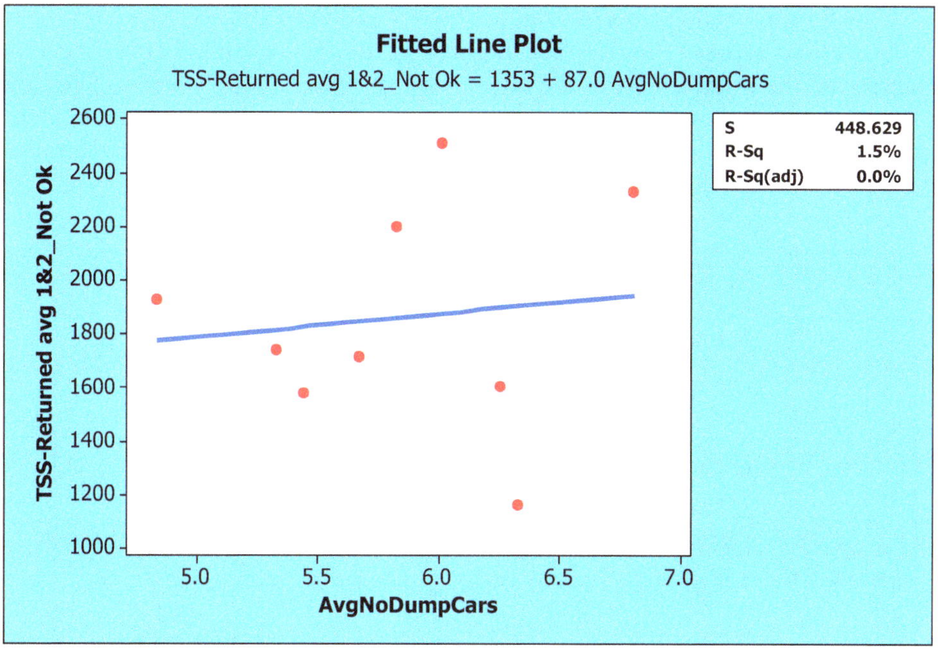

Figure 9.7. Fitted line plot between the average TSS of return water to RSTs 1 and 2 and the average number of dump cars from dust catchers; with an R^2 value of 1.5%.

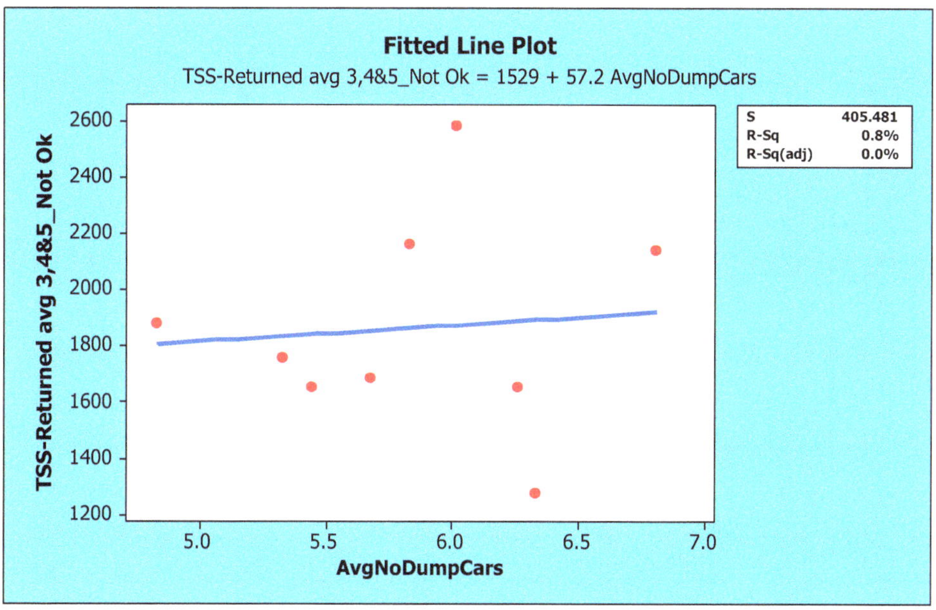

Figure 9.8. Fitted line plot between the average TSS of return water to RSTs 3, 4 and 5 and the average number of dump cars from dust catchers; with an R^2 value of 0.8%.

All the three fitted line plots (see figures 9.6, 9.7 and 9.8) show that there is little or no correlation between the average number of dump cars filled from the dust catchers and the average TSS of the return water as well as supply water. Thus, the dust catchers were eliminated from the scope of this project, and only the closed loop circuit of water supply to the scrubbers of GCPs 1–6 was taken into consideration

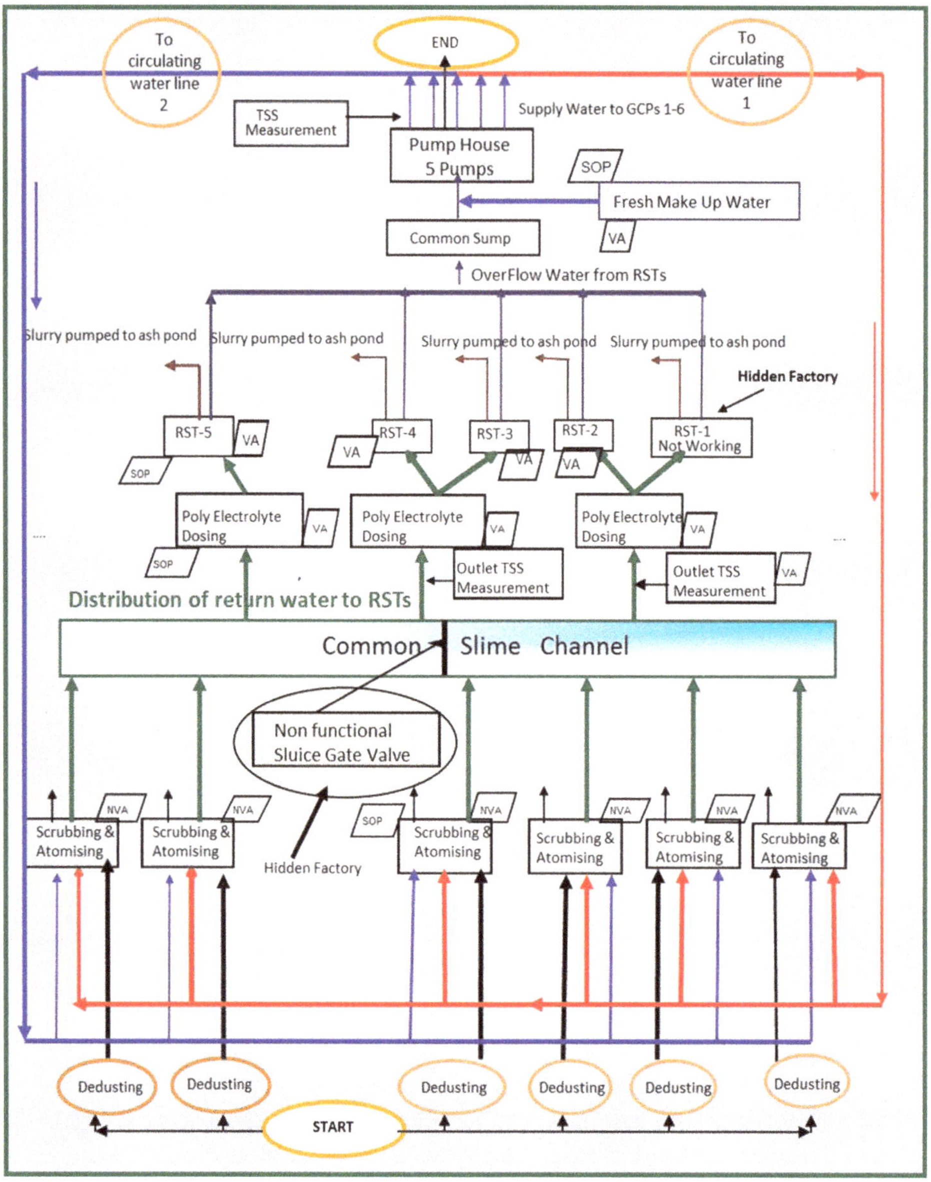

Figure 9.9. Process flow diagram of the closed loop treatment circuit for GCP 1–6 supply water for the removal of total suspended solids.

for identifying critical independent variables that could significantly affect the dependent variable, namely the TSS of supply water, in the case of our project.

Process flow diagram (PFD): Through PFD, we created a visual representation of the closed loop process flow of the supply of water to the GCP scrubbers, exit of dust laden return water to the common slime channel, its distribution to different RSTs, settling and removal of the dust, addition of make-up water to the overflow water and pumping back of the supply water to the scrubbers. The PFD helped us to identify loops, hidden factories and non-value adding activities in the process (see figure 9.9).

A process is a combination of sequential or closed loop activities (or sub-processes) that culminate in an outcome in the form of a product or service. All activities in a processes can be categorized, in six sigma parlance, into the broad categories of:

1. *Value added*: This activity in the process adds form, function and value to the end product or service and is of value to the customer.
2. *Non-value added*: This activity in the process does not add any form, function or value to the end product or service and can be eliminated.
3. *Non-value added but necessary*: This activity by itself may not add any value but is necessary in the final sequence of activities in order to add value to the final product, or may be deemed as value added for another concurrent, dependent or symbiotic sub-process that leads to the final product or service.

A hidden factory refers to the activities or sub-processes that reduce the quality or efficiency of an operation or business process, and may result in waste or bad work as they remain hidden under the day-to-day operations.

The hidden factories can be in the form of:

a. Production loss.
b. Compromised quality.
c. Reduced availability of equipment.
d. Delayed delivery.

Identifying the non-value added activities and the hidden factories can help to address the operational issues that can adversely affect the quality of the process outcome.

Mapping of the process flow diagram of the closed loop treatment circuit for the supply water of BF GCPs 1–6 for removal of TSS helped to identify two hidden factories that were adversely affecting the TSS levels in the supply water.

The two existing hidden factories identified were:

a. The jamming of a valve plate in the common slime channel that was causing uneven distribution of water into the RSTs, thereby increasing the water load in RSTs 1 and 2 beyond their capacities, and reducing the water load in RSTs 3, 4 and 5. This was in turn adversely affecting the TSS settling capability of the RSTs 1 and 2.
b. Further, it was also found that RST 1 was not working properly, thereby again impacting the removal of TSS from the return slime water.

Table 9.1. Input–output worksheet.

					Process Ouputs/Features				Process Inputs(s)				
S.No.	Process Step	VA/ NVA	SOP	Equipment	Output	Feature (Y)	Specification	Capability	Input(X)	Controlable/ Noise	Feature	Specification	Test(s) / Analysis
1	Scrubbing &Atomizing the raw BF gas	VA	SOP available(GCP water level etc)	Scrubber & Automizer	Clean BF Gas	Dust Content	≤5mg/m3		Semi Clean gas	C	Dust	≤15gm/m3	No test
2			SOP for Nozzle Flushing	Scrubber & Automizer	Return Water from GCPs	Suspended Solids	TSS<1000 ppm		Supply water	C	1) Suspended solids 2) Pressure 3) Volume	1) TSS<100 ppm 2) 4.5 Kg/cm2 3) 600 m3/hr.	Measurements done
3	Distribution of return water to RST1,2 & 3, 4,5	VA		Slime Channel & One no. of sluice gate valve	Return water to RST1,2 & 3,4,5	Volume	m3/hr		Return water from GCP's	C	a) Volume	No specs	No Analysis
4	Poly Electrolyte dosing in channels to RSTs at 3 points	VA	SOP for dosing available	Feeding drum and pipe	Controlled volume of polyelectrolyte solution in return water channels to RST's	Flow	100 gms/ 3000 litres/8 hrs.		Polyelectrolyte solution in water	C	1) Flow 2) Ratio of Poly-electrolyte to water 3) Quality of polyelectrolyte	1) m3/hr based on rate at which drum gets emptied 2) 100 gms/3000 litres	No test
5	Distribution of return water to 4 nos of radial settling tanks	VA		Control valves	Regulated water flow to each RST	Volume & Suspended Solids	RST 1&2- 3500 M3/hr. RST 3,4,5 - 2200 M3/hr.		Return water from slime channel	C	a) Flow	m3/hr	TSS test is done
6	Settling of suspended solids from return water in RSTs	VA	SOP available	Radial Settling Tanks with Rotating arms (4 nos.)	Clear supply water with low TSS	Suspended Solids	<100 ppm		a)Regulated volume of water after throttling and Polyelectrolyte dosing	C	Suspended solids	TSS <1000 ppm	No test
7	Removal of settled slurry from RSTs to ash pond	VA	SOP available	1)Two Nos.of slurry pumps for each RST (1 working + 1 stand by)	Slurry pumped to ash pond	Density	No specs		Return water from slime channel with settled slurry after PolyElectrolyte Dosing	C	Suspended solids	No specs	visual inspection
8	Collection of clear supply water in suction chambers			Suction Chamber	Clear supply water with low TSS	Suspended solids & Flow	TSS		Clear overflow water from RSTs	C	Suspended solids	TSS	No test done
9	Addition of fresh make-up water to clear water from RSTs	VA		Pumps & Piping system	Fresh water supply to pump house	Suspended solids & Flow	TSS = 10-20 ppm, Flow = 300-400m3/hr		Fresh Water from Maroda Reservoir	C	Suspended solids	TSS=10-20 ppm	Test done
10	Pumping of clear supply water to GCPs 1-6		SOP for process available	Pumps and piping (2 working + 3 standby)	Clear supply water to GCPs	Water Pressure,Flow	Delivery pressure 7kg/cm2, Flow=700m3/hr		clear water from suction chamber + fresh make up water	C	water head	3.5 - 5metres	Measurement done
11	Supply of clear water through pipes (Lines 1 &2) to GCPs 1-6			Piping system (2 nos.)	Clear Water with low TSS	Pressure & Suspended Solids	Pressure - 5.5 Kg/cm2 TSS <100 ppm		Clear supply water thru pipes	C	1) Pressure 2) Suspended Solids	1) 7 Kg/cm2 2) TSS < 100 ppm	1) Pressure measurement done 2) TSS test done

Input–output (I/O) worksheet: An I/O worksheet was constructed to provide details of the inputs and outputs identified in the PFD. Through a spreadsheet in the form of an I/O worksheet we listed steps from PFD and identified process steps, process input/output, specification and VA or NVA. The output of this I/O worksheet was directly fed into the cause and effect matrix and FMEA. (See table 9.1 for the I/O worksheet.)

Fishbone diagram: The fishbone diagram was made after a brainstorming session with the extended team members. Individuals from the water management department up to the supervisory grades were also included. (See figure 9.10.)

Cause and effect matrix: The team members prepared the cause and effect matrix by taking inputs from all process steps from the I/O worksheet as well as from the fishbone diagram. It was found that inputs from man (personnel) and machine were the most frequent and critical. With these as the starting point, we proceeded further to carry out the failure mode and effects analysis (FMEA) based on both the I/O sheet and fishbone diagram. (See table 9.2 for the cause and effect matrix.)

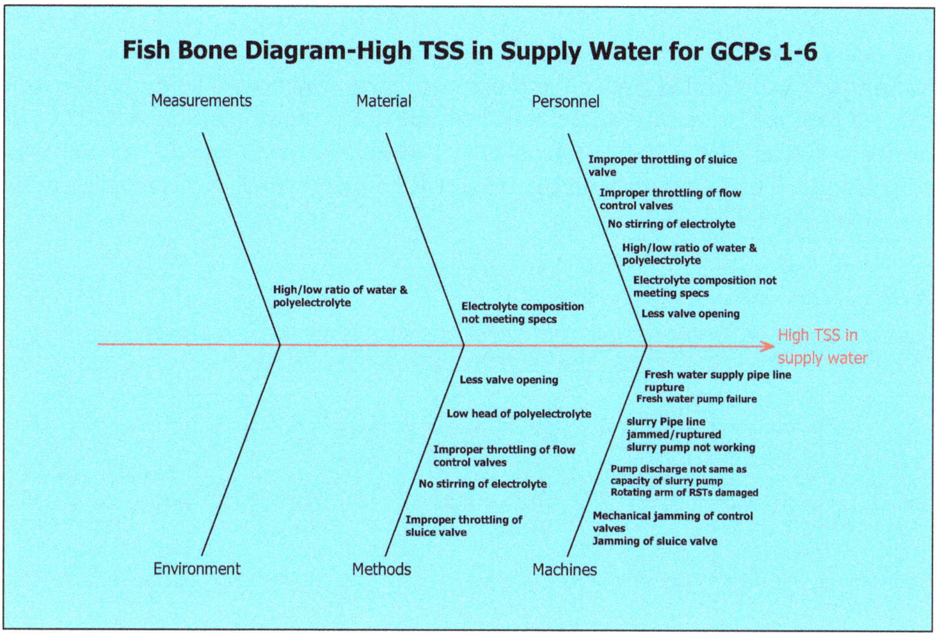

Figure 9.10. Fishbone diagram to identify possible causes for high TSS in the supply water.

Table 9.2. Cause and effect matrix.

	Customer Priority	10	
Process Step	Process Input	HIGH TSS	Total
Scrubbing & Atomizing the raw BF gas	Semi Clean gas	0	0
	Supply water	1	10
	Nozzles	0	0
Distribution of return water to RSTs 1, 2 & 3, 4, 5	Return water from GCP's	0	0
	Sluice Valve	9	90
Poly Electrolyte dosing in channels to RSTs at 3 points	Head	9	90
	Regulation of valve	9	90
	Quality of Poly.elec.	9	90
	Quantity of Poly Elctrolyte	9	90
Distribution of return water to 4 nos of radial settling tanks	Regulation of individual valves for controled flow of water	9	90
Settling of suspended solids from return water in RSTs	a)Regulated water after throttling and Polyelectrolyte dosing	0	0
	Rotating Arm	9	90
Removal of settled slurry from RSTs to ash pond	Return water from slime channel with settled slurry after Polyelectrolyte Dosing	0	0
	Pumping system pipes	9	90
	Pumps	9	90
Collection of clear supply water in suction chambers	Clear overflow water from RSTs	0	0
Addition of fresh make-up water to clear water from RSTs	Water intake from reservoir	3	30
Pumping of clear supply water to GCPs 1-6	Water stored in suction chamber	0	0
Supply of clear water through pipes (Lines 1 &2) to GCPs 1-6	Clear supply water thru pipes	0	0
	Total	0	
	LSL		
	TGT		
	USL		

Failure modes and effects analysis: The input for FMEA was taken from the detailed map, cause and effects matrix and fishbone diagram. The important processes and their outputs were tabled under the column process function. These outputs were analyzed for their modes of failure and their effects on the primary metric. The risk priority numbers (RPNs) were calculated, according to their severity, occurrences and detection. The processes with the highest RPNs were taken for further analysis. (See table 9.3 for FMEA.)

Table 9.3. Failure modes and effects analysis.

Process Function	Potential Failure Mode	Potential Effects of Failure	SEV	Potential Cause(s)/ Mechanism(s) of Failure	OCC	Current Process Controls	DET	RPN	Recommended Action(s)
The highest value process steps from the C&E matrix.	In what ways might the process potentially fail to meet the process requirements and/or design intent?	What is the effect of each failure mode on the outputs and/or customer requirements? The customer could be the next operation, subsequent operations, another division or the end user.	How Severe is the effect to the cusotmer?	How can the failure occur? Describe in terms of something that can be corrected or controlled. Be specific. Try identify the causes that directly impacts the failure mode, i.e., root causes.	How often does the cause or failure mode occur?	What are the existing controls and procedures (inspection and test) that either prevent failure mode from occurring or detect the failure should it occur? **Should include an SOP number.**	How well can you detect cause or FM?	SEV x OCC x DET	What are the actions for reducing the occurrence, or improving detection, or for identifying the root cause if it is unknown? **Should have actions only on high RPN's or easy fixes.**
Distribution of return water to RST1, 2 &3, 4,5	Uneven Distribution of water	High TSS	7	Mechanical jamming of Valve	7	Manual Inspection	9	441	1) Lubrication of valve 2) Frequent operation of valve 3) Valve lubrication schedule to be prepared
			7	Improper Throttling	9	No control	10	630	1) SOP to be prepared 2) Operator to be trained
			8	Deposit of mud and slabs in slime channel	9	No control	5	360	1) Prepare maintenance checklist, maintenance schedule, housekeeping schedule
Poly Electrolyte dosing in channels to RSTs at 3 points	No settling of suspended solids	High TSS	6	Low head of Poly electrolyte	4	Manual Inspection	2	48	
			6	Less Valve opening	4	Manual Inspection	2	48	
			6	Electrolyte chemical composition not as per spec	1	Chemical composition testing	1	6	
			6	No stirring of electrolyte	10	No control	10	600	1) stirring with compressed air 2)Stirring to be included in SOP
			8	High/Low ratio of water & poly electrolyte	10	ratio of mixing defined	9	720	1)Ratio of polyelectrolyte to water to be increased 2) New SOP to be prepared
Distribution of return water to 4 nos. of radial settling tanks	No/less settling of dust	Will result in High TSS	7	Mechanical jamming of Valve	7	Manual Inspection	9	441	1) Lubrication of valve 2) Frequent operation of valve 3) Valve lubrication schedule to be prepared
			7	Improper Throttling	9	No control	10	630	1) SOP to be prepared 2) Operator to be trained
Settling of suspended solids from return water in RSTs	No/less settling of dust	Will result in High TSS	7	Rotating Arm Damaged	5	Manual Inspection	5	175	
Removal of settled slurry from RSTs to ash pond	Accumulation of Dust in RST	Will result in High TSS	8	Pump not working	3	Operator monitors the working of pumps	3	72	
			8	Pump discharge not upto capacity of pump	7	Manual checking of flow of slurry	10	490	1) Check pump discharge 2) Install higher capaciy pump, if required 3) Check slurry pipeline condition 4) Clean slurry pipeline
			8	Pipe line jammed/Ruptured	2	Manual Inspection	3	48	
Addition of fresh make-up water to clear water from RSTs	Supply water will have slightly high amount of Dust	Will result in slightly High TSS	7	Pump failure	1	Operator monitors the working of pumps	3	21	
			7	Supply pipe line rupture	1	Manual Inspection	3	21	

Table 9.4. Why-why for improper throttling of valve.

Improper throttling of valve		
Question		Answer
What is your final action?		Proper regulation of throttling of valve
After regulation of throttling, is the distribution of water proper?		Yes
Why	**Answer**	**Action**
Why was throttling not proper?	Lack of SOP for throttling.	SOP to be made.
	Lack of knowledge of the operator.	Train the operator.
		Prepare a training manual for operators.

9.3.3 Analyze and improvement

Why-why analysis: Why-why analysis of each process with the highest RPNs was carried out to identify the root cause and eliminate or reduce it to the minimum extent possible. The why-why analysis sheets of each cause is as shown in tables 9.4–9.9.

Based on the inputs from FMEA and why-why analysis, certain process improvements in the form of quick wins were carried out immediately. These were:

1. Elimination of the hidden factory: The broken valve plate obstructing the homogeneous flow of return water in the common slime channel was removed, thereby facilitating optimum distribution of return water to all the RSTs.
2. Cleaning of the common slime water channel: This also facilitated the optimum distribution of return water to all the RSTs
3. Increasing the ratio of polyelectrolyte to water in the polyelectrolyte emulsion: Ratio of polyelectrolyte to water was increased from 100 g/ 3000 l/ 8 h to 250 g/6000 l/ 8 h in RSTs 3–5 and 100 g/3000 l/ 8 h to 400 g/ 10 000 l/ 8 h in RSTs 1–2, to improve the sedimentation rate of the suspended solids.
4. Lubrication and throttling of distribution valves: Lubrication of valves was carried out to ensure full throttling of the valves in each distribution channel to each RST so that the full flow of the return water to each RST could be ensured. This also resulted in correct distribution of return water to each RST.

After bringing about the above mentioned improvements, data were collected for the TSS of return water to RSTs 1 and 2 and RSTs 3, 4 and 5 as well as the TSS of the supply water at the pump house outlet one day per week, over a period of two months. On the basis of the new data on the TSS of the supply water, capability analysis of the process was carried out, as shown in figure 9.11. With Cp and Pp of more than 1, the process capability has been improved.

Table 9.5. Why-why for the incorrect ratio of water and polyelectrolyte.

Incorrect ratio of water and polyelectrolyte		
Question		Answer
What is your final action?		Ratio of polyelectrolyte and water increased.
After increasing the ratio, is the settling of dust particles okay?		Yes
Why	**Answer**	**Action**
Why was the ratio of polyelectrolyte to water increased?	It was felt that the ratio of polyelectrolyte dose to water was low and inadequate.	Regulated increase in polyelectrolyte dosing ratio.
Why was the ratio of polyelectrolyte to water low and inadequate?	The exact ratio of polyelectrolyte to water was not known to the operator.	Ratio of polyelectrolyte and water told to the operator.
	Higher polyelectrolyte dosing leads to higher settling of dust which in turn leads to choking of the RSTs.	Recommended installation of higher capacity slurry pumps and cleaning of slurry pipeline.
Why was the exact ratio of polyelectrolyte to water not known to the operator?	No SOP existing for polyelectrolyte dosing.	SOP to be prepared.

Table 9.6. Why-why for mechanical jamming of valve.

Mechanical jamming of valve		
Question		Answer
What is your final action?		Removal of jamming.
After removal of jamming, is valve ok?		Yes
Why	**Answer**	**Action**
Why did you remove jamming?	The valve was jammed and not operating.	Lubricating the valve.
Why was the valve jammed?	Less frequent operation.	Frequent operation of the valve is needed.
	No lubrication/less lubrication.	Frequent lubrication of the valve is required.
Why was valve not operated frequently?	No standard schedule was made for the operating valve.	Make operation schedule.
Why was no schedule made for valve operation?	The importance of frequent valve operation was not known to the working group.	Impart information to working group about the frequent operation of valve.
Why was valve not lubricated frequently?	No schedule was made for lubricating the valve.	Prepare the valve lubrication schedule.

Table 9.7. Why-why for the deposits of mud and slabs in the slime channel.

Deposits of mud and slabs in slime channel		
Question		Answer
What is your final action?		Cleaning of the slime channel.
After cleaning, is the distribution of water proper?		Yes
Why	**Answer**	**Action**
Why was the slime channel filled with mud and slabs?	Because of slabs falling into the slime channel.	Removal of slabs from the slime channel.
	Because of a broken valve piece stuck in the channel.	Broken valve piece removed.
Why did slabs fall into the slime channel?	Slabs were old and broken.	Replacement of slabs done.
	Broken slabs were not replaced.	Replacement of slabs done.
Why were broken slabs not replaced?	Not much importance given to housekeeping.	Periodic housekeeping to be done.
Why was a broken valve piece stuck in the channel?	A valve was broken during operation and was not removed.	Prevent breaking of valves during operation.
Why was the valve broken during operation and not removed?	The valve was not maintained properly	Proper maintenance of the valve to be done.
	Removal of broken valve piece not considered necessary.	Attention to be paid to removal of broken parts in future.
Why was the valve not maintained properly?	NO SOP for valve maintenance.	Prepare a maintenance schedule and maintenance checklist.
Why was the removal of the broken valve piece not considered important?	The housekeeping was not proper.	Prepare housekeeping schedule.

Box plot analysis was also carried out for three sets of TSS data:

a. Average TSS of supply water for the period December 2011 to March 2012, when the TSS value was meeting minimum requirements: TSS average supplied—ok.
b. Average TSS of supply water for the period April 2012 to December 2012, when TSS values were not meeting minimum requirements: TSS average supplied—not ok.
c. Average TSS of the supply water for the period April 2013 to May 2013, after carrying out the six sigma improvements: TSS average supplied—improved.

Table 9.8. Why-why for no stirring of electrolyte emulsion.

No stirring of electrolyte emulsion		
Question		Answer
What is your final action?		Stirring of electrolyte with compressed air.
After stirring, is the settling of dust particles okay?		Yes.
Why	**Answer**	**Action**
Why was stirring not done?	A lack of facilities for supply of compressed air.	Air supply provided and stirring being done.
	Less importance given to stirring.	Operator asked to monitor stirring of polyelectrolyte with compressed air.
Why was less importance given to stirring?	Operator unaware of the operational importance of stirring.	Importance of stirring to achieve homogeneous polyelectrolyte solution communicated to operator.
Why was the operator unaware of the operational importance of stirring?	Stirring not included in the SOP for polyelectrolyte dosing.	Include stirring in SOP.

Table 9.9. Why-why for insufficient pump discharge.

Pump discharge not sufficient		
Question		Answer
What is your final action?		Check existing pump discharge/install higher capacity pump.
After checking existing pump discharge, is it okay?		Yes.
Why	**Answer**	**Action**
Why did you check the pump capacity?	Settled slurry in the RSTs is not being removed completely, affecting the efficiency of the RSTs.	Check discharge of slurry pumps.
Why is the settled slurry not being removed completely?	Existing pump is old.	Recommend installation of higher capacity slurry pumps in RSTs.
	Slurry pipeline may be choked.	Pipeline condition checked.
Why is slurry pipeline choked?	Not cleaned regularly.	Slurry pipeline to be cleaned.
Why is slurry pipeline not cleaned regularly?	NO SOP for cleaning of slurry pipeline.	Prepare SOP for cleaning of slurry pipeline.

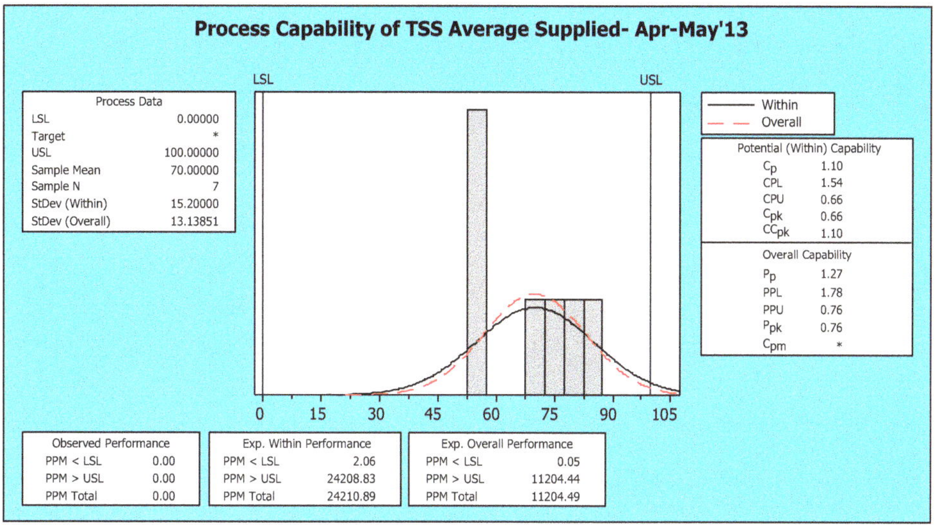

Figure 9.11. Capability analysis of TSS average supplied for the period April–May 2013—post improvement.

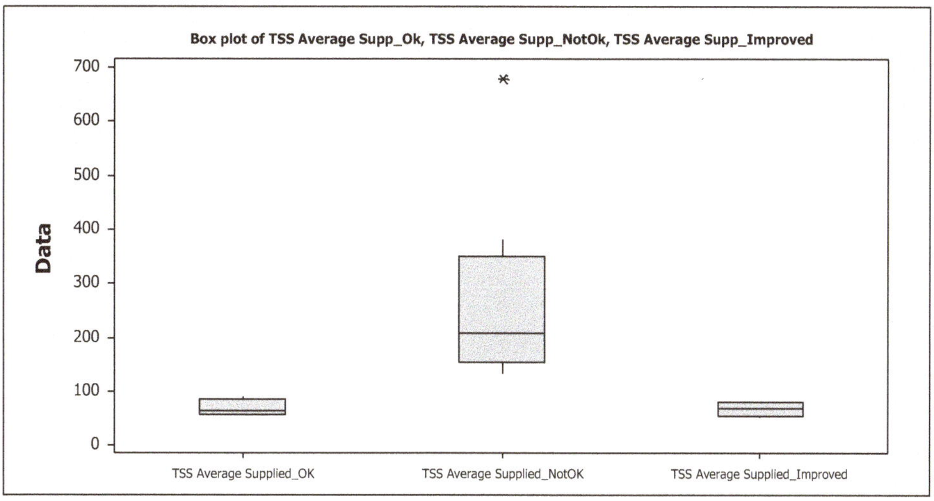

Figure 9.12. Box plot analysis of the TSS of the supply water: historical, skewed and improved.

The result is as depicted in figure 9.12.

The probability plot of the improved TSS of the supplied water (April–May 2013) compared to the historical data (December 11–March 2012) also shows that the process variations have been minimized and brought within the specification limit, at 95% CI (see figure 9.13). The process variations have been minimized and brought within the specification limit, at 95% CI.

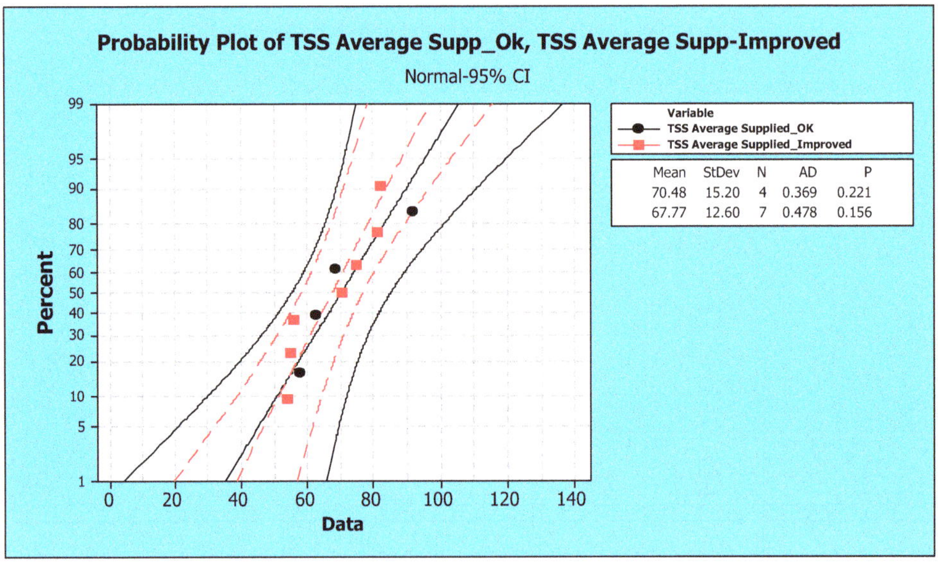

Figure 9.13. Probability plot of TSS average supplied: historical versus improved.

9.3.4 Control

Various control measures were adopted and practiced for maintaining the processes and the same activities are evaluated for reducing breakdowns. The control plan is shown in table 9.10.

9.4 Conclusion

9.4.1 Financial benefits

The financial benefits forecast, on account of savings in terms of reduced breakdown and reduced replacement of spares, as validated by the finance department, was 0.992 crores/annum and the actual savings from March 2013 to May 2013 were Rs. 24.8 lakhs.

9.4.2 Non-financial benefits

The non-financial benefits have been in the form of reduced suspended solids (TSS) of less than 100 ppm consistently since April 2013. Other consequent non-financial benefits have been as follows (the domino effect):

1. Reduced choking of nozzles/throttle assemblies, U-seals, and drain pots of u-seals as well as gas lines.
2. Visible reduction in the wearing out/erosion of scrubber shell plates.
3. Fewer unscheduled/emergency breakdowns/shutdowns in GCPs 1–6, due to BF gas leakages.
4. Renewed motivation in the workforce, due to the elimination of non-value adding activities towards correcting repetitive unscheduled/sudden breakdowns.

Table 9.10. Control plan for improved TSS in supply water.

Date:	**Revision:**
Product: Return water from GCPs 1-6	
Process: Removal of suspended solids from return water	

Process Step	What's Controlled?	Input or Output?	Spec. Limits/ Requirements	Measurement Method	Sample Size	Frequency	Who/What Measures	Where Recorded	Decision Rule/ Corrective Action	SOP's
Cleaning of slime channel	Proper Distribution of return water to RSTs	Input	To maintain water level 2 feet below top of channel	Manual	Entire slime Channel	Once in six months	Operator of WMD/EMD	Water Chemical lab & Pump House control room log book.	**Manual cleaning of slime channel to be done every six months**	**Manual cleaning of slime channel to be done every six months till water level is below 2 feet from top**
Poly Electrolyte dosing in channels to RSTs at 3 points	Ratio of water and polyelectrolyte powder	Input	250 gram poly /3000 litre water/8 hrs	Manual	3000 litres of water	Every 8 hrs	Operator of WMD Pump House	WMD Pump House control room log book.	Continue the specification ratio	Mixture of stated specification to be prepared and dosed in each shift
Distribution of return water to RST1,2 &3, 4,5	Operation of sluice valve for proper distribution of water to each RSTs	Input	% opening	Manual	Five valves	Once in a month	Operator of WMD	WMD Pump House control room log book.	Regulation of valves till TSS comes down to normal limits	Keep valves open as per level of water in slime channel
Distribution of return water to RST1,2 &3, 4,5	Lubrication of suice valves	Input	Free operation of valves	Manual	Five valves	Once in a month	Maintenance section of WMD	WMD Pump House control room log book.	Lubrication of valves till valve operation is smooth	Application of grease on mating parts of the valves
Removal of settled slurry from RSTs to ash pond	Discharge of slurry pump	Output	Pump capacity in 72 M3/hr.	Current flow measurement- 26 Amp.	Five outlets	continuous	Ultrasonic Flowmeter (to be installed)	CLR in pumphouse control room	To be installed and if discharge rate is found less, then repair/replacement of pump to be planned	Range of current flow- 18 to 22 Amp. Current flow should not be less than 18 Amp
Shifting of slurry to the centre of RSTs	Wear and tear of rotating arm	Output	No damage in steel structures of the arm	Visual Inspection	1 Rotating Arm	Every year	WMD Maint	WMD Maint. log book.	Repair/replacement of damaged parts	Check for no existence of old damaged structures

Approved by:	
Approved by:	
Approved by:	

5. The lifespan of the throttle assemblies (septum valves), nozzles, gate valves and overflow throttles of the scrubber and atomizers will increase.
6. The BF gas leakage problem will be reduced. The safety of GCP operators as well as the area surrounding GCPs will be ensured.
7. Better removal of dust from BF gas, thereby causing less choking of valves and burners at the consumer end, leading to reduced specific fuel consumption and better furnace productivity.

IOP Publishing

Modern Optimization Methods for Science, Engineering and Technology

G R Sinha

Chapter 10

Performance evaluation and measures

K A Venkatesh and N Pushkala

Measuring performance is an art and identifying the right tool to evaluate the performance is another Herculean task. Philosophically, anything and everything is perception. Evaluating the performance in terms of the perception of the promoters, the internal customers/employees or the end users are different tasks. This goal of this chapter is to introduce a tool which could address them all. In this chapter, we introduce an elegant non-parametric tool, called data envelopment analysis (DEA), in which the choice of inputs and outputs is based on the needs of the evaluator. Among the many other tools, the growth of DEA literature is incomparable. DEA is an extension of linear programming problems, in which one has to solve a number of linear programming problems in order to obtain relative efficiency scores. Initially this chapter introduces the various paradigms in performance measurement for decision making and ends with a discussion of the tool R in addressing DEA.

10.1 Performance measurement models

For success in any environment, the most important aspect is sustainability. Efficiency is the only parameter which makes an individual or organization sustainable. Many methods and models are available in the literature and in practice. Most of them are parametric ones, that is, mundane ones. These methods and models may not be suitable from the perspective of the observer. Needlessly, the observer may not be part of the system; sometimes the observer may be a one-time user of the system. Are there any scientific ways to measure the efficiency? This chapter will address this question and discuss various models and methods to measure the performance. Generally these models are used to assess the intended system to make a decision in order to attain the mission/goal. Optimization techniques are the most common tools used to make the best choice among the available options in order to make appropriate decisions. The decision making process can be achieved by deploying any of the following three approaches, or hybrid approaches: descriptive analysis, prescriptive analysis and normative analysis.

Further, the decision making process can be classified based on single or multiple criteria. Multiple-criterion decision making problems (MCDM) can be classified as multiple attribute decision making problems (MADM) and multiple objective decision making problems (MODM). MADM evolved from the Bernoulli utility theory. More precisely, multiple attribute utility theory (MAUT) and out-ranking methods (such as ELECTRE and PROMETHEE) are the two subcategories of MADM. Fuzzy MADM has been introduced to deal with situations of uncertainty.

The objective of a multi-objective decision making problem is to attain an optimal solution for the given set of (contradictory) objectives subject to the set of constraints. Hence, to associate an optimization technique with MODM is the natural choice. The foremost aim in these models is to convert the given multiple objectives into a weighted single objective function, otherwise pareto solutions must be obtained. Evolutionary algorithms, genetic algorithms, particle swarm optimization and vector optimization are the popular models of MCDM.

Single criteria decision making problems are modeled as linear programming problems to obtain the desired optimal solution. In this chapter, we present DEA as a tool to study performance measurement. In addition to DEA, this chapter deals with the analytical hierarchy process (AHP) and fuzzy AHP.

10.1.1 Fuzzy sets

This section deals with the fundamentals of fuzzy sets and operations on the fuzzy sets.

Definitions

1. Let X be a set. The *fuzzy set* A' is given by $A' = \{(a, \mu_A(x))|x \in X\}$, where $\mu_{A'}(x){:}X \to [0, 1]$ is the membership function.
2. The *support of the fuzzy set* A' is given by $\mathrm{supp}(A') = \{x \in X|\mu_{A'}(x) > 0\}$.
3. The α − cut of a fuzzy set A' is defined as $A'(\alpha) = \{x \in X|\mu_{A'}(x) \geqslant \alpha\}$, $\forall\ \alpha\epsilon[0, 1]$.
4. The height of $\mathcal{A}'$ is given by $h(\mathcal{A}') = \sup_{x\in X} \mu_A(x)$. If $h(A') = 1$, then the fuzzy set is said to be *normal*, otherwise it is called *subnormal*.

Operations

The fundamental arithmetic operations addition, subtraction, multiplication and division (/) are obtained based on the extension principle and α − cut arithmetic.

Extension principle

Let a', b' be any two fuzzy numbers and c be a particular event. The fundamental operations are defined as:

1. $\mu_{a'+b'}(c) = \mathrm{supp}_{x,y\in X}\{\mathrm{mini}(a'(x), b'(y)|x + y = c\}$.
2. $\mu_{a'-b'}(c) = \mathrm{supp}_{x,y\in X}\{\mathrm{mini}(a'(x), b'(y)|x - y = c\}$.

3. $\mu_{a'Xb'}(c) = \text{supp}_{x,y \in X}\{\text{mini}(a'(x), b'(y)|xXy = c\}$.
4. $\mu_{a'/b'}(c) = \text{supp}_{x,y \in X}\{\text{mini}(a'(x), b'(y)|x/y = c\}$.

α – cut **arithmetic**

Let r, s, t be the infimum, the mode and the superimum of the two fuzzy numbers $a' = [a^r, a^s, a^t]$ and $b' = [b^r, b^s, b^t]$, respectively. Using α – cut, one could define the fundamental operations as:

1. $a'(\alpha) + b'(\alpha) = [a^r(\alpha) + b^r(\alpha), a^t(\alpha) + b^t(\alpha)]$.
2. $a'(\alpha) - b'(\alpha) = [a^r(\alpha) - b^s(\alpha), a^t(\alpha) - b^r(\alpha)]$.
3. $a'(\alpha)Xb'(\alpha) = [A, B]$
   where $A = \text{mini}\{a^r, b^r, a^s, b^s, a^t, b^t\}$,
   $B = \text{mini}\{a^r, b^r, a^s, b^s, a^t, b^t\}$.
4. $\frac{a'(\alpha)}{b'(\alpha)} = [a^r(\alpha), a^t(\alpha)]\left[\frac{1}{b^t(\alpha)}, \frac{1}{b^r(\alpha)}\right]$.

10.2 AHP and fuzzy AHP

AHP is one of the MADM methods [1]. The various steps involved in AHP are:

1. Identify the nature of the problem.
2. Construct the hierarchy at all possible levels.
3. Construct pairwise judgement matrix at all levels.
4. Synthesization.
5. Consistency test.
6. Comparisons at all levels.
7. Develop overall rankings to identify the best alternatives.

Let a_1 a_2, …, a_n be the attributes and the relative importance is given by w_1, w_2, …, w_n and the pairwise comparison is given in the matrix form, called an association matrix

$$A = \begin{bmatrix} a_{11} & \cdots & a_{1n} \\ \vdots & \ddots & \vdots \\ a_{n1} & \cdots & a_{nn} \end{bmatrix},$$

where $a_{ij} = 1/a_{ji}$ and $a_{ij} \cdot a_{jk} = a_{ik}$. Generally the ratio w_i/w_j is unknown. The problem of AHP is to determine $a_{ij} = w_i/w_j$. Let W be the weight matrix given as

$$A = \begin{bmatrix} w_1/w_1 & \cdots & w_1/w_n \\ \vdots & \ddots & \vdots \\ w_n/w_1 & \cdots & w_n/w_n \end{bmatrix},$$

where $w = [w_1, w_2, .., w_n]^T$ is a column vector. The system can be written as $(W - nI)w = 0$. One could employ an eigenvalue method to solve the system $(W - nI)w = 0$, then obtain the comparative weights by finding the largest eigenvalue λ of A in magnitude. For the consistency test we introduce two more indices called the consistency index (CI) and consistency ratio (CR). The index CI is given by CI = $(\lambda - n)/(n - 1)$ and CR is given by CR = CI/random CI.

10.2.1 Fuzzy AHP

To capture the vague information from a real world problem, one has to deploy fuzzy numbers/linguistic variables. Triangular fuzzy numbers were incorporated in AHP by van Laaroven and Pedrycz in a pairwise comparison matrix. Subsequently many deployed various fuzzy numbers with AHP and the theory and models of FAHP have increased.

The geometric mean method is one of the popular methods to address FHAP. The pairwise comparisons are represented by linguistic variables using fuzzy numbers; let us define the fuzzy pairwise comparison matrix $A = [a'_{ij}]$ of order n, where n is the number of attributes with $a_{ij} \odot a_{ji} = 1$; and $a_{ij} = \frac{w_i}{w_j}$.

The weights can be computed by geometric mean method as $w'_i = s_i \odot (s_1 \oplus s_2 \ldots \oplus s_n)^{-1}$, where $s_i = (a_{i1} + a_{i2} + \ldots + a_{in})^{\frac{1}{n}}$.

One of the limitations of this method is that it does not consider whether the sum of the weights is equal to unity. To overcome this situation, one could deploy optimization techniques to obtain the weights. The major disadvantage of the geometric mean method is the increasing length of the irrational fuzzy interval, due to the multiplication of fuzzy numbers. A linear programming method to address the FAHP is presented.

10.2.2 Linear programming method

Given with a fuzzy positive reciprocal matrix $A = [a'_{ij}]$ of order n, the pairwise judgement can be described in the fuzzy interval $[Il_{ij}, Fu_{ij}]$ as

$$Il_{ij} \leqslant \frac{w_i}{w_j} \leqslant Fu_{ij}.$$

For the given level of α − cut,

$$Il_{ij}(\alpha) \leqslant \frac{w_i}{w_j} \leqslant Fu_{ij}(\alpha), \ \forall \ \alpha \in [0, 1]. \tag{10.1}$$

The inequality (10.1) can be written as

$$Il_{ij}(\alpha)w_j + w_i \leqslant 0, \text{ and}$$

$$w_i - Fu_{ij}(\alpha)w_j \leqslant 0.$$

The above two inequalities can be written in matrix notion as $Sw \leqslant 0$, where S is the matrix of order $2m \times n$.

The linear membership function is used to measure the consistency test towards the interval judgement

$$\mu_t(S_t w) = \begin{cases} 1 - \dfrac{S_t w}{T_t} & \text{if } S_t w \leqslant T_t \\ 0, & \text{otherwise} \end{cases},$$

where T_t is the parameter of tolerance.

The optimal weights are

$$\mu_D(w) = \max_w\{\mu_1[\text{mini}_{1\leqslant t\leqslant m}[\mu_1(S_1, w), \mu_2(S_2, w), \ldots, \mu_m(S_m, w)] \\ |w_1 + w_2 + \ldots + w_n = 1]\}.$$

The linear programming model corresponding to FAHP is given by

$$Z = \text{maximize } \theta.$$

Subject to the constraints

$$\theta \leqslant 1 - \frac{S_t w}{T_t}$$

$$w_1 + w_2 + \ldots + w_n = 1, \quad w_1 > 0, \quad 1 \leqslant i \leqslant n \,\&\, 1 \leqslant t \leqslant 2m.$$

Here, θ is the degree of satisfaction.

10.3 Performance measurement in the production approach

Performance evaluation is the most fundamental thing to benchmark. A systematic comparison of the performance of the system against others is popularly known as benchmarking. By comparing one's own performance with other firms, one can help to identify the position of the firm and opportunities to learn toward improvement. In general, efficiency is defined as the ratio between the output and the input. This section introduces necessary definitions for performance evaluation.

Definitions

1. For the given input i, the system produces an output o then production possibility set is given by $T = \{(i, o)|i \text{ can produce } o\}$. That is, for the amount of input i, the system/firm produces the amount of output o. Note that the inputs may be resources such as raw material and human resources, and outputs may be the services, number of transactions and so on.
2. Farrell input efficiency tells us the amount of reduction in input i that is able to produce the same amount of output o. It can be defined as $F_{ie} = \min\{k|ki \text{ can produce } o\}$, where $k \geqslant 0$.
3. Farrell out efficiency tells us the amount of increase in the output o, for the same amount of input i, and is given by $F_{oe} = \max\{l|i \text{ can produce } lo\}$, where $l \geqslant 0$.
4. Let (i_1, o_1) and (i_2, o_2) be the members of the production possibility set (i_2, o_2) dominates (i_1, o_1) iff $i_2 \leqslant i_1$ and $o_2 \geqslant o_1$ and $(i_1, o_1) \neq (i_2, o_2)$.
5. A (i, o) is efficient in the production possibility set iff $\exists$ no member $(i_l, o_l) \in T$ dominates (i, o).

Let us assume that we are interested in studying the performance of an organization among its r peers and denote the firm of interest as the tth firm. The tth firm produced m outputs $o_t = (o_{t1}, o_{t2}, \ldots, o_{tm})$ using n inputs $i_t = (i_{t1}, i_{t2}, \ldots, i_{tn})$, the production possibility set for the tth firm is given by $T_t = \{(o_t, i_t) \in \mathbb{R}^n X \mathbb{R}^m | i_t \text{ can produce } o_t\}$.

Using Farrell efficieny scores, the ranking of organizations is determined. There are other methods to find efficiency scores based on distance, called the Shepard distance. In the literature there are methods based on the reduction of input and expansion output simultaneously to obtain the efficiency scores. The efficiency obtained based on Farrell is called technical efficiency (TE). If we replace the input vector with a cost vector and the efficiency computation is carried out, the resulting computation is called cost efficiency (CE). Cost efficiency is choosing the right candidate with minimal cost. The other efficiency is allocative efficiency (AE) and it can be computed as AE = CE/TE. In the same way one could define revenue efficiency and profit efficiency.

10.3.1 Free disposability hull

Consider the production possibility set $T = (i, o)|i$ can produce $o\}$, where with the given inputs $\boldsymbol{i}$, o outputs can be produced. It is also true that one could produce the same output o with more than i inputs. This idea is called *free disposability of input* and it can be translated as whenever (i, o) is in T and $i' \geqslant i$ then (i', o) is in T.

In the same way, free disposability of output can be defined and it is given as: whenever (i, o) is in T and $o' \leqslant o$ then (i, o') is in T. The combination of free disposability of input and free disposability of output along with the production set of feasibility for the observed set is called the *free disposability hull* (FDH).

Convexity: The production possibility set T is said to be convex if the line segment joining any two points in T lies entirely inside T. Let (i, o) and (i_1, o_1) be any two members in T and λ $(0 < \lambda < 1)$ be a scalar such that $\lambda(i, o) + (1 - \lambda)(i_1, o_1) \epsilon T$.

Returns to the scale: Let $T = \{(i, o)|i$ can produce $o\}$ be the production possibility set and be a scalar. If for any possible combination (i, o) can be scaled down or scaled up for the given scalar λ, that is if $(i, o) \in T$ then $\lambda(i, o) \in T$.

Based the value of λ, various returns to scale are classified as:

1. Constant returns to the scale (CRS), if $\lambda \geqslant 0$.
2. Decreasing returns to the scale (DRS), if $1 \geqslant \lambda \geqslant 0$.
3. Increasing returns to the scale (IRS), if $\lambda \geqslant 1$.

Based on the need, one could construct the production possibility set in terms of production function or cost function with CRS, IRS, DRS, FDH and VRS.

10.4 Data envelopment analysis

DEA is one of the non-parametric methods for studying the performance of any firm and was introduced by Charnes, Cooper and Rhodes (CCR) in 1978, and in 1984 it was refined by Banker, Charnes and Cooper (BCC). This is an extended application of the linear programming model. The firms/entities whose performance is to be studied are called decision making units (DMU). The major advantage of the DEA model is that it can handle multiple inputs and multiple outputs. DEA models assume that all considered DMUs are efficient and able to compute a single score of efficiency for all DMUs, unlike in ratio analysis.

Based on the requirements, DEA can be input oriented or output oriented with any of the returns to scale. The various returns to scale are as given in figures 10.1–10.4.

This chapter only deals with CRS and VRS and is input oriented as well as output oriented.

10.4.1 CCR model

Now we will present the most basic model, popularly known as CCR DEA. The underlying principle of this model is to create a piecewise linear convex frontier in such way that the frontier envelops the assumed input and output data, as per the set objective function [2]. From the generated linear piecewise frontiers, efficiency scores are computed by a set of linear programs. The decision making unit (DMU) is the firm, which has the capability of transforming the inputs into outputs. Suppose we have n firms, each producing n distinct outputs from m distinct inputs. We have to compute the relative efficiency score of all considered DMUs. The scenario is modeled as a fractional linear program with the objective of maximizing the ratio of the sum of weighted outputs to the sum of weighted inputs subject to the constraints of non-decreasing weights. As it is a maximization of the ratio, the relative efficiency score can be at most 1. That is, to find the optimal weights so that the efficiency score is maximized.

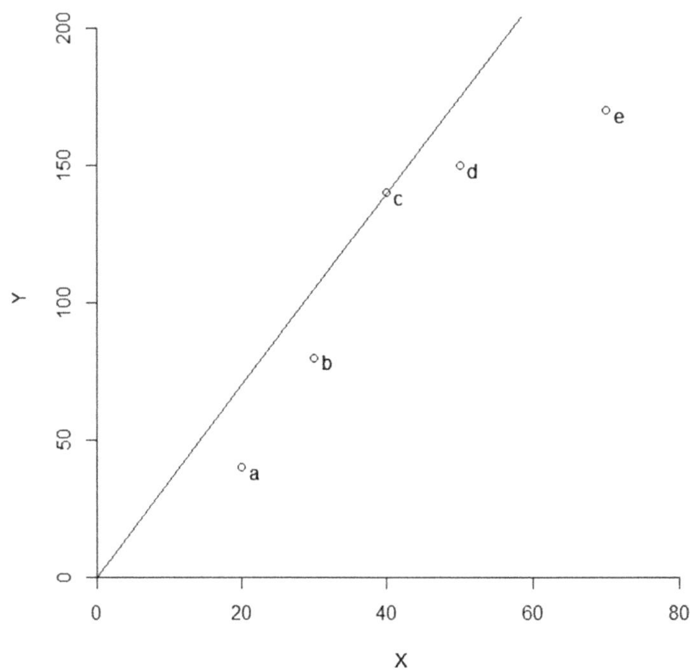

Figure 10.1. CRS.

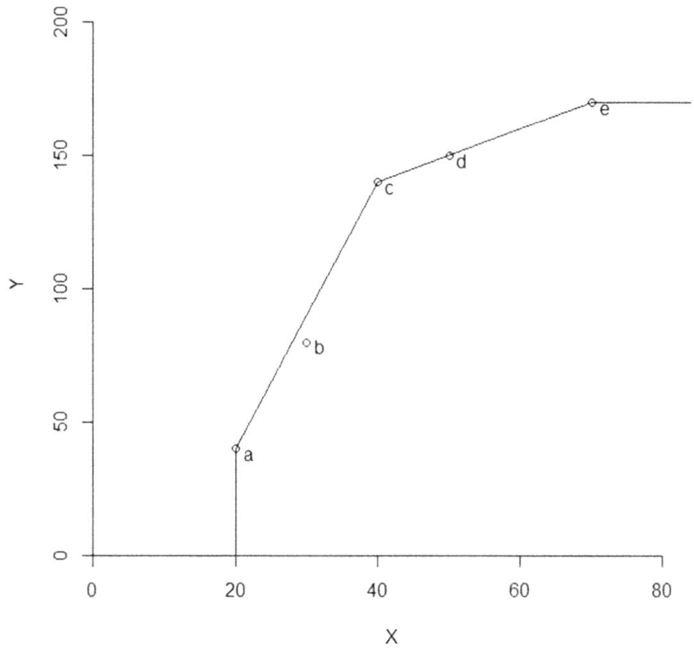

Figure 10.2. VRS.

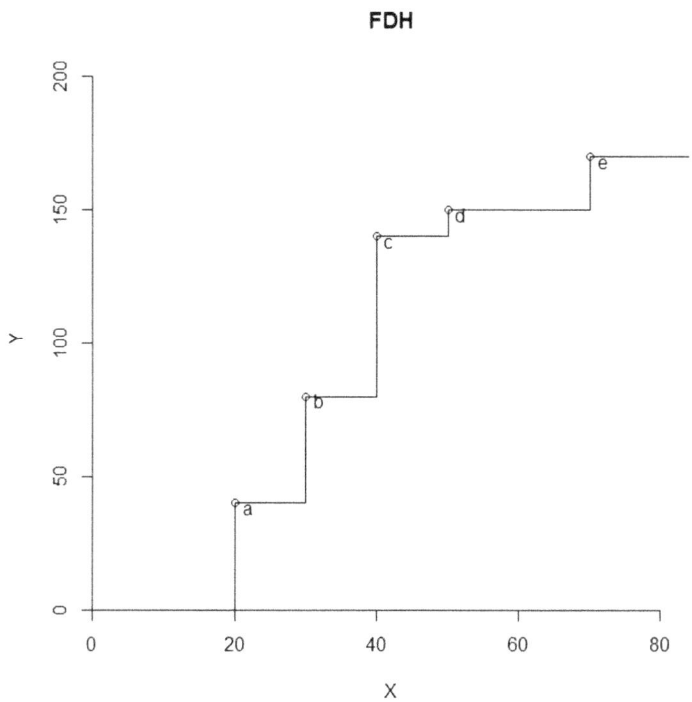

Figure 10.3. FDH.

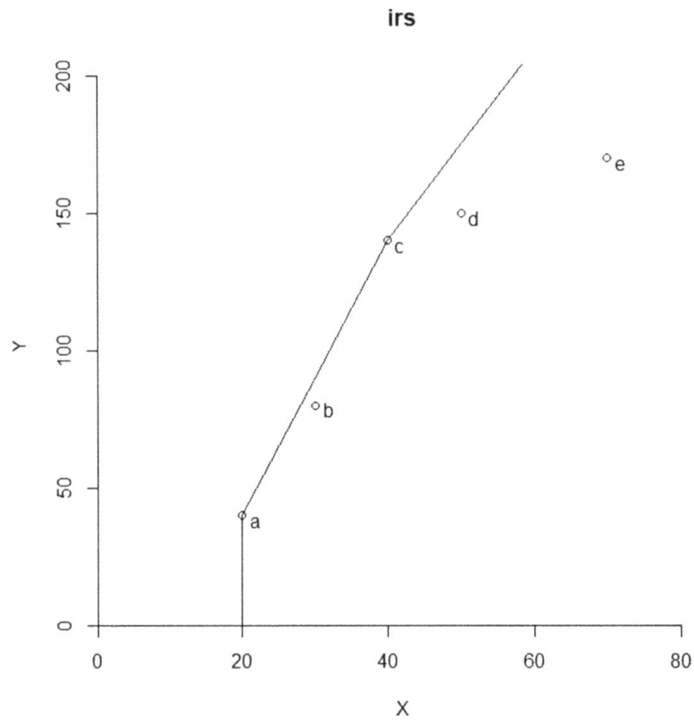

Figure 10.4. IRS.

The corresponding mathematical problem can be written as

$$\text{Max}_{u,\,v}z = \frac{u_1y_{1,0} + u_2y_{2,0} + \ldots.+u_ny_{n,0}}{v_1x_{1,0} + v_2x_{2,0} + \ldots+v_mx_{m,0}} \tag{10.2}$$

subject to the constraints

$$\frac{u_1y_{1,j} + u_2y_{2,j} + \ldots.+u_ny_{n,j}}{v_1x_{1,j} + v_2x_{2,j} + \ldots+v_mx_{m,j}} \leqslant 1$$

$$u_i \geqslant 0\,(1 \leqslant i \leqslant n) \text{ and } v_j(1 \leqslant j \leqslant m),$$

where y_{ik} denotes the output k produced by DMU i, x_{ij} denotes the input j used by DMU i, and u_i and v_j are weights supplied to outputs and inputs, respectively. The objective of this fractional linear programming problem (10.2) is to obtain the weights u_i, v_j which could maximize the ratio DMU_0; the DMU_0 is being evaluated.

The model (10.2) can be converted to a multiplier model using CCR transformation and is

$$\text{Max}_{\text{u},v}z = u_1y_{1,j} + u_2y_{2,j} + \ldots.+u_\text{n}y_{n,j} \tag{10.3}$$

subject to the constraints

$$v_1x_{1,j} + v_2x_{2,j} + \ldots+v_mx_{m,j} = 1$$

$$u_1y_{1,j} + u_2y_{2,j} + \ldots.+u_ny_{n,j} \leqslant v_1x_{1,j} + v_2x_{2,j} + \ldots+v_mx_{m,j}$$

$$u_i \geqslant 0(1 \leqslant i \leqslant n) \text{ and } v_j(1 \leqslant j \leqslant m).$$

The relative efficiency scores of all DMUs can be obtained by solving model (10.3) n (the number of DMUs) times. In optimization problems, we are accustomed to talking about the dual of the model. Here, the dual of this model is called the envelopment form. The model (10.2) is just the envelopment model and assumes constant returns to scale (CRS), but imposing one more condition, that the sum of lambdas is equal to 1, paves the way for the BCC model with a variable scale to returns (VRS). All DMUs are operating in an optimal environment, and then one should consider the CRS. In the majority of reports, this CRS assumption is violated due to improper competition, and the variety of prevailing regulations and restrictions. This led to a new measure of modified technical efficiency called super-efficiency (SE). Hence the estimated efficiency scores based on VRS are at least equal to the efficiency scores based on CRS.

Another important issue in the DEA approach needs to be discussed: the orientation, input orientation or output orientation. While measuring the technical efficiency, the efficiency measure improves the efficiency by the reduction in the input quantities, whereas the output quantities are unaltered. The technical efficiency for the cost frontier concides with this approach. The output orientation is just the reversal of the input orientation that improves the efficiency scores by proportional reduction of the output quantities, without altering the input quantities; this approach works well for the revenue frontier. The choice of orientation has a very minor influence on the efficiency scores, even though there is a strong correlation between both orientations [3]. The profitability falls under technical efficiency and hence it can be measured with both orientations. Importantly, the profit efficiency does not need to fall between 0 and 1 (sometimes, the profit could be more than 100% and at times it could be negative as well). Whether the profit is maximum or normalized is identified by the parameter return on expenditures. If the return of expenditure is minimal or negative, it indicates the inefficiency of the firm. Both input and output orientation models identify identical frontiers, more precisely both orientation models fetch the same set of efficient DMUs.

Example 10.1. Multiple input and multiple output.
Consider a chain of restaurants; the promoter of the group wishes to measure the performance of every branch. He/she decides the input and output to measure, given in table 10.1.

Let us translate this problem into LPP.
For DMU1

$$\text{Max}Z = 9u_1 + 4u_2 + 16u_3,$$

subject to

Table 10.1. Input and output.

DMU	Input 1	Input 2	Output 1	Output 2	Output 3
DMU1	5	14	9	4	16
DMU2	8	15	5	7	10
DMU3	7	12	4	9	13

$$5v_1 + 14v_2 = 1$$

$$9u_1 + 4u_2 + 16u_3 - 5v_1 - 14v_2 \leqslant 0$$

$$5u_1 + 7u_2 + 10u_3 - 8v_1 - 15v_2 \leqslant 0$$

$$4u_1 + 7u_2 + 13u_3 - 7v_1 - 12v_2 \leqslant 0$$

$$u_1, u_2, u_3, v_1, v_2 \geqslant 0.$$

For DMU2

$$\text{Max}, Z = 5u_1 + 7u_2 + 10u_3$$

subject to

$$8u_1 + 15u_2 = 1$$

$$5u_1 + 7u_2 + 10u_3 - 8v_1 - 15v_2 \leqslant 0$$

$$9u_1 + 4u_2 + 16u_3 - 5v_1 - 14v_2 \leqslant 0$$

$$4u_1 + 7u_2 + 13u_3 - 7v_1 - 12v_2 \leqslant 0$$

$$u_1, u_2, u_3, v_1, v_2 \geqslant 0.$$

For DMU3,

$$\text{Max } Z = 4v_1 + 7v_2 + 13v_3,$$

subject to

$$7u_1 + 12u_2 = 1$$

$$4u_1 + 7u_2 + 13u_3 - 7v_1 - 12v_2 \leqslant 0$$

$$9u_1 + 4u_2 + 16u_3 - 5v_1 - 14v_2 \leqslant 0$$

$$5u_1 + 7u_2 + 10u_3 - 8v_1 - 15v_2 \leqslant 0$$

$$u_1, u_2, u_3, v_1, v_2 \geqslant 0.$$

One could use any tool to solve the above linear programming and find the efficiency scores of DMU1, DMU2 and DMU3.

CCR efficiency: A DMU is efficient in the context of CCR if $z = 1$, that is there exists at least one optimal solution for non-negative u and v, otherwise the DMU is inefficient.

Duality plays a vital role in linear programming problems. We have k DMUs under consideration, which use m inputs to produce n outputs. The dual of the model (10.3) can be expressed for a real number z^* and a non-negative vector $\lambda = (\lambda_1, \lambda_2, \dots, \lambda_k)$ as

$$z^* = \text{mini}\, z. \tag{10.4}$$

Subject to

$$\sum_{j=1}^{k} x_{ij}\lambda_j \leqslant z x_{i0.}\ 1 \leqslant i \leqslant n$$

$$\sum_{j=1}^{k} y_{rj}\lambda_j \leqslant y_{r0}.\ \ 1 \leqslant r \leqslant m$$

$$\lambda_j \geqslant 0, \forall j.$$

The dual model (10.4) is feasible if $z^* = 1, \lambda_j^* = 1$ and $\lambda_0 = 0$, for $j \neq 0$. This process has to be repeated as many times as the number of DMUs.

Note that the optimal solutions occur on the boundary points and not all boundary points are efficient. In such a scenario, we employ the modified model (10.5) as

$$\text{Max} \sum_{i=1}^{n} s_i^- + \sum_{r=1}^{s} s_r^+, \tag{10.5}$$

subject to

$$\sum_{j=1}^{k} x_{ij}\boldsymbol{\lambda}_j + s_i^- = z x_{i0}, \quad 1 \leqslant i \leqslant n$$

$$\sum_{j=1}^{k} y_{rj}\boldsymbol{\lambda}_j - s_r^+ = y_{r0}, \quad 1 \leqslant r \leqslant m$$

$$\boldsymbol{\lambda}_j \geqslant 0,\ 1 \leqslant j \leqslant k$$

$$s_i^- \geqslant 0,\ 1 \leqslant i \leqslant m$$

$$s_r^+ \geqslant 0,\ 1 \leqslant i \leqslant n,$$

where s_r^+ and s_i^- are slack variables.

DEA efficiency: Any DMU is efficient if and only if $z^* = 1$ and all slack variables are zero.

Now we will solve example 10.1 using input oriented and output oriented methods and present the efficiency scores for three DMUs. The efficiency scores of DMU1, DMU2 and DMU3 are, 1, 0.7733 and 1, respectively. Clearly DMU2 is inefficient.

The models given in (10.2) to (10.5) are CCR—i.e. input oriented. In the same manner, one can define the output oriented models.

10.4.2 BCC model

In the BCC model convexity is assumed. The BCC model was introduced by Banker, Charnes and Cooper to evaluate the efficiency of the DMU_0. The convexity assumption is the additional feature of BCC compared to CCR.

The corresponding LPP is given as

$$z^* = \text{mini}\, z, \tag{10.6}$$

subject to

$$\sum_{j=1}^{k} x_{ij}\lambda_j \leqslant z x_{i0},\quad 1 \leqslant i \leqslant n$$

$$\sum_{j=1}^{k} y_{rj}\lambda_j \leqslant y_{r0},\quad 1 \leqslant r \leqslant m$$

$$\sum_{j=1}^{k} \lambda_j = 1$$

$$\lambda_j \geqslant 0,\ \forall j.$$

BCC efficient: Any DMU is BCC efficient if $z = 1$, that is there exists one optimal solution for non-negative u and v, otherwise the DMU is inefficient.

As we did in the CCR model, in the case of an inefficient boundary point we deploy the following model:

$$\text{Max} \sum_{i=1}^{n} s_i^- + \sum_{r=1}^{s} s_r^+, \tag{10.7}$$

subject to

$$\sum_{j=1}^{k} x_{ij}\lambda_j + s_i^- = z x_{i0},\quad 1 \leqslant i \leqslant n$$

$$\sum_{j=1}^{k} y_{rj}\lambda_j - s_r^+ = y_{r0},\quad 1 \leqslant r \leqslant m$$

$$\sum_{j=1}^{k} \lambda_j = 1$$

$$\lambda_j \geqslant 0,\ 1 \leqslant j \leqslant k$$

$$s_i^- \geqslant 0,\ 1 \leqslant i \leqslant m$$

$$s_r^+ \geqslant 0,\ 1 \leqslant i \leqslant n,$$

where $s_r^+$and s_i^- are slack variables.

The BCC output oriented model is given as

$$w^* = \max w, \tag{10.8}$$

subject to

$$\sum_{j=1}^{k} x_{ij}\lambda_j \leqslant x_{i0},\quad 1 \leqslant i \leqslant n$$

$$\sum_{j=1}^{k} y_{rj}\lambda_j \geqslant wy_{r0},\quad 1 \leqslant r \leqslant m$$

$$\sum_{j=1}^{k} \lambda_j = 1$$

$$\lambda_j \geqslant 0,\ \forall j.$$

Example 10.2. We are interested in studying the performance of 27 public sectors banks in India for the year 2017. The inputs are cash-in-hand (CH), money at call and short notice (MCS), advances (ADV) and investments in India (II) and outputs are interest income (IIc) and business per employee (BPE). The input and output are given in table 10.2.

Of these 27 banks, only 12 are fully efficient. The largest bank in India, SBI, is also inefficient in the year 2016, whereas under VRS, only four banks are inefficient. The computed efficiency scores based on CRS-input oriented, VRS-input oriented and the scale efficiency are given in table 10.3.

10.4.3 Other models

A variety of DEA models is available. To name a few, we have the slack based measure DEA, network DEA, stochastic DEA, fuzzy DEA and so on. In some cases, one may not have either inputs or outputs, in such cases, use unity as inputs/outputs and adopt the chosen model to obtain the efficiency scores [4–7].

Table 10.2. Inputs and outputs of DMUs.

Banks	CH	MCS	II	ADV	IIC	BPE
Allahabad Bank	6137.051	46 378.76	556 579.193	1560 695.2	188 849.46	148.5
Andhra Bank	8847.228	0	537 209.13	416 888.62	176 346.75	166.5
Bank of Baroda	37 564.451	353 920.31	1108 851.704	2289 771.6	440 612.77	168
Bank of India	26 310.955	217 519.64	1136 107.609	3538 093.2	417 964.7	179.6
Bank of Maharashtra	8189.764	0	362 308.724	299 714.37	130 529.86	180.2
Bharatiya Mahila Bank Ltd	56.178	1350	4788.136	5.841	1558.457	32.9
Canara Bank	20 118.673	66 393.138	1413 138.326	2879 455.5	440 221.35	144.462
Central Bank of India	16 584.535	12 000	888 200.49	768 034.42	258 878.97	119.478
Corporation Bank	12 213.76	6000	632 800.927	805 599.41	194 112.37	187.9
Dena Bank	5975.075	0	352 262.182	390 862.62	106 457.34	146.2
Idbi Bank Limited	15 856.281	14 437.12	989 994.323	1983 065.7	280 431.02	251.8
Indian Bank	5371.302	62.942	505 146.666	297 525.23	162 437.84	153.1
Indian Overseas Bank	18 253.872	57 568.519	757 189.041	758 587.35	235 172.95	124.1
Oriental Bank of Commerce	6947.381	1656.375	656 578.381	403 696.29	200 587.1	168.873
Punjab and Sind Bank	2507.299	8000	276 450.374	149 244.64	87 443.409	162
Punjab National Bank	22 238.461	4690	1525 444.026	3357 959.2	474 243.5	135.9
Syndicate Bank	8945.607	112 963.74	678 464.959	813 299.55	231 977.8	146.1
Uco Bank	5838.033	45 147.936	818 859.854	593 404.56	185 609.74	126.8
Union Bank of India	11 248.14	3619.76	877 728.066	3937 164	321 988.01	155.1
United Bank of India	5588.093	21 000	447 233.834	115 839.1	99 366.709	123.7
Vijaya Bank	5545.97	301.612	418 424.895	172 765.8	120 835.79	145.7
State Bank of Bik and Jai	5025.53	0	247 823.746	502 362.9	95 924.728	121.5
State Bank of Hyderabad	5191.821	7006.436	380 075.95	694 043.33	141 872.07	147.1
State Bank of India	150 809.19	124 570.23	4376 605.631	9719 560.1	1636 853.1	141.1
State Bank of Mysore	3882.42	0	201 239.554	393 592.22	71 277.658	118.3
State Bank of Patiala	2693.868	0	309 170.205	302 991.46	104 570.99	128.8
State Bank of Travancore	5879.217	11 496.507	360 618.29	407 926.73	96 088.79	117.6

Table 10.3. Various efficiency scores of DMUs.

Banks	VRS EFF	CRS EFF	SCALE EFF
Allahabad Bank	1	0.956 817	0.956 817
Andhra Bank	1	1	1
Bank of Baroda (BOB)	1	1	1
Bank of India (BOI)	1	0.945 135	0.945 135
Bank of Maharashtra	1	1	1
Bharatiya Mahila Bank Ltd	1	1	1
Canara Bank (CB)	1	0.831 885	0.831 885
Central Bank of India (CBI)	1	0.822 517	0.822 517
Corporation Bank (CRB)	0.932 689	0.835 85	0.896 172
Dena Bank (DB)	0.860 194	0.858 416	0.997 934
Idbi Bank Limited	1	0.749 013	0.749 013
Indian Bank (IB)	1	1	1
Indian Overseas Bank (IOB)	0.935 99	0.849 889	0.908 011
Oriental Bank of Commerce	1	0.945 505	0.945 505
Punjab and Sind Bank (PSB)	1	1	1
Punjab National Bank (PNB)	1	0.847 056	0.847 056
Syndicate Bank	1	0.957 495	0.957 495
Uco Bank	1	0.836 404	0.836 404
Union Bank of India (UNBI)	1	1	1
United Bank of India (UBI)	1	0.946 651	0.946 651
Vijaya Bank	1	1	1
State Bank of Bik and Jai	1	1	1
State Bank of Hyderabad (SBH)	1	1	1
State Bank of India (SBI)	1	0.963 954	0.963 954
State Bank of Mysore (SBM)	1	1	1
State Bank of Patiala (SBP)	1	1	1
State Bank of Travancore (SBT)	0.745 526	0.743 562	0.997 366

10.5 R as a tool for DEA

R is an open source tool for statistical computing. Because of its capabilities, it is a popular and powerful tool. This section will introduce R for solving DEA problems in an elementary manner. R can be downloaded from https://www.r-project.org/.

Store your input and output as a .csv file and type the following in R-prompt: First install the Benchmarking package then type: require (Benchmarking):

1. To call your file to R

   Dea_data <- read.csv(file.choose( ))

2. To verify the data and number of DMUs

   head(Dea_data) // to show the first six rows of your data

   dim(Dea_data) // to show the number of DMUs

3. Arrange your input as well as output in matrix form
   xImat<-cbind(Dea_datacol_name1, Dea_datacol_name2, ..., Dea_data $col_namek)
   yOmat<-cbind(Dea_datacol_name1, Dea_datacol_name2, ..., Dea_data $col_namel)
4. To compute the efficiency scores
   CRS and input oriented
   deaCrsIn<- dea(xImat, yOmat, RTS='crs',ORIENTATION = 'IN')
   eff(deaCrsIn)
   pears(deaCrsIn)// list the pears
   deaCrsIn$sx and deaCrsIn$sy // list the slacks in inputx, outputy
5. To get the output in csv file
   write.csv(eff(deaCrsIn), file='eff_scores.csv')

References

[1] Venkatesh K A, Pushkala N and Mahamayi J 2017 IFA - Off-Balance Sheet Items and Performance Evaluation of Public and Private sector banks in India: A DEA approach CFA Institute https://arx.cfa/post/IFA-Off-Balance-Sheet-Items-and-Performance-Evaluation-of-Public-and-Private-sector-banks-in-India-A-DEA-approach-4968.html

[2] Cooper W, Seiford L M and Tone K 2007 *Data Envelopment Analysis: A Comprehensive Text with Models, Applications, References and DEA-Solver Software* (Berlin: Springer)

[3] Coelli T and Perelman S 1999 A comparison of parametric and non-parametric distance functions: with application to European railways *Eur. J. Operation. Res.* **117** 326–39

[4] Behr A 2015 *Production and Efficient Analysis with R* (Berlin: Springer)

[5] Ray S C *DEA, Theory and Techniques for Economics and Operations Research* (Cambridge: Cambridge University Press)

[6] AliEmrouznijad M T *Performance Management with Fuzzy Data Envelopment Analysis* (Berlin: Springer)

[7] Bogetoft P and Otto L 2011 *Benchmarking with DEA, SFA and R* (Berlin: Springer)

Chapter 11

Evolutionary techniques in the design of PID controllers

Santosh Desai, Rajendra Prasad and Shankru Guggari

Currently, the world is witnessing great changes in almost all areas, including the field of system engineering, with steeply increasing system complexity resulting in outsized systems. Generally these systems are best described by a large number of differential or difference equations to form a mathematical model and ease the process of analysis, simulation, design and control. However, the said task is not as easy as it seems to be. It is very grueling, sometimes not feasible and also proves to be a costly affair because of what may be called 'the curse of dimensionality'. Hence model order reduction has been borne out of the necessity to provide simplified models, thereby addressing the effects of higher dimensional models and allowing us to cope with the emergence of present day technology.

Today, various methods are being proposed by authors [1–9] and are available in the literature. However, choosing the right method is one aspect which needs to be considered. The area of application may also have an effect in addition to satisfying the design requirements. Finally, this results in the need for approximations (low order). Hence there is a need for a suitable technique to combat/improve the drawbacks of the prevailing methods by developing/proposing new reduction techniques satisfying the need of the hour. The proposed methods are not only limited to continuous time systems, but can also be applied to systems presented in discrete time. The same is justified by applying the proposed technique on several numerical models. The outcomes are compared to other well-known methods when subjected to specific test inputs. However, these simulations were performed to obtain the behavior of the system in open loop configuration, but in most practical cases some sort of controller always exists to control the system behavior. The design of such a controller becomes a crucial task, in particular when the plant size is very large. In that case, the size of the controller also increases, thereby resulting in

doi:10.1088/978-0-7503-2404-5ch11

complicated and costly design. Apart from this, more computational time, and difficulties in analysis, simulation and understanding of the system arise. Hence, there is a need for a suitable lower order controller which can be derived from a higher order controller, preserving the crucial dynamics of the original. Furthermore, the derived reduced controller will be able to control the original higher order system satisfactorily. This has resulted in the application of order reduction methods to address controller reduction problems. In this chapter, the design of a PID controller is achieved with the aid of evolutionary techniques such as PSO and BBBC. Further, these techniques are extended to fine-tune the controller parameters, namely λ (integral order) and μ (derivative order) in the fractional order proportional integral derivative (FOPID) controller structure.

11.1 PID controller

As per the available literature, authors have proposed different methods [1–9]. But, the choice of method also plays a major role. The suitability of the method chosen depends on meeting the design requirements without sacrificing accuracy. This results in approximations of lower order. The PID controller $G_c(s)$ design connected in series with an uncontrolled plant $G_p(s)$ is of prime importance in this chapter. The behavior of the overall closed loop system designed has to be stable enough to drive under unity feedback conditions. Also, the responses of the proposed system and the reference model should match each other in the time domain. Figure 11.1 indicates two approaches [10–12] that are being considered here for the design of the controller, namely plant/process reduction and controller reduction.

The process reduction method comprises reducing the original plant $G_p(s)$ to $R_p(s)$. Then a suitable controller $R_c(s)$ is designed and placed in series with $R_p(s)$, as shown in figure 11.2. Further, the closed loop response of $R_{\mathrm{CL}}(s)$ is obtained assuming unity feedback. The block diagram in figure 11.2 depicts the original, reduced controller configurations using the so called reference model $M(s)$. This method is also referred to as the direct approach.

In the second (controller reduction) approach, the (higher order) controller $G_c(s)$ is designed on the basis of the higher order original uncontrolled plant $G_p(s)$. Then, the transfer function of the closed loop configuration $G_{\mathrm{CL}}(s)$ is reduced

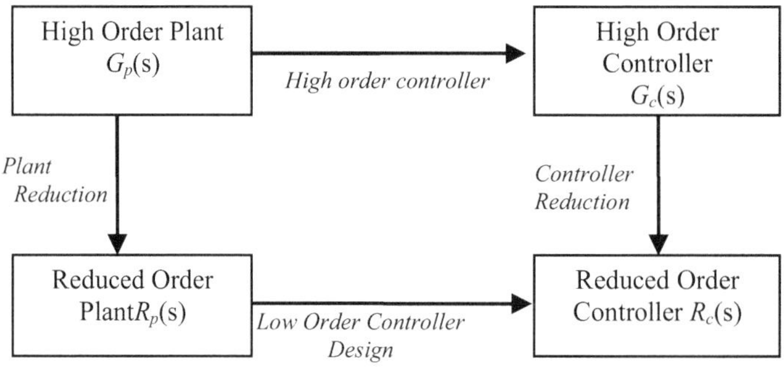

Figure 11.1. Controller design approaches.

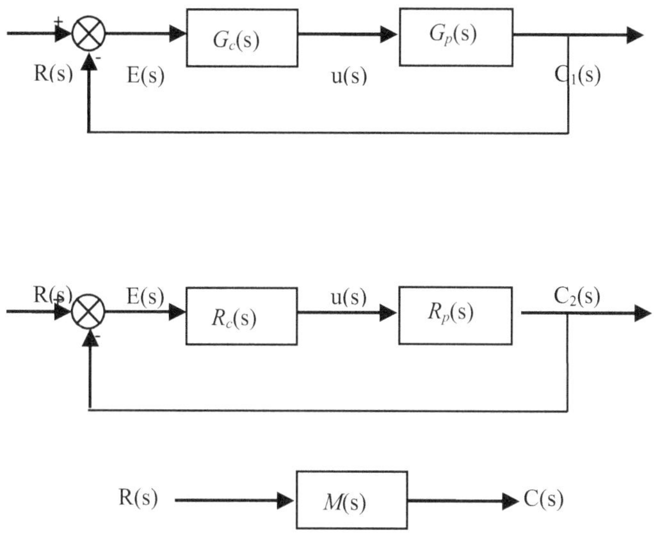

Figure 11.2. Original and reduced controller configuration with reference model.

appropriately with unity feedback and compared to $R_{CL}(s)$. The procedure carried out in the controller reduction approach is also referred to as an indirect approach. Here, the reduction process is carried out at the last stages and the possibility of error propagation is ruled out. In the direct approach the issue of error creeps in since the reduction is carried out during the initial stage of design.

In the present study, both direct and indirect design approaches have been considered. Another optimization technique based on big bang big crunch (BBBC) theory [13] is being used to tune the parameters of the PID controller for the said purpose. This approach has yielded better results when compared to the HNA assisted approach. In addition to this PSO, another evolutionary method, is also used for optimizing the design parameters. The evolutionary technique under discussion helps in searching for and selecting the best possible set of parameters which satisfies the design requirements. When the design task is completed, the closed loop responses are then compared with the reference model $M(s)$. The reference model $M(s)$, also called the specification model or standard model, is the desired closed loop system. To conclude, $M(s)$ meets all the desired performance specifications and acts as the basis for comparison.

11.1.1 Design procedure

Figure 11.1 indicates the two approaches of the controller design [10]. The first step involves the design of the higher order controller as per the plant followed by reducing to a lesser order. The closed loop behavior of the controller (higher order) with the plant and the controller (lower order) with the plant are compared with that of the reference model. Approximate model matching in the Pade sense is used to obtain the controller parameters. The controllers (full order and reduced order) are then compared in terms of their performance, as shown in figure 11.2.

11.1.1.1 Plant reduction and controller design (direct method)

The following steps are followed for design purposes based on the Pade sense.

Step 1. A reference model $M(s)$ is obtained based on time/frequency/complex domain specifications of the plant $G_p(s)$. The controlled system response (with unity feedback) in a closed loop is similar to $M(s)$.

Considering the given original plant $G_p(s)$, reference model $M(s)$ represented by

$$G_p(s) = \frac{a_o + a_1 s + \dots + a_m s^m}{b_o + b_1 s + b_2 s^2 + \dots + s^n}; \quad m < n \tag{11.1}$$

$$M(s) = \frac{g_o + g_1 s + \dots + g_u s^u}{h_o + h_1 s + h_2 s^2 + \dots + h_v s^v}; \quad u < v. \tag{11.2}$$

Step 2. Obtain the equivalent open loop specification model from the reference model $M(s)$ using

$$\tilde{M}(s) = \frac{M(s)}{1 - M(s)}. \tag{11.3}$$

Step 3. Consider the structure of the controller given by

$$G_c(s) = \frac{p_o + p_1 s + \dots + p_k s^k}{q_o + q_1 s + q_2 s^2 + \dots + q_j s^j}; \quad k < j. \tag{11.4}$$

Step 4. Compare the response of the system with the reference model to obtain the unknown controller parameters by

$$G_c(s)G_p(s) = \tilde{M}(s)$$
or
$$G_c(s) = \frac{\tilde{M}(s)}{G_p(s)} = \sum_{i=0}^{\infty} e_i s^i, \tag{11.5}$$

where $e_i s$ = coefficients of power series expansion ($s = 0$).

Equate (11.4) and (11.5) in the Pade sense to determine unknown control parameters $_{pi}$ and q_i:

$$\begin{aligned}
p_0 &= q_0 e_0 \\
p_1 &= q_0 e_1 + q_1 e_0 \\
p_2 &= q_0 e_2 + q_1 e_1 + q_2 e_0 \\
&\vdots \\
p_i &= q_0 e_i + q_1 e_{i-1} + \dots + q_1 e_0 \\
0 &= q_0 e_{i+1} + q_1 e_1 + \dots + q_{i+1} e_0 \\
&\vdots \\
0 &= q_0 e_{i+j} + q_1 e_{i+j-1} + \dots + q_j e_i.
\end{aligned} \tag{11.6}$$

The controller with the desired structure is obtained by solving (11.6).

Step 5. Obtain the closed loop transfer function using

$$G_{\mathrm{CL}}(s) = \frac{G_c(s)G_p(s)}{1 + G_c(s)G_p(s)}. \tag{11.7}$$

Step 6. $R_P(s)$ is obtained from $G_P(s)$ using the reduction method. Steps 4 and 5 are repeated and the transfer function (closed loop) for $R_{\mathrm{CL}}(s)$ (reduced order model) is given by

$$R_{\mathrm{CL}}(s) = \frac{R_c(s)R_p(s)}{1 + R_c(s)R_p(s)}. \tag{11.8}$$

11.1.1.2 Indirect approach: controller design and reduction

Here, the original plant $G_p(s)$ is considered and a high order controller $G_c(s)$ is designed. The closed loop transfer function $G_{\mathrm{CL}}(s)$ is determined considering feedback as unity. Finally, $G_{\mathrm{CL}}(s)$ is reduced to obtain the reduced closed loop transfer function $R_{\mathrm{CL}}(s)$.

11.1.2 Method 1: PID controller design using PSO

PSO, as a global optimization algorithm, has gained popularity in both academia and industry. Its simplicity and intuitiveness, ability to handle both discrete and continuous variables, quite easy implementation, and exploration of the majority of the problem space are its advantages. As a result, it is being used to solve optimization problems related to engineering systems. Another main attraction of PSO is that it works well for any dimensional problem. Hence, it is used to find the optimum for functions. These functions include single, multi-objectives and may be nonlinear or linear. Although, the problem exists in local minima, it is uncomplicated to code and understand its most basic form. Conceptually, PSO and the genetic algorithm (GA) are similar and are easier to implement. To conclude, PSO is comparable to and may be better than GA [14, 15].

The PSO algorithm was originally introduced by Kennedy and Eberhart in 1995 [16]. A swarm can be formally defined as a group of mobile agents that communicate with each other directly or indirectly [17]. Since its inception, many problems in various engineering fields have benefited because of its simple computation and information sharing within the algorithm, as it derives its internal communications from the social behavior of individuals (particles). A multi-dimensional optimization problem [18] is represented by each individual (particle). However, the appropriateness of the solution is based on the function which is related to an optimization problem.

The process of PSO begins by initializing the population of particles, randomly positioned across the search range with an initial random velocity having values not greater than a certain percentage of the search space in each direction. Each particle (candidate solution) is conveyed as a location within the problem search space. This is established by reviving its velocity at consistent interims towards both the positions pbest [19] (personal best) and gbest (global best) found by the entire swarm. pbest and gbest are iteratively updated for each particle, until a better or more dominant solution (in terms of fitness) is found. This process continues, until the maximum iterations are reached or specified criteria are met. The equation governing the movement of the particles of the swarm is

$$v_i = v_i + c_1 r_1 (p_i - x_i) + c_2 r_2 (p_g - x_i), \tag{11.9}$$

where
v_i is the velocity vector of the ith particle,
x_i is the position vector of the ith particle,
$_{pi}$ is the n-dimensional personal best of the ith particle found from initialization,
p_g is the n-dimensional global best of the swarm found from initialization,
c_1 is the cognitive acceleration coefficient and
c_2 is the social acceleration coefficient

and the position is updated using

$$x_i = x_i + v_i. \tag{11.10}$$

The classical version of the PSO algorithm defined by velocity update equations (11.9) and (11.10) inherits a weakness that can be nullified by inertia weight w or constriction coefficient χ. The method of introducing inertia weight was first introduced by Shi and Eberhart [20]. The modified velocity equation is given by

$$v_i = w^* v_i + c_1 r_1 (p_i - x_i) + c_2 r_2 (p_g - x_i). \tag{11.11}$$

According to Eberhart and Shi [21], the optimal strategy is to initially set w to 0.9 and reduce it linearly to 0.4, allowing initial exploration followed by acceleration toward an improved global optimum.

The problems in velocity update equations (11.9) and (11.10) were addressed by Clerc [15] by introducing constriction coefficient χ so as to result in

$$v_i = \chi[v_i + c_1 r_1 (p_i - x_i) + c_2 r_2 (p_g - x_i)], \tag{11.12}$$

where χ is computed as

$$\chi = \frac{2}{|2 - \phi - \sqrt{\phi(\phi - 4)}|}, \tag{11.13}$$

where $\phi = c_1 + c_2$, $\phi > 4$.

The velocity equation forms the main component of the PSO algorithm and is unlikely to become stuck in the local minima [22]. The basic PSO algorithm is as follows:

Step 1. [Start.] Initialize the position, velocity and the personal best of each particle in the swarm at random.

Step 2. [Evaluate fitness value.] For each iteration, the particles will move progressively into the solution space. The action involves updating the particle velocity and its movement, and assessing the fitness function for the current position.

Step 3. [Compare fitness function.] The fitness function of the current position and gbest are compared. Revisit the above stages for the whole particles.

Step 4. [Maximum iteration.] Continue until the iteration limit or until the termination criteria are reached. Stop, and return the best solution gbest, otherwise update w and go to the next iteration.

Step 5. [Loop.] Go to step 2, evaluate fitness value.

The flowchart as shown in figure 11.3 indicates the process flow of PSO. Table 11.1 gives the typical parameters used for PSO in the present study.

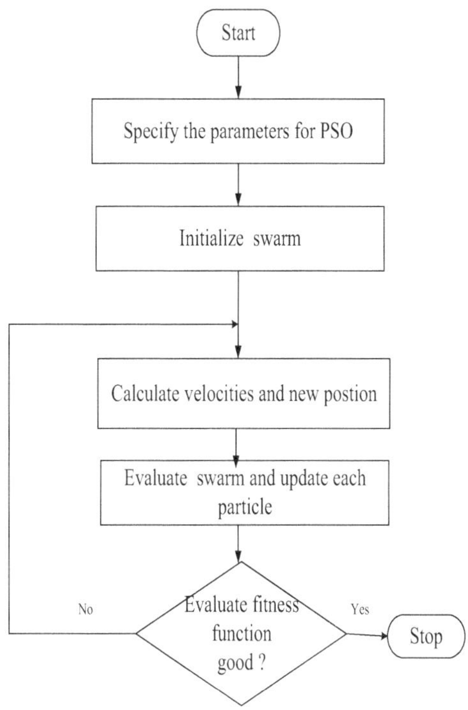

Figure 11.3. PSO optimization process.

Table 11.1. Typical parameters used for PSO.

Parameters	Value
Swarm size	20
Maximum generations	100
c_1, c_2	2, 2
w_{start}, w_{end}	0.9, 0.4

11.1.2.1 Illustrative examples

Numerical examples are presented to illustrate both indirect and direct methods of PID controller design. The design parameters are optimized using PSO and solved in detail for the first example. Later, the results are compared to other methods.

11.1.2.1.1 Direct method

Example 11.1. A regulator problem is considered and its transfer function and reference model are represented as [23]

$$G_p(s) = \frac{s^5 + 8s^4 + 20s^3 + 16s^2 + 3s + 2}{2s^6 + 36.6s^5 + 204.8s^4 + 419s^3 + 311.8s^2 + 67.2s + 4}$$

$$M(s) = \frac{0.023s + 0.0121}{s^2 + 0.21s + 0.0121}.$$

Step 1. Consider $M(s)$ and determine the equivalent open loop transfer function using (11.3):

$$\tilde{M}(s) = \frac{0.023s^3 + 0.01693s^2 + 0.002819s + 0.0001464}{s^4 + 0.397s^3 + 0.05137s^2 + 0.002263s}.$$

Step 2. Let the desired controller be according to (11.5) and this is given by

$$G_c(s) = \frac{\tilde{M}(s)}{G_p(s)} = \frac{1}{s}(0.064707 + 0.767859s + 0.801795s^2 - 4.681159s^3 + \ldots).$$

Step 3. Take the PID controller structure as

$$G_c(s) = K_1 + \frac{K_2}{s} + K_3 s = \frac{K_1 s + K_2 + K_3 s^2}{s}.$$

Step 4. $G_c(s)$ is compared to the power series expansion to obtain the parameters K_1, K_2 and K_3. This results in the PID controller

$$G_c(s) = \frac{0.064707 + 0.767859s + 0.801795s^2}{s}.$$

Step 5. Obtain $G_{CL}(s)$ using (11.7)

$$G_{CL}(s) = \frac{0.8228s^7 + 7.349s^6 + 22.66s^5 + 29.02s^4 + 16.03s^3 + 4.981s^2 + 1.728s + 0.1294}{1.823s^7 + 25.65s^6 + 125.1s^5 + 238.5s^4 + 172s^3 + 38.58s^2 + 3.728s + 0.1294}.$$

Step 6. Apply the PSO technique to attain a reduced order model (second) from $G_P(s)$. This is achieved by minimizing the ISE between the $G_p(s)$ and $R_p(s)$ using the formula

$$I = \int_{\infty}^{0} [y(t) - y_r(t)]^2 \, dt, \tag{11.14}$$

where $y(t)$ = unit step response of $G_p(s)$ and $y_r(t)$ = unit step response of $R_p(s)$. Thus

$$R_{\text{ppso}}(s) = \frac{0.02555s + 0.01036}{s^2 + 0.4756s + 0.01036}.$$

Step 7. Similarly using GA and HNA, the reduced system will be

$$R_{\text{pga}}(s) = \frac{0.01414s + 0.009369}{s^2 + 0.1436s + 0.009369}$$
$$R_{\text{phna}}(s) = \frac{0.0113s + 0.0736}{s^2 + 0.1436s + 0.009369}.$$

Step 8. Now, following step 4 in 11.1.1.1, the controller structure will be

$$R_{\text{cga}}(s) = \frac{\tilde{M}(s)}{R_{\text{pga}}(s)}$$
$$= \frac{1}{s}(0.06471 + 0.714036s + 1.89926s^2 + \ldots).$$

Step 9. Considering the PID controller structure as

$$R_c(s) = K_1 + \frac{K_2}{s} + K_3 s$$

or

$$R_c(s) = \frac{K_1 s + K_2 + K_3 s^2}{s}.$$

Step 10. The values of K_1, K_2 and K_3 are obtained with the comparison of power series coefficients:

$$R_{\text{cga}}(s) = \frac{0.06471 + 0.714036s + 1.89926s^2}{s}.$$

Step 11. Obtain the transfer function of $R_{\text{CLPSO}}(s)$, $R_{\text{CLGA}}(s)$ $R_{\text{CLHNA}}(s)$ as per step 6 in 11.1.1.1

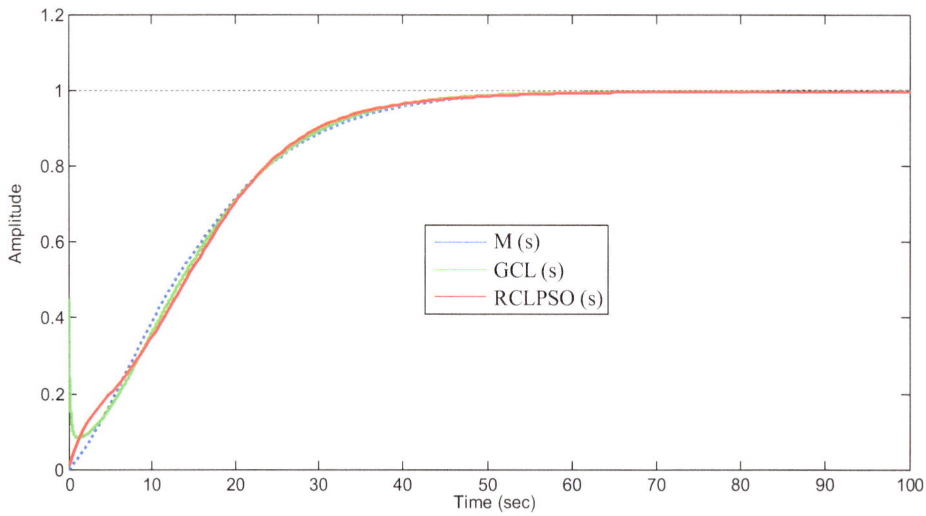

Figure 11.4. Comparison of step responses.

Table 11.2. Qualitative comparison of original and reduced order plants in terms of time domain specifications for example 11.1.

System	Rise time t_r (s)	Peak overshoot % M_p	Settling time t_s (s)
$M(s)$	28.1	0.003 83	46.2
$G_{CL}(s)$	22.3	0	45.1
$R_{CLPSO}(s)$	28.8	0	43.6
$R_{CLGA}(s)$	28	0	45.3
$R_{CLHNA}(s)$	30	0.95	42.1

$$R_{CLPSO}(s) = \frac{0.04853s^3 + 0.03861s^2 + 0.00933s + 0.0006703}{1.049s^3 + 0.2142s^2 + 0.01969s + 0.0006703}$$
$$R_{CLGA}(s) = \frac{0.01165s^3 + 0.07664s^2 + 0.005604s + 0.004762}{1.012s^3 + 0.417s^2 + 0.0792s + 0.004762}$$
$$R_{CLHNA}(s) = \frac{0.05177s^3 + 0.0445s^2 + 0.007102s + 0.0006152}{1.052s^3 + 0.1885s^2 + 0.0167s + 0.0006152}.$$

The unit step responses obtained are compared in figure 11.4. All the responses are found to be comparable. Table 11.2 shows the qualitative comparison of the original and reduced order plants in terms of time domain specifications.

11.1.2.1.2 Indirect method

Example 11.2. Consider a stable original practical system [24] along with a reference model represented by

$$G_p(s) = \frac{248.05s^4 + 1483.3s^3 + 91931s^2 + 468730s + 634950}{s^6 + 26.24s^5 + 1363.1s^4 + 26803s^3 + 326900s^2 + 859170s + 528050}$$

$$M(s) = \frac{4}{s^2 + 4s + 4}.$$

Step 1. Considering $M(s)$, the equivalent transfer function in an open loop configuration is obtained using (11.3)

$$\tilde{M}(s) = \frac{4s^2 + 16s + 16}{s^4 + 8s^3 + 20s^2 + 16s}.$$

Step 2. According to (11.5), the controller transfer function is given as

$$\tilde{M}(s) = G_c(s)G_p(s)$$

or

$$G_c(s) = \frac{\tilde{M}(s)}{G_p(s)}$$

$$= \frac{1}{s}(0.8316 + 0.5313s - 0.2841s^2 + 0.1159s^3 + \ldots).$$

Step 3. Choose the controller structure as

$$G_c(s) = \frac{k(1 + k_1 s)}{s(1 + k_2 s)}.$$

Step 4. Comparing the power series expansion coefficients with controller structure results in

$$K = 0.8316, \quad K_1 = 1.1735, \quad K_2 = 0.5347$$

and $G_c(s)$ for the original plant will be

$$G_c(s) = \frac{0.976s + 0.8316}{0.5347s^2 + s}.$$

Step 5. Then $G_{CL}(s)$ will be

$$G_{CL}(s) = \frac{242.5s^5 + 1656s^4 + 9.11 \times 10^4 s^3 + 5.347 \times 10^5 s^2 + 1.011 \times 10^6 s + 5.28 \times 10^5}{0.5363s^8 + 15.07s^7 + 757.3s^6 + 1.598 \times 10^4 s^5 + 2.038 \times 10^5 s^4 + 8.788 \times 10^5 s^3 + 1.677 \times 10^6 s^2 + 1.539 \times 10^6 s + 5.28 \times 10^5}.$$

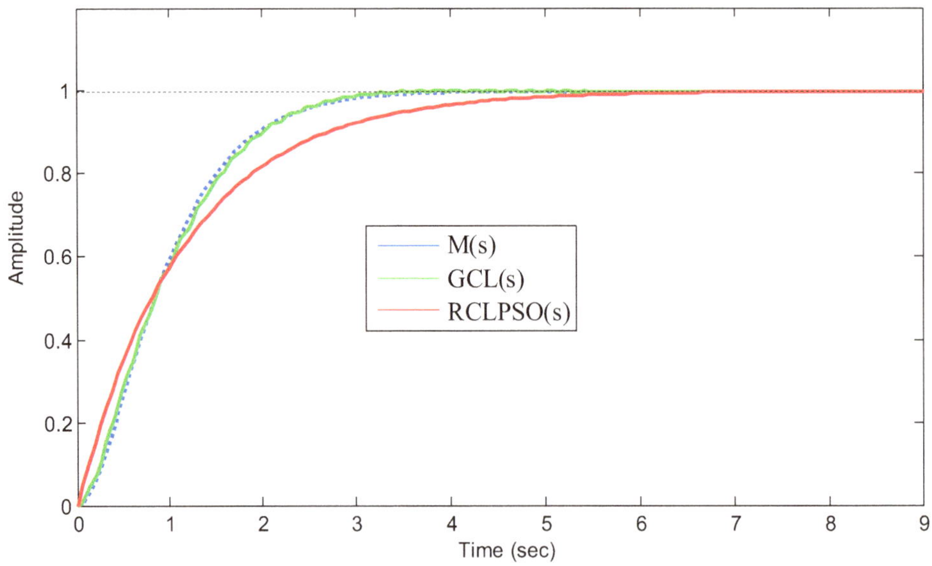

Figure 11.5. Comparison of step responses.

Table 11.3. Qualitative comparison of original and reduced order plants in terms of transient response parameters for example 11.2.

System	Rise time t_r (s)	Peak overshoot % M_p	Settling time t_s (s)
$M(s)$	1.68	0	2.92
$G_{\mathrm{CL}}(s)$	1.76	0.239	2.81
$R_{\mathrm{CLPSO}}(s)$	2.57	0	4.57
$R_{\mathrm{CLGA}}(s)$	1.82	0.158	2.79
$R_{\mathrm{CLHNA}}(s)$	2.57	0.02	4.6

Step 6. The third order reduced model using PSO is obtained by reducing the transfer function (closed loop) by minimizing (11.14):

$$R_{\mathrm{CLPSO}}(s) = \frac{0.8622s^2 + 2.05s + 0.9609}{s^3 + 3.258s^2 + 3.172s + 0.9609}$$

$$R_{\mathrm{CLGA}}(s) = \frac{0.4844s^2 + 2.393s + 1.674}{s^3 + 3.233s^2 + 4.045s + 1.674}$$

$$R_{\mathrm{CLHNA}}(s) = \frac{0.9633s^2 + 3.88s + 1014}{1.176s^3 + 1.404s^2 + 1190s + 1013}.$$

The comparison of the step responses of $M(s)$, $G_{\mathrm{CL}}(s)$ and $R_{\mathrm{CLPSO}}(s)$ is depicted in figure 11.5. These responses are found to be comparable in terms of transient response specifications with the responses from other methods in table 11.3. It can be concluded that the result obtained by PSO is comparable.

11.1.3 Method 2: PID controller design using BBBC

Techniques based on the theory of evolution of the universe, such as BBBC, are found to be useful for optimization [13]. In the present section, the BBBC based method is utilized to optimize the values of the PID controller parameters. It is seen that this method is better/comparable to the conventional methods in the following solved examples.

11.1.3.1 Illustrative examples

An example is being considered to illustrate that BBBC outperforms PSO, GA and HNA methods. The same is verified by two methods (indirect and direct) of PID controller design by comparing in terms of time domain specifications.

11.1.3.1.1 Direct method

Example 11.3. Consider the regulator problem in example 11.1 in section 11.1.2.1.1

$$G_P(s) = \frac{s^5 + 8s^4 + 20s^3 + 16s^2 + 3s + 2}{2s^6 + 36.6s^5 + 204.8s^4 + 419s^3 + 311.8s^2 + 67.2s + 4}$$
$$M(s) = \frac{0.023s + 0.0121}{s^2 + 0.21s + 0.0121}.$$

Step 1. Following steps 1 to 5 of example 11.1 in section 11.1.2.1.1, we obtain

$$G_{\mathrm{CL}}(s) = \frac{0.8228s^7 + 7.349s^6 + 22.66s^5 + 29.02s^4 + 16.03s^3 + 4.981s^2 + 1.728s + 0.1294}{1.823s^7 + 25.65s^6 + 125.1s^5 + 238.5s^4 + 172s^3 + 38.58s^2 + 3.728s + 0.1294}.$$

Step 2. The second order model is obtained after the original system is reduced using the proposed BBBC by minimizing (11.14) to obtain

$$R_{\mathrm{PBBBCOA}}(s) = \frac{0.0233s + 0.01176}{s^2 + 0.2035s + 0.01176}.$$

Step 3. Similarly, using PSO, GA and HNA, the reduced transfer function is

$$R_{\mathrm{ppso}}(s) = \frac{0.02555s + 0.01036}{s^2 + 0.4756s + 0.01036}$$
$$R_{\mathrm{pga}}(s) = \frac{0.01414s + 0.009369}{s^2 + 0.1436s + 0.009369}$$
$$R_{\mathrm{phna}}(s) = \frac{0.0113s + 0.0736}{s^2 + 0.1436s + 0.009369}.$$

Step 4. Now, the reduced controller will be

$$R_{\text{CBBBC}}(s) = \frac{\tilde{M}(s)}{R_{\text{PBBBC}}(s)}$$
$$= \frac{1}{s}(0.06191 + 0.76252s + 0.5764s^2 - \ldots).$$

Step 5. Take the PID controller structure as

$$R_c(s) = K_1 + \frac{K_2}{s} + K_3 s$$

or

$$R_c(s) = \frac{K_1 s + K_2 + K_3 s^2}{s}.$$

Step 6. K_1, K_2 and K_3 of the controller are obtained using power series coefficients

$$R_{\text{CBBBC}}(s) = \frac{0.06191 + 0.7625s + 0.5764s^2}{s}.$$

Step 7. Second order model is obtained by reducing $G_p(s)$ and $G_C(s)$ (PSO, HNA, BBBC, GA) in closed loop results as

$$R_{\text{CLBBBC}}(s) = \frac{0.01343s^3 + 0.02455s^2 + 0.01041s + 0.000728}{1.013s^3 + 0.2281s^2 + 0.02217s + 0.000728}$$
$$R_{\text{CLPSO}}(s) = \frac{0.04853s^3 + 0.03861s^2 + 0.00933s + 0.0006703}{1.049s^3 + 0.2142s^2 + 0.01969s + 0.0006703}$$
$$R_{\text{CLGA}}(s) = \frac{0.01165s^3 + 0.07664s^2 + 0.005604s + 0.004762}{1.012s^3 + 0.417s^2 + 0.0792s + 0.004762}$$
$$R_{\text{CLHNA}}(s) = \frac{0.05177s^3 + 0.0445s^2 + 0.007102s + 0.0006152}{1.052s^3 + 0.1885s^2 + 0.0167s + 0.0006152}.$$

The response of $M(s)$, $G_{\text{CL}}(s)$ and $R_{\text{CLBBBC}}(s)$ for a given step input is plotted in figure 11.6. Further, the outcomes are compared with other methods as listed in table 11.4.

11.1.3.1.2 Indirect method

Example 11.4. A sixth order stable practical system (example 11.2 in section 11.1.2.1.2) is considered:

$$G_p(s) = \frac{248.05s^4 + 1483.3s^3 + 91931s^2 + 468730s + 634950}{s^6 + 26.24s^5 + 1363.1s^4 + 26803s^3 + 326900s^2 + 859170s + 528050}$$
$$M(s) = \frac{4}{s^2 + 4s + 4}.$$

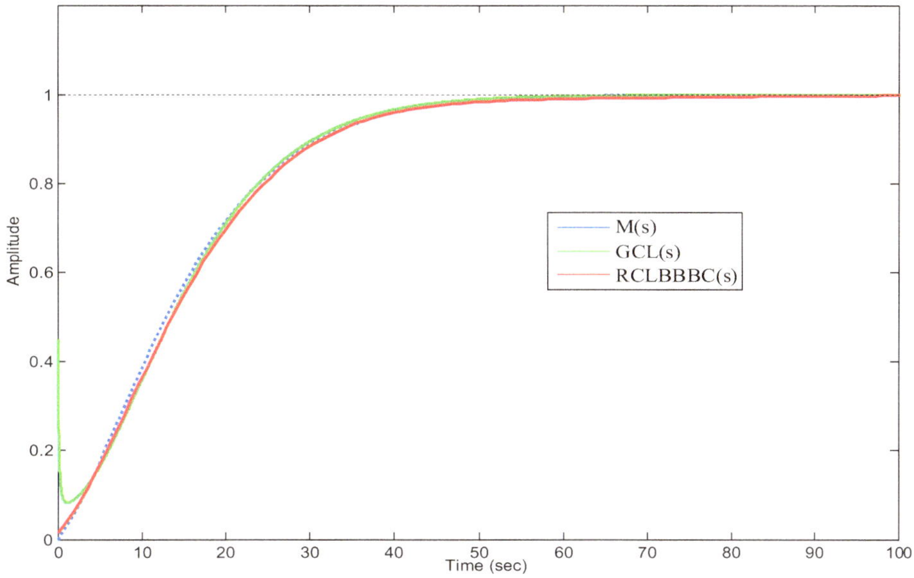

Figure 11.6. Comparison of step responses.

Table 11.4. Qualitative comparison of original and reduced order plants in terms of transient response parameters for example 11.3.

System	Rise time t_r (s)	Peak overshoot % M_p	Settling time t_s (s)
$M(s)$	28.1	0.003 83	46.2
$G_{\mathrm{CL}}(s)$	22.3	0	45.1
$R_{\mathrm{CLBBBC}}(s)$	28.2	0	47.5
$R_{\mathrm{CLPSO}}(s)$	28.8	0	43.6
$R_{\mathrm{CLGA}}(s)$	28	0	45.3
$R_{\mathrm{CLHNA}}(s)$	30	0.95	42.1

Step 1. Following the same steps 1 to 5 of example 11.2 in section 11.1.2.1.2, we obtain the closed loop transfer function $G_{\mathrm{CL}}(s)$ as

$$G_{\mathrm{CL}}(s) = \frac{242.5s^5 + 1656s^4 + 9.11 \times 10^4 s^3 + 5.347 \times 10^5 s^2 + 1.011 \times 10^6 s + 5.28 \times 10^5}{0.5363s^8 + 15.07s^7 + 757.3s^6 + 1.598 \times 10^4 s^5 + 2.038 \times 10^5 s^4 + 8.788 \times 10^5 s^3 + 1.677 \times 10^6 s^2 + 1.539 \times 10^6 s + 5.28 \times 10^5}.$$

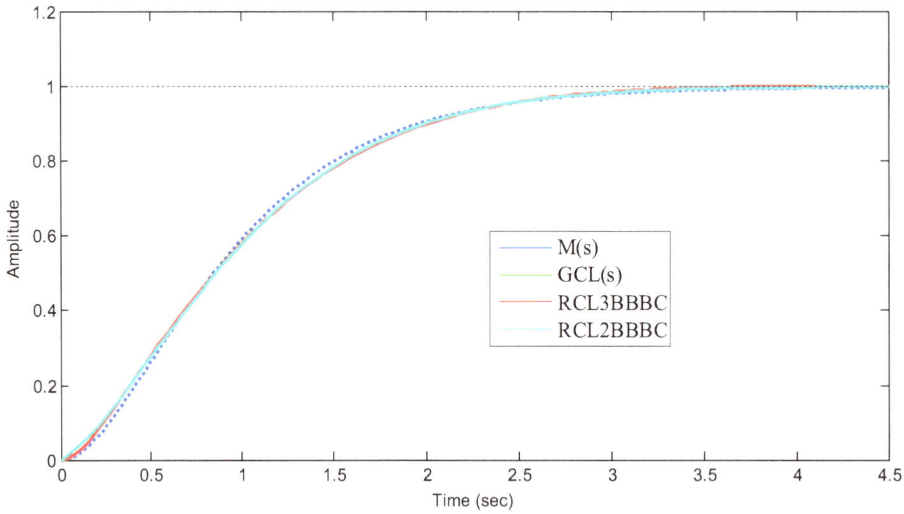

Figure 11.7. Comparison of step responses.

Step 2. $G_{\mathrm{CL}}(s)$ is reduced to third and second order using BBBC by minimizing (11.14) and is given by

$$R_{\mathrm{CL3BBBC}}(s) = \frac{-0.04996s^2 + 7.323s + 25.83}{s^3 + 14.13s^2 + 33.04s + 25.83}$$
$$R_{\mathrm{CL2BBBC}}(s) = \frac{0.2917s + 2.797}{s^2 + 3.083s + 2.797}.$$

Step 3. The reduced third order systems obtained by PSO, GA and HNA are

$$R_{\mathrm{CL3PSO}}(s) = \frac{0.8622s^2 + 2.05s + 0.9609}{s^3 + 3.258s^2 + 3.172s + 0.9609}$$
$$R_{\mathrm{CL3GA}}(s) = \frac{0.4844s^2 + 2.393s + 1.674}{s^3 + 3.233s^2 + 4.045s + 1.674}$$
$$R_{\mathrm{CL3HNA}}(s) = \frac{0.9633s^2 + 3.88s + 1014}{1.176s^3 + 1.404s^2 + 1190s + 1013}.$$

The comparison of the step responses of $M(s)$, $G_{\mathrm{CL}}(s)$, $R_{\mathrm{CL3BBBC}}(s)$ and $R_{\mathrm{CL2BBBC}}(s)$ is depicted in figure 11.7. It is seen that $G_{\mathrm{CL}}(s)$, $R_{\mathrm{CL3BBBC}}(s)$ and $R_{\mathrm{CL2BBBC}}(s)$ almost overlap each other and are competitive. The same can be concluded by examining table 11.5, which compares the responses from other methods in terms of transient domain specifications. It can be concluded that the result obtained by BBBC is outstanding.

Table 11.5. Qualitative comparison of original and reduced order plants in terms of transient response parameters.

System	Rise time t_r (s)	Peak overshoot % M_p	Settling time t_s (s)
$M(s)$	1.68	0	2.92
$G_{CL}(s)$	1.76	0.239	2.81
$R_{CL3BBBC}(s)$	1.76	0.213	2.81
$R_{CL2BBBC}(s)$	1.76	0.0583	2.84
$R_{CLPSO}(s)$	2.57	0	4.57
$R_{CLGA}(s)$	1.82	0.158	2.79
$R_{CLHNA}(s)$	2.57	0.02	4.6

11.2 FOPID controller

PID controllers have become quite popular because of their various advantages [25]. However, it is observed that there is scope for better tuning of PID controllers [26]. In the recent past, different tuning methods have been proposed and this is still a challenging task [27]. Here, a FOPID controller is being proposed, providing some sort of solution to the problems under consideration. Both FOPID and PID controllers are quite similar. The difference lies in the introduction of additional flexibility in design. This is accomplished due to the presence of derivative λ and integral μ in fractional order and results in better gain variations and phase characteristics. To conclude, fractional order controllers have proved to be quite powerful in robust control design.

The FOPID controller proposed [28] is applicable to linear systems [29] and this can be termed as a drawback. However, due to various tuning methods [29, 30], FOPID controllers are being used widely. In this regard, the use of evolutionary techniques such as PSO [31] and GA [32, 33] in tuning has proved to be fruitful. As a result, FOPID controllers based on evolutionary techniques have gained importance when compared to conventional methods.

The $PI^{\lambda}D^{\mu}$, also known as the FOPID controller, was dealt with in both the time [28] and frequency [30] domains. The FOPID controller generally takes the form

$$G_{FOPID}(s) = \frac{K_D s^{\lambda+\mu} + K_P s^{\lambda} + K_I}{s^{\lambda}}, \quad (11.15)$$

where λ, μ are real numbers (positive), and K_p, K_I, K_D represent the proportional, integral and differential gains, respectively.

The FOPID controller provides flexibility in the design of the PID control as shown by the shaded portion in figure 11.8. The values of λ and μ define the four control coordinates, i.e. P, PI, PD and PID of the classical PID. This results in generalizing the conventional PID controller (integer order) and expands it from point to plane.

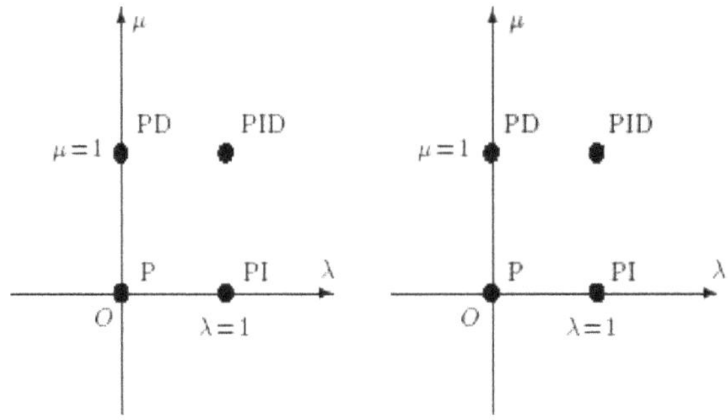

Figure 11.8. Fractional order PID and the classical PID (integer and fractional order).

11.2.1 Statement of the problem

The original nth order LTI-SISO system is taken into account represented by the transfer function

$$G_p(s) = \frac{\sum_{j=1}^{n} b_j s^{j-1}}{\sum_{i=1}^{m} a_i s^{i-1}}, \tag{11.16}$$

where b_j, a_i are scalar constants, n is the order of the numerator and m is the order of the denominator polynomial. The FOPID controller $G_{\text{FOPID}}(s)$ in the form of (11.15) is the objective of this proposal. Later, the response of $G_p(s)$ connected in series with $G_{\text{FOPID}}(s)$ (with unity feedback) injected with unit step input is obtained. The closed loop response meets the specified time domain parameters, i.e. M_P, T_S, T_r, steady state error (SSE) and integral square error (ISE).

11.2.2 BBBC aided tuning of FOPID controller parameters

The introduction of evolutionary based tuning methods has changed the way problems are being solved. It is noted that these methods are successful in almost all areas and hence have become popular amongst researchers. BBBC [34–37] is one such evolutionary technique being used for optimizing design parameters [13]. Here, BBBC is used to optimize the values of the FOPID controller for a given uncontrolled original plant $G_P(s)$. This is accomplished by tuning the areas of search space through the interaction of individuals and has proven to be fruitful. The response of the system obtained in closed loop (unity feedback) is stable, satisfying time domain specifications.

11.2.3 Illustrative examples

Example 11.5. A third order original system [32] represented by $G_p(s)$ is considered. It is desired to design a suitable FOPID controller to provide better M_P, T_S, SSE and ISE when injected with a unit step signal:

$$G_p(s) = \frac{40}{2s^3 + 10s^2 + 82s + 10}.$$

Step 1. Assume the structure of the FOPID controller as

$$G_{\text{FOPID}}(s) = \frac{K_D s^{\lambda+\mu} + K_P s^{\lambda} + K_I}{s^{\lambda}}.$$

Step 2. Minimize the fitness function E using BBBC:

$$E = \int_0^{\infty} [y(t) - y_r(t)]^2 \, \mathrm{d}t, \tag{11.17}$$

where $y(t)$ = the unit step response of the higher order model and $y_r(t)$ = unit step response of the reduced order model at time t, and the peak overshoot is given by

$$M_p = \exp^{\left(\frac{-\sigma_d \pi}{\omega_d}\right)} 100, \tag{11.18}$$

where, $\sigma_d \pm j\omega_d$ represents the dominant pole position.

Step 3. This results in

$$G_{\text{FOPIDBBBC}}(s) = \frac{5.3896 s^{0.9+1.1} + 27.0406 s^{0.9} + 19.6417}{s^{0.9}}.$$

Step 4. The transfer function in the open loop configuration is given by

$$\begin{aligned} G_{\text{OL}}(s) &= G_P(s) G_{\text{FOPIDBBBC}}(s) \\ &= \left(\frac{40}{2s^3 + 10s^2 + 82s + 10}\right)\left(\frac{5.3896 s^{\lambda+\mu} + 27.0406 s^{\lambda} + 19.6417}{s^{\lambda}}\right). \end{aligned}$$

Step 5. The transfer function in the closed loop configuration is given by

$$\begin{aligned} G_{\text{CLBBBC}}(s) &= \frac{G_p(s) G_{\text{FOPIDBBBC}}(s)}{1 + G_p(s) G_{\text{FOPIDBBBC}}(s)} \\ &= \frac{215.584 s^2 + 1081.624 s^{0.9} + 785.668}{2s^{3.9} + 10s^{2.9} + 215.584 s^2 + 82 s^{1.9} + 1091.624 s^{0.9} + 785.668}. \end{aligned}$$

Step 6. The FOPID controller tuned with the aid of GA [32] is

$$G_{\text{FOPIDGA}}(s) = \frac{0.36015 s^{\lambda+\mu} + 6 s^{\lambda} + 12.24}{s^{\lambda}}.$$

Step 7. The transfer function in the closed loop configuration (for $\lambda = 0.9$, $\mu = 1.1$) is given by

$$G_{\text{CLGA}}(s) = \frac{14.40s^2 + 240s^{0.9} + 489.6}{2s^{3.9} + 10s^{2.9} + 14.406s^2 + 82s^{1.9} + 250s^{0.9} + 489.6}.$$

Step 8. $G_{\text{CLBBBCRED}}(s)$ is obtained with the aid of BBBC

$$G_{\text{CLBBBCRED}}(s) = \frac{4.858s^2 + 3.25 \times 10^{-5}s + 0.07416}{s^3 + 4.859s^2 + 0.0153s + 0.07416}.$$

The unit step responses of $G_P(s)$, $G_{\text{CLBBBC}}(s)$, $G_{\text{CLBBBCRED}}(s)$, $G_{\text{CLGA}}(s)$ [32] ($\lambda = 0.9$, $\mu = 1.1$) are compared, as shown in figure 11.9. It is noted that the method proposed with the aid of BBBC yielded improved results compared to GA. Figure 11.10 indicates the unit step response of the integer PID controller. Further, the fractional order PID controller ($\lambda = 0.9$, $\mu = 1.1$) exhibits better results. The responses of $G_{\text{CLBBBC}}(s)$ (various values of λ, μ) when subjected to a unit step signal are indicated in figures 11.11–11.17. Table 11.6 shows a comparison of T_S, M_P, ISE and integral absolute time error (IATE) obtained for various values of λ and μ.

Figure 11.18 illustrates the path traced by λ, μ for a population size of 75; the number of iterations is 60 along with 0.75 as the reduction rate. The values obtained in Matlab (2011b) during every iteration during the optimization process using BBBC are considered for plotting the path.

Here, another approach to optimally tune the FOPID parameters using BBBC [34–37] is being proposed. The results yielded are compared to GA [32]. The values

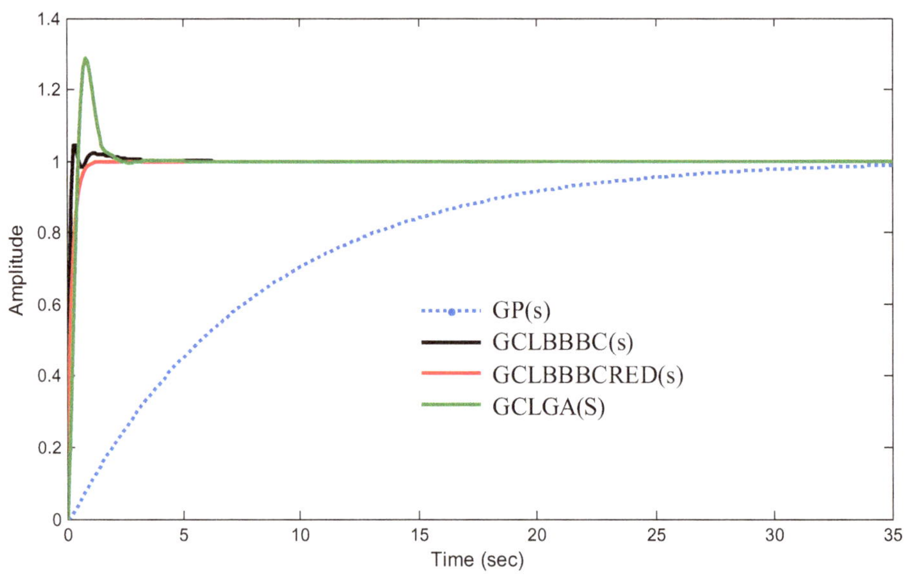

Figure 11.9. Comparison of unit step responses (FOPIDBBBC and FOPIDGA controllers for $\lambda = 0.9$ and $\mu = 1.1$).

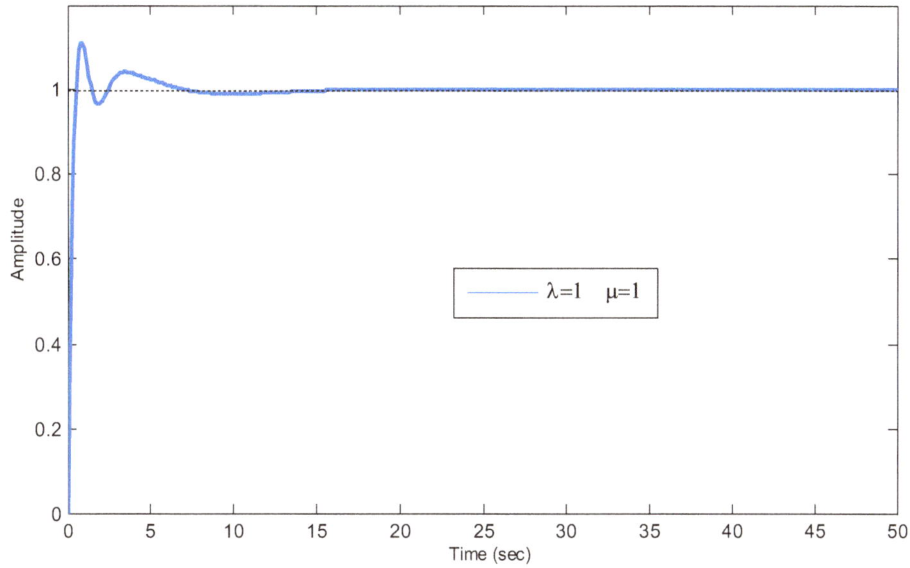

Figure 11.10. Unit step response of FOPID controller (λ, μ = 1).

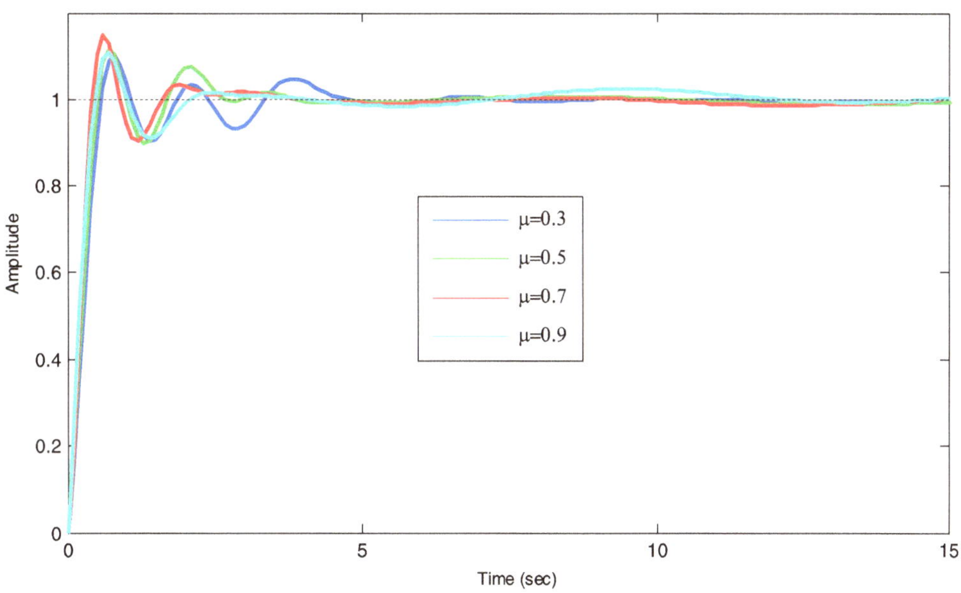

Figure 11.11. Unit step response of FOPID controller ($\mu < 1$).

of M_p, T_s, ISE and ITAE are being taken into account for the qualitative analysis and it is observed that the proposed technique has performed comparatively better. Also, the FOPID controller yielded good results compared to the PID controller (integer mode). Further, researchers can continue tuning the values of λ, μ up to two or three decimal digits for better responses.

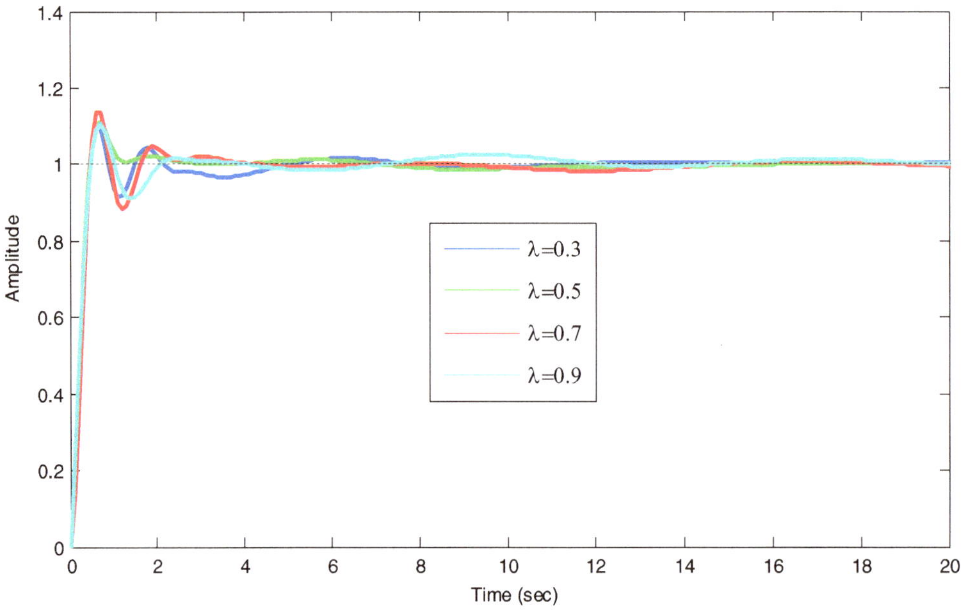

Figure 11.12. Unit step response for FOPID controller λ ($\lambda < 1$).

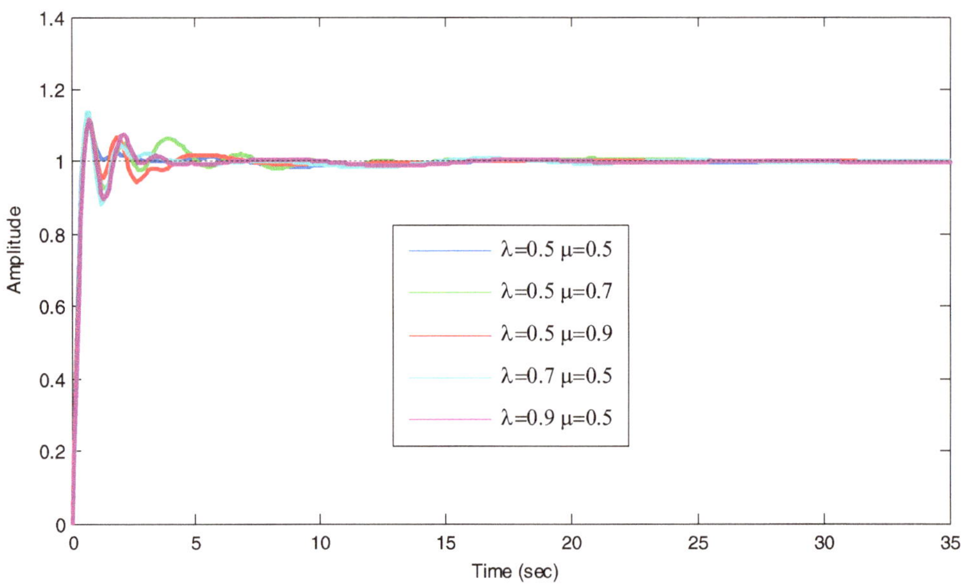

Figure 11.13. Unit step response for FOPID controller ($\lambda < 1$, $\mu < 1$).

11.3 Conclusion

In this chapter, the task of designing both PID and FOPID controllers has been accomplished successfully. In the case of the PID controller, both direct and indirect methods of controller design are considered in this work. The reduction of the closed

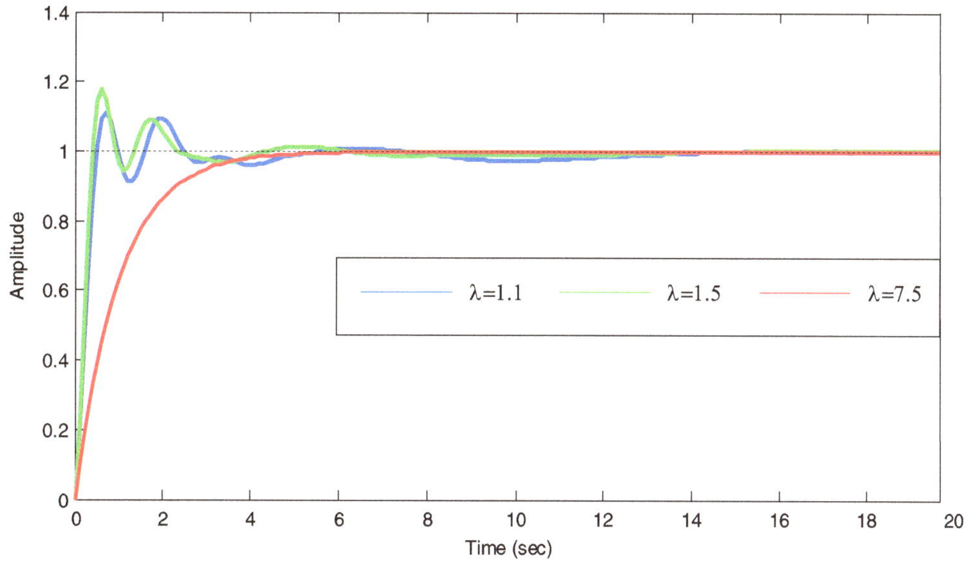

Figure 11.14. Unit step response for FOPID controller ($\lambda > 1$).

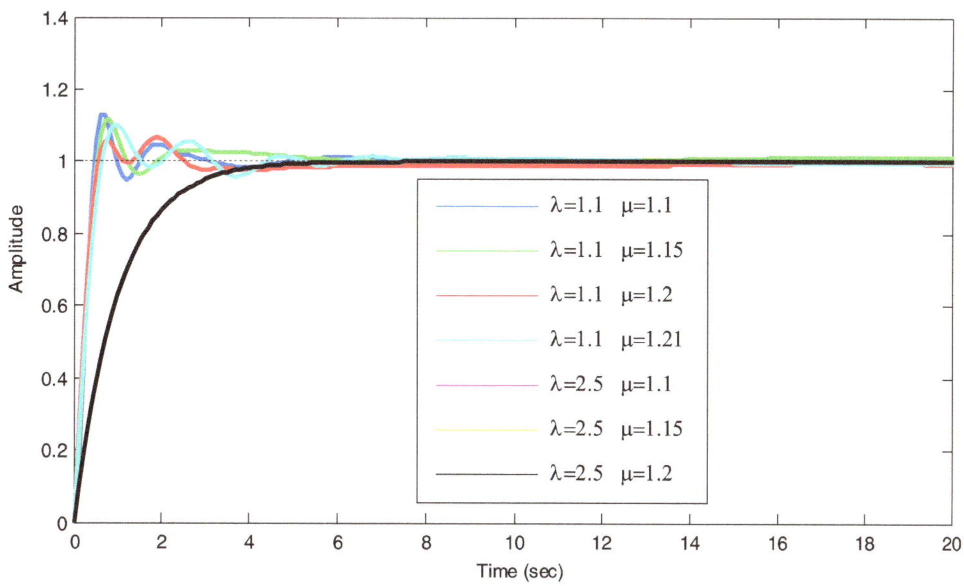

Figure 11.15. Unit step response for FOPID controller ($\lambda > 1$, $\mu > 1$).

loop system is then performed using PSO and BBBC by reducing the error between the reference model and the reduced model. Later, unknown controller parameters are found. The suitability of the proposed methods is justified by solving numerical examples from the literature available. The step responses illustrate the value of the proposed method.

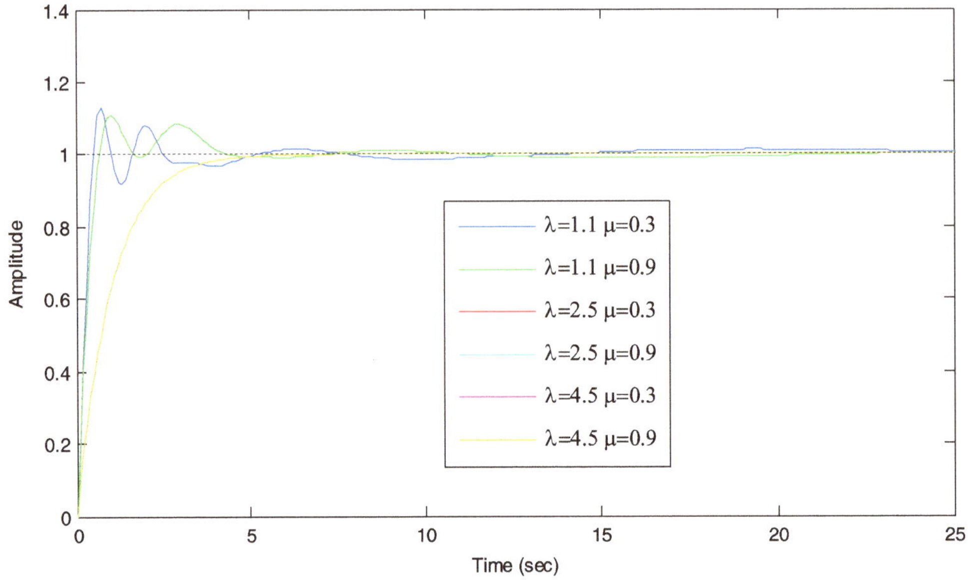

Figure 11.16. Unit step response for FOPID controller ($\lambda > 1$, $\mu < 1$).

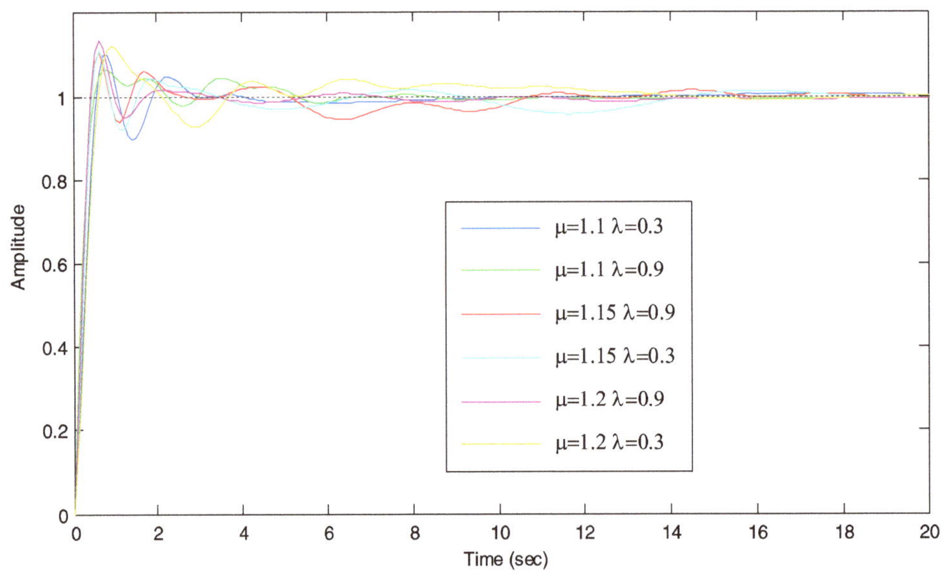

Figure 11.17. Unit step response for FOPID controller ($\lambda < 1$, $\mu > 1$).

In the case of the FOPID controller, the values of λ and μ are tuned by using BBBC and then the results are compared to the existing technique. Qualitative comparison in terms of settling time, peak overshoot, ISE and IATE are tabulated apart from step responses to justify the proposed method.

Table 11.6. Effect of variations in λ and μ on time domain specifications.

		Values obtained by BBBC (proposed)				Values obtained by GA [32]			
λ	μ	Settling time T_s	Peak overshoot M_p	ISE	IATE	Settling time T_s	Peak overshoot M_p	ISE	IATE
1	1	5.4627	11.3171	1.9246	36.3443	23.65	47.15	2.350	53.70
1	0.3	4.3278	10.0349	2.4982	33.1984	41.742	46.0245	2.889	123.7
1	0.5	2.5002	11.6848	2.2635	23.6048	34.903	46.9016	2.659	89.85
1	0.7	2.2034	14.3641	2.0940	18.2592	25.293	47.2256	2.505	67.85
1	0.9	10.4159	10.6855	1.9705	35.1073	24.148	47.3067	2.405	56.97
0.3	1	4.2231	10.8044	2.1812	29.7341	27.228	37.3379	2.243	68.56
0.5	1	1.9948	11.3069	2.0830	23.9988	24.263	40.0857	2.253	56.97
0.7	1	3.2528	14.6011	2.3424	28.4420	24.031	42.9715	2.291	54.11
0.9	1	10.4159	10.6855	1.9705	35.1073	23.632	45.8391	2.335	53.58
0.5	0.5	1.9548	11.4000	1.9830	24.1121	35.404	38.7205	2.543	95.84
0.5	0.7	6.8865	12.2687	2.2163	38.9573	26.266	39.7392	2.397	72.19
0.5	0.9	3.7937	11.5378	2.1081	23.6011	24.703	40.1304	2.300	60.52
0.7	0.5	3.3538	14.8011	2.3436	28.4520	28.661	42.1173	2.589	92.45
1.1	0.5	11.1732	11.2030	2.2012	24.1625	23.4050	48.5314	2.381	54.04
1.5	0.5	3.8488	0.9096	2.0339	23.2997	27.4584	53.5798	2.475	56.14
7.5	0.5	3.9123	0.1060	5.5167	11.0425	37.8834	141.1292	5.643	121.3
1.1	1.1	2.4884	13.5817	2.2104	18.5385	22.925	45.8945	2.142	49.08
1.1	1.15	4.4859	11.8160	2.2241	502 059	22.096	36.124	1.438	38.42
1.1	1.2	4.5773	6.7630	1.9964	70.8083	13.319	44.7987	0.2839	13.54
1.1	1.21	4.1228	10.1151	2.2047	19.7048	20.628	93.2290	0.8556	30.17
2.5	1.1	3.9123	0.0	5.5167	11.0425	25.475	60.6300	2.441	57.84
2.5	1.15	3.9120	0.0	5.5180	11.0200	24.189	46.7059	1.658	44.59
2.5	1.2	3.9200	0.0	5.5200	11.0100	17.147	44.2505	0.4092	21.84
1.1	0.3	4.5326	12.9902	2.2375	36.9991	41.357	47.7478	2.918	123.5
1.1	0.9	3.9314	10.7146	2.3835	28.9681	23.860	48.6390	2.428	57.27
2.5	0.9	3.9123	0.0	5.5167	11.0425	30.375	64.9865	2.764	67.17
4.5	0.3	3.9122	0.0	5.5169	11.0430	40.946	102.1544	4.574	135.4
0.3	1.1	2.6811	10.5548	2.2649	22.5673	26.564	36.0403	2.021	63.14
0.9	1.1	4.9125	6.7860	1.6417	23.2709	23.366	43.4405	2.098	48.6
0.3	1.15	13.3011	11.1522	2.0382	47.0520	25.260	30.5292	1.344	48.47
0.9	1.15	10.1522	10.8518	2.0804	27.5545	22.705	34.3791	1.403	38.11
0.3	1.2	1.5322	13.6581	0.1837	2.98015	4.8701	47.5611	0.1939	5.328
0.9	1.2	1.4958	12.2436	0.2507	5.36410	13.042	45.1063	0.2591	11.7

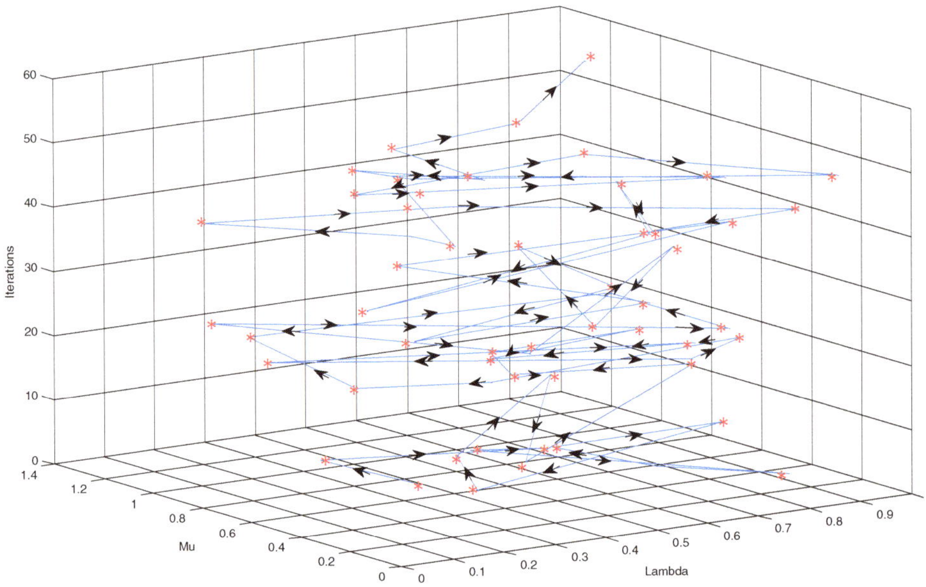

Figure 11.18. Path traced by λ and μ.

References

[1] Aguirre L A 1991 New algorithm for closed-loop model matching *Electron. Lett.* **27** 2260–2

[2] Bandyopadhyay B, Unbehauen H and Patre B M 1998 A new algorithm for compensator design for higher-order system via reduced model *Automatica* **34** 917–20

[3] Goddard P J and Glover K 1998 Controller approximation: approaches for preserving H∞ performance *IEEE Trans. Autom. Control* **43** 858–71

[4] Mukherjee S 1997 Design of compensators using reduced order models *Int. Conf. on Computer Applications in Electrical Engineering, Recent Advances (CERA-97)* (Roorkee: Indian Institute of Technology Roorkee), pp 529–36

[5] Pal J and Sarkar P 2002 An algebraic method for controller design in delta domain *Int. Conf. on Computer Applications in Electrical Engineering, Recent Advances (CERA-01)* (Roorkee: Indian Institute of Technology Roorkee), pp 441–9

[6] Prasad R, Pal J and Pant A K 1990 Controller design using reduced order models *14th National Systems Conf. (NSC-1990)* (Aligarh: Aligarh Muslim University), pp 182–6

[7] Wang G, Sreeram V and Liu W Q 2001 Performance preserving controller reduction via additive perturbation of the closed-loop transfer function *IEEE Trans. Autom. Control* **46** 771–5

[8] Ramesh K, Nirmalkumar A and Gurusamy G 2009 Design of discrete controller via novel model order reduction technique *Int J Elect Power Eng.* **3** 163–8

[9] Hemmati R, Boroujeni S M S, Delafkar H and Boroujeni A S 2011 PID controller adjustment using PSO for multi area load frequency control *Aust. J. Basic Appl. Sci.* **5** 295–302

[10] Anderson B D O and Liu Y 1989 Controller reduction: concepts and approaches *IEEE Trans. Autom. Control* **34** 802–12

[11] Wilson D A and Mishra R N 1979 Design of low order estimators using reduced models *Int. J. Control* **29** 447–56

[12] Yousuff A and Skelton R E 1984 A note on balanced controller reduction *IEEE Trans. Autom. Control* **29** 254–7

[13] Erol O K and Eksin I 2006 New optimization method: Big Bang–Big Crunch *Adv. Eng. Software* **37** 106–11

[14] Hassan R, Cohanim B, de Weck O and V G 2005 A comparison of particle swarm optimization and the genetic algorithm *46th AIAA/ASME/ASCE/AHS/ASC Structures, Structural Dynamics and Materials Conf. (Austin, TX)*

[15] Clerc M 1999 The swarm and the queen: towards a deterministic and adaptive particle swarm optimization *ICEC 1999* (Washington, DC) pp 1951–7

[16] Kennedy J and Eberhart R 1995 Particle swarm optimization *Proc. IEEE Int. Conf. on Neural Networks* **vol 4** pp 1942–8

[17] Engelbrecht A P 2007 *Computational Intelligence: An Introduction* 2nd edn. (Chichester: Wiley)

[18] Kennedy J, Eberhart R C and Shi Y 2001 *Swarm Intelligence* vol 10 (San Francisco, CA: Morgan Kaufmann)

[19] Bahrepour M, Mahdipour E, Cheloi R and Yaghoobi M 2009 SUPER-SAPSO: a new SA-based PSO algorithm *Applic. Soft Comput.* **58** 423–30

[20] Shi Y and Eberhart R C 1998 A modified particle swarm optimizer *IEEE Int. Conf. on Evolutionary Computation* (Piscataway, NJ: IEEE), pp 69–73

[21] Eberhart R C and Shi Y 2000 Comparing inertia weights and constriction factors in particle swarm optimization *Congress on Evolutionary Computation 2000 (San Diego, CA)* pp 84–8

[22] Clerc M 2006 *Particle Swarm Optimization* (London: ISTE)

[23] Jamshidi M 1983 *Large Scale Systems Modelling and Control Series* vol 9 (Amsterdam: North Holland)

[24] Prasad R 1989 Analysis and design of control systems using reduced order models *PhD Thesis* University of Roorkee, India

[25] Valerio D and da Costa J S 2006 Tuning-rules for fractional PID controller *Fract. Diff. Appl.* **2** 28–33

[26] Abu-Al-Nadi D, Othman M K A and Zaer S A-H 2011 Reduced order modeling of linear MIMO systems using particle swarm optimization *The Seventh Int. Conf. on Autonomic and Autonomous Systems (ICAS 2011)*

[27] Zhang J, Wang N and Wang S 2004 A developed method of tuning PID controllers with fuzzy rules for integrating processes *2004 American Control Conf. (Boston, MA)* pp 1109–14

[28] Podlubny I 1999 Fractional-order systems and PID controllers *IEEE Trans. Autom. Control* **44** 208–14

[29] Caponetto R, Fortuna L and Porto D 2004 A new tuning strategy for a non-integer order PID controller *First IFAC Workshop on Fractional Differentiation and its Applications (Bordeaux)*

[30] Petras I 1999 The fractional order controllers: methods for their synthesis and application *J. Electr. Eng.* **50** 284–8

[31] Cao J-Y and Cao B-G 2006 Design of fractional order controller based on particle swarm optimization *Int. J. Control Autom. Syst.* **4** 775–81

[32] Padhee S, Gautam A, Singh Y and Kaur G 2011 A novel evolutionary tuning method for fractional order PID controller *Int. J. Soft Comput. Eng. (IJSCE)* **1** 1–9

[33] Saptarshi D, Indranil P, Shantanu D and Amitava G 2012 Improved model reduction and tuning of fractional-order $PI^{\lambda}D^{\mu}$ controllers for analytical rule extraction with genetic programming *ISA Trans.* **51** 237–61

[34] Desai S R and Prasad R 2012 PID controller design using BB-BCOA optimized reduced order model *IJCA Special Issue on Advanced Computing and Communication Technologies for HPC Applications ACCTHPCA* **5** 32–7

[35] Desai S R and Prasad R 2013 A novel order diminution of LTI systems using big bang big crunch optimization and routh approximation *Appl. Math. Modell.* **37** 8016–28

[36] Desai S R and Prasad R 2014 Novel technique of optimizing FOPID controller parameters using BBBC for higher order system *IETE J Res.* **60** 211–7

[37] Desai S R and Prasad R 2013 A new approach to order reduction using stability equation and big bang big crunch optimization *Syst. Sci. Control Eng.* **2013** 20–7

IOP Publishing

Modern Optimization Methods for Science, Engineering and Technology

G R Sinha

Chapter 12

A variational approach to substantial efficiency for linear multi-objective optimization problems with implications for market problems

Glenn Harris and Sien Deng

Market problems can often be modeled using multi-objective optimization and solved by finding properly efficient solutions. However, even when the problems are solved with a properly efficient solution, sometimes the markets in these problems can be manipulated by individuals to receive substantial gains at the expense of the integrity of the market. For this reason it may be desirable to find solutions which avoid this potential exploitation. Substantially efficient solutions, which are an extension of properly efficient solutions, do just that. Using variational analysis and by constructing novel examples, substantially efficient solutions and their topology are explored in the context of linear multi-objective optimization problems. What is discovered is that, in general, these problems do not necessarily have substantially efficient solutions but there are ways to identify them. Furthermore, while substantially efficient solutions are difficult to guarantee, it is possible to ensure their existence under demanding constraints. The implications of this work are a more principled choice for solutions to market problems which preserve market stability and the recognition of potential market misuse when those choices are not available.

12.1 Introduction

Often in problems of engineering, industry, economics, and the like, decisions are made concerning optimizing multiple objectives. However, these objectives are not always measured on a common scale. For example, a genuine analysis of a company's profit, worker happiness, public image and time spent on a project cannot all necessarily be put in terms of a single currency. These are multi-objective optimization problems and are a very important area in optimization.

doi:10.1088/978-0-7503-2404-5ch12

Deciding on a solution for most multi-objective optimization problems can be difficult because there is often no choice that is clearly a minimum in every objective. An efficient solution is one for which no other solution is better in performance in at least one objective while also being better or equal in performance in all other objectives. For the most part, only efficient solutions are considered for decision makers in multi-objective optimization problems. So another difficulty arises for a decision maker since there is often more than one efficient solution to a problem. Some efficient solutions are anomalous in that the ratio, called the trade-off, between an objective that stands to gain and any objective that stands to lose by changing the solution to some other solution can be made arbitrarily large. An aversion to that type of solution is understandable as the decision maker may feel unsatisfied knowing there is a potentially worthwhile trade-off available. Proper efficient solutions are efficient solutions that are not anomalous in that way as they require that for any other solution there is a boundary on the trade-off between any improving objective and at least one diminishing objective.

Only achieving improper efficiency would leave the ability to change to some other solution where every component of the objective function that stands to do better can do so at a negligible loss of all the components that stand to do worse. One can imagine a situation where multiple parties have come together with a problem to solve. This could be businesses looking to increase profits, nation states looking to increase their currency value, or engineers looking to design a product. Each party has their own objective function that assigns an output to a resource allocation, such as regional control, cash input, labor distribution or concentration of material. A properly efficient solution seems like it would be reasonable because if a petition was put forth to change to some other solution, at least one party would stand to lose proportionally to someone who would stand to gain. The party that stood to lose may consider it unfair and attempt to put a stop to the change in resource allocation. So, the allocation would not change from the properly efficient solution. However, resource allocation is not always entirely held in the hands of people who have a vested interest in a particular objective. This leads to the real world issue that a properly efficient solution may not address, that is, market manipulation. Some mechanisms of manipulation that could be employed in this way are in currency exchanges or stock option markets.

Assuming multiple countries' currency value could be modeled by assigning objective functions to each country and certain states of affairs (such as policies, resource holdings, etc) as inputs for those functions, all the members of the global community may want to collectively maximize their country's currency value. A state of affairs resulting in a properly efficient solution may not be enough to avoid a sort of currency exchange manipulation that has the potential for seizing the whole market or inflating currency values. Imagine three countries who have in some way achieved a state of affairs which is properly efficient. This means that any changes that would increase the value of one state's currency would decrease at least one other proportionately. This would give some governing body that determines exchange rates for currencies helpful information to adequately set market prices. However, being only properly efficient still leaves open the possibility of a relatively nontrivial increase in the value of one currency traded for a negligible decrease of value of one of the other two currencies. An intrepid investor may spot this possibility and, with the advent of instant trading and the lightning speed at which information travels in the modern age, see an opportunity for exploitation. Evidence of this speed goes as far back as 2004 when electronic currency trading took place at two centralized electronic broker systems between parties in under a second [1]. In order to see how this manipulation might be achieved, let f_1, f_2 and f_3 be the objective functions of countries 1, 2 and 3, respectively, and consider this a maximization problem. Let x be the properly efficient solution at which the system is currently. Let $\{y_n\}_{n\in\mathbb{N}}$ be some other collection of efficient solutions such that $f_1(x) < f_1(y_n)$, $f_2(x) > f_2(y_n)$ and $f_3(x) > f_3(y_n)$, but $\frac{f_1(y_n) - f_1(x)}{f_3(x) - f_3(y_n)} > n$. Even if as $n \to \infty$ the gain in the first objective is minuscule, the loss of the other is that much smaller.

The investor could profit by exchanging a large sum of currency from country 3's to country 1's currency, then implement actions that would ensure the state of affairs y_n occurs. They would then change their currency back to country 3's. They could, depending on the size of the sum of money they exchanged and the state y_n they aimed for, achieve practically any level of profit they wanted. Not only that, but given that computer programs can be implemented to carry out instant transactions repeatedly over a short period of time, it could be the case that this operation can be carried out multiple times over the course of seconds or minutes pushing between the states y_n and x. In between those swaps, they would be purchasing and selling currencies between the two countries resulting in profit for the investor. This could occur repeatedly until no one is willing to buy or sell their currency anymore, effectively skewing the global currency market driving prices up and seizing up the system that would otherwise be needed for international investment. This may not even be immediately detectable as y_n may be a state very near x. Knowing when the state of affairs does not have a bounded trade-off between all pairs of contravening objectives helps to determine what sorts of regulation (fines/fees) to apply to block the misuse. Either that or the currency brokers could know when a sort of extra premium should be charged for a currency exchange between countries whose marginal trade-off is not bounded in order to stop investors from taking advantage.

Another realistic example, which one could call an 'option scheme', is rooted in the stock market. An option is a finance derivative contract which gives the holder the liberty (but is not compelled, or forced) to buy or sell an asset at a given price by a certain date. A savvy business entity could purchase and sell stock options appropriately and then nudge the market one way or the other to make significant gains in one stock to negligible losses in another. So a person could sell the option to buy from the company that stands to lose negligibly beneath that loss. They could also short a stock knowing full well what they will do later will not change the stock values notably, then use those funds to buy the option to sell the stocks of the company that stands to gain a noticeable amount (in relation to the negligible loss). Admittedly, these options may be harder to come by, but it is still a possibility that a large or unlimited number of options may be offered in certain circumstances. Indeed, in 2014 the total dollar value of the Chicago Board Options Exchange, a powerful options trading place, was roughly 570 billion dollars [2] and in 2016 it was roughly 666 billion dollars [3]. In addition to options, the scheme may be applicable to other derivitive products as well. There seems to be an almost unlimited potential for derivatives in the form of credit default swaps, with an overall market value of between 45 and 62 trillion dollars in 2007 [4], with a large portion of them responsible for exacerbating the housing market crisis in 2008. Moreover when these swaps are on securitized debt they can be exercised for partial losses and do not require an entire collapse [4]. That is an important point for the scheme because the exchanges are on marginal amounts, but high in volume.

As long as there is a customer, one can sell the option to buy a stock. Such an option scheme would probably require more than diminutive gains, because they may not trigger the stock option to sell the rising stock. They can still be small though, such as $0.10 or $0.02 but perhaps not $0.00003. It may also be difficult for the exploiter to manipulate stock values, but it is not impossible. Companies can misinform or withhold information, as in the case of Enron projecting estimated profits up to 20 years out, even though there were obvious concerns about those profits [5]. Commodities holders and bankers can withhold or saturate stored commodities or capital, resulting in altered valuation of stocks. These are only some of the ways people could contribute to revaluing stocks, which could be considered as components of the vectors in the feasible set of solutions. While purchasing stocks can be done manually and valued individually by humans, many stocks are evaluated using models. These values are recalculated over the day instantly by computers. That in conjunction with actions on the options being instantly transacted (within milliseconds) with the modern stock systems [6] makes it possible to repeatedly take advantage of the option scheme. This could be considered a misuse of stock options that could degrade trust in the market system resulting in stagnation or depression. Not knowing this potential for certain solutions means that stock options may not be valued correctly.

These examples show that there are many situations when properly efficient solutions to multi-objective problems are not enough. In multi-objective market problems it is more desirable to have a solution whose trade-off is bound for all pairs of objectives that have opposing outcomes when switching to another solution. This is where substantial efficiency enters. Substantial efficiency is a natural extension of proper efficiency, requiring the trade-off bound apply to all the objectives that stand to suffer.

The sorts of machinations given above make knowing when a solution to a problem is or is not substantially efficient of interest to many groups. Investors, business people and fortune seekers may want to know when a solution to a problem is not substantially efficient in order to exploit that to potentially build a treasure trove. Governing bodies and institutions may want to ensure that a solution to a problem they are solving is a substantially efficient solution in order to avoid such actions by said investors. Knowing when the current state is not substantially efficient could result in a governing body taxing such option transactions, making the option scheme impossible. Or in the currency case, a governing body could make small changes to the value of a currency, or note that more premiums should be charged for a currency change to currency brokers to avoid such manipulations. Either way, knowing if a solution to a market problem is substantially efficient may reduce the chance that the market can be exploited. Further, engineers can appreciate situations where such marginal trade-offs being bound between all objectives would be preferable in their optimization problems.

The aspiration of this chapter is to inquire about and then examine the prevalence and nature of substantially efficient solutions for linear multi-objective optimization problems. This is done through exploring many examples and expanding the current framework of results that exist. It will be shown that such solutions do not commonly exist. What this means for market problems is that regulators and administrators must be ever-vigilant and on guard for market manipulators. For engineers, it means that many solutions to their problems still provide the potential for another solution where the trade-off between certain objectives is net positive, giving them more to consider when decision making.

The new results are summarized as follows, starting with the simplest type of linear multi-objective problem (LMOP) where the solutions come from a one-dimensional space, all the efficient solutions must be substantially efficient (SE) as well. This leads one to wonder if that is always the case, but an example is then given showing that efficiency on its own does not imply substantial efficiency in LMOPs. A further example is then given, showing that LMOPs do not even necessarily have a substantially efficient solution. Since SE solutions are not always available, some necessary conditions are sought out to determine what kinds of further restrictions may need to be placed on problems to ensure an SE solution. A bounded feasible set of solutions and a relation between the tangent cone at a point with the objectives turn out to be necessary conditions. An example shows they are not sufficient however, so another approach is attempted. A solution being properly efficient for every restriction of the LMOP to pairs of objectives makes that solution SE. This finally leads to a very specific type of problem where an SE solution can be guaranteed. Seeing how difficult it is to ensure an SE solution exists for a problem, the authors turn to the topology of the collection of SE solutions to help find them when they do exist. This culminates in two theorems, one relating the collection of bounds of SE solutions being bounded themselves to the closure of the set of SE solutions. The other theorem shows that the location of SE solutions is on the boundary of the feasible collection of solutions for a very wide range of problems.

This work is split into five sections. Section 2 will give the necessary background for understanding what follows. Section 3 will introduce substantially efficient solutions and attempt to explain the current status of research into these types of solutions. Section 4 will provide new results and illuminating examples discovered by the authors for the linear case of problems. Section 5 will conclude with some final remarks and some suggestions for future research.

12.2 Background

The information that is contained in this section is a collection of terminology that is required for understanding the later sections. Everything that follows will be using real Euclidean vector spaces. The following notation is for describing vectors and comparing them, along with some topological notation to be used.

Notation 12.1. For vectors $x = (x_1, \dots, x_n) \in \mathbb{R}^n$ and $y = (y_1, \dots, y_n) \in \mathbb{R}^n$, $x(i) = x_i$, and

by $x = 0$ it is meant that $x = (0, 0, \dots, 0)$,
by $x = y$ it is meant that $x_i = y_i$ for all i,
by $x < y$ it is meant that $x_i < y_i$ for all i,
by $x \leqslant y$ it is meant that $x_i \leqslant y_i$ for all i and $x_j < y_j$ for some j.

For a space X, let X° denote the interior of X.

The fundamental structure of a problem and solutions to those problems are as follows.

Definition 12.1. Let $f(x) = (f_1(x), \dots, f_m(x))$ be a vector-valued function and $X \subseteq \mathbb{R}^n$. Consider the problem

$$\begin{aligned} (P) \quad &\text{Minimize: } f(x) \\ &\text{Subject to : } x \in X. \end{aligned}$$

The set X contains what are called the *feasible solutions* for (P). A feasible solution is known as an *efficient solution* to (P) when there is no other $y \in X$ such that $f(y) \leqslant f(x)$ and $f(y) \neq f(x)$. Solving (P) is understood to be the process of identifying all the points in X that are efficient.

Introduced by Pareto, efficiency made its début in [7] and was modernized in [8] by Koopmans. It was further formalized in the form given here by Kuhn and Tucker in [9]. The particular type of problem considered in this work is the linear case of problem (P). ◊

Definition 12.2. Consider the problem

$$\begin{aligned} (L) \quad &\text{Minimize: } f(x) := (m_1 \cdot x + b_1, \dots, m_m \cdot x + b_m) \\ &\text{Subject to: } x \in X = \{y \in \mathbb{R}^n | Ay \leqslant c\}, \\ \text{where} \quad &\{m_i\}_{i=1}^m \subset \mathbb{R}^n, \{b_i\}_{i=1}^m \subset \mathbb{R}, \\ &r, m, n \in \mathbb{N}, c \in \mathbb{R}^r, \text{ and } A \in \mathbb{R}^{r \times n}. \end{aligned}$$

This type of problem is called a *linear multi-objective optimization problem* or LMOP. Here m_i will be referred to as the *direction* of the objective function f_i. ◊

To rephrase, an LMOP is a multi-variate multi-objective function where each objective component is a linear function and the feasible space is the intersection of closed half-spaces. Since the intersection of half-spaces is convex and linear functions are convex, this problem is automatically a convex problem. It may be assumed that $b_i = 0$ for all i as these constant terms do not affect the subsequent results. Moving on, a special type of efficiency called proper effiency was introduced by Geoffrion in [10].

Definition 12.3. A properly efficient solution of (*P*) is properly efficient solution x $\in X$ for which there exists a scalar $M > 0$ such that for each i, if $y \in X$ and $f_i(y) < f_i(x)$ then there exists an index j for which $f_j(y) > f_j(x)$ and

$$\frac{f_i(x) - f_i(y)}{f_j(y) - f_j(x)} \leqslant M.$$

The quotient on the left may be referred to as the *trade-off quotient* from here on out. Efficient solutions that are not properly efficient are simply called *improperly efficient.* ◊

Solutions that are properly efficient cannot do any 'better' in any single component without doing proportionally 'worse' in at least one other component. Proper efficiency has been shown to be of interest to many in the optimization community and within industry as well. For instance, in [11] Seinfeld and McBride state that proper efficiency help to avoid anomalous solutions in their problem solving process that could arise when just any efficient solution would be chosen. Another example, in [12] Belenson and Kapur explicitly state that an efficient solution on its own may not be desirable and instead opt for properly efficient solutions as they avoid some anomalies as well. In [13], Borwein characterized proper efficient points in terms of tangent cones. In [14], Benson and Morin developed some necessary and sufficient conditions for the presence of properly efficient solutions. Then they related those conditions to the stability of specified single-objective maximization problems. Also, in [15] Wendell and Lee involved proper efficient points in their generalization of results from LMOPs to non-linear cases relying on duality. Further, in [16], Benson furthered Borwein's work by defining properly efficient points in terms of projection cones. In [17], Liu extends the concept of proper efficiency to ϵ-proper efficiency. In [18], Jiang and Deng extend the concept of proper efficiency to α-proper efficiency. These are just a few of the many considerations relevant to the concept of proper efficiency. The area has garnered much more attention than the record provided.

A well known fact following [14] corollary 1 or [18] corollary 3.2 is the following.

Proposition 12.1 *In an LMOP, every efficient solution is properly efficient.*

Some helpful notation is provided to distinguish the different sets of solutions.

Notation 12.2. Let $E_{(P)} := \{x \in X: x \text{ is efficient for the problem } (P)\}$ and let $P_{(P)} := \{x \in X: x \text{is properly efficient for the problem } (P)\}$.

The following definition of a tangent cone, which can be thought of as all the vectors away from x into X, can be found in [19].

Definition 12.4. The tangent cone $T_X(x)$ of a vector x in X is defined by $w \in T_X(x)$ if

$$w = \lim_{k\to\infty} (x_k - x)/\lambda_k$$

for some sequence $\{x_k\}_{k\in\mathbb{N}} \subseteq X$ with $x_k \to x$ and some scalar sequence $\lambda_k \searrow 0$. ◊

This was a short overview of what is necessary for what follows. Good references that can be used for further variational analysis and optimization are [20] and [19]. Further information concerning multi-objective optimization is located in [21] and [22].

12.3 A review of substantial efficiency

The concept of proper efficiency can be extended to include a boundary on more than just one trade-off quotient. That extension leads to the concept introduced by Kaliszewski in [23].

Definition 12.5. The solution $x \in X$ is *substantially efficient* or SE if it is efficient and there exists some real $M > 0$ such that for each f_i and $y \in X$ satisfying $f_i(y) < f_i(x)$, then when $f_j(x) < f_j(y)$,

$$\frac{f_i(x) - f_i(y)}{f_j(y) - f_j(x)} \leqslant M.$$

Let $S_{(P)} := \{x \in X\colon x \text{ be substantially efficient for the problem } (P)\}$. ◊

Solutions that are substantially efficient have the property that that one cannot do any 'better' in any single component without doing proportionally 'worse' in all components that stand to suffer.

The definition was again stated in [24] under the name 'strongly properly efficient'. In this paper Khaledian, Khorram and Soleimani-Damaneh rediscovered the concept of substantial efficiency, giving examples and studying the notion in different senses. One was in the sense of Geoffrion, which is the sense presented in this paper. The other was in the sense of Benson using cones.

Some more work was done in [25] where Pourkarimi and Karimi provided two characterizations for substantial efficiency, one rooted in a scalar problem and the other based on a stability concept. They also introduced the concept of quasi-substantial efficiency and presented two similar characterizations. Quasi-substantially efficient solutions can have varying orders, with each order determining the type of bound on the trade-off quotients.

A fact that will be used, that is adapted from [23, p 135] is the following.

Proposition 12.2. *For any problem (P), if the number of efficient solutions is finite then all the efficient solutions are substantially efficient.*

The following are trivial facts based on the definitions needed before moving forward to the new results.

Proposition 12.3. *If the problem (P) has only two objective functions (i.e.* $f(x) = \{f_1(x), f_2(x)\}$*) then any properly efficient solution is actually substantially efficient.*
Remark. It is apparent that $S_{(P)} \subseteq P_{(P)} \subseteq E_{(P)}$ for any problem (P).

12.4 New results and examples

Engineers and economists will want to know the conditions for which SE solutions exist. The first proposition relates substantially efficient solutions with all efficient solutions in the mono-variable case as it is a simple type of LMOP.

Proposition 12.4. *If the LMOP (L) has* $X \subseteq \mathbb{R}$, *then every efficient solution is actually substantially efficient.*

Proof. Fix an efficient solution x^*. Note that for each $i = 1, \ldots, m$ that $f_i(x) = m_i x + b_i$. So for any fixed i if $x \in X \subseteq \mathbb{R}$ is a feasible solution and $j \in \{1, \ldots, m\}$ for which $f_i(x^*) > f_i(x)$ and $f_j(x) > f_j(x^*)$ note that $m_j \neq 0$ and $x \neq x^*$ or else $f_j(x) > f_j(x^*)$ is impossible. Observe that

$$\begin{aligned}\frac{f_i(x^*) - f_i(x)}{f_j(x) - f_j(x^*)} &= \frac{m_i x^* - b_i - m_i x + b_i}{m_j x - b_j - m_j x^* + b_j}\\ &= \frac{m_i(x^* - x)}{m_j(x - x^*)} \leqslant \frac{-m_i}{m_j}.\end{aligned} \tag{12.1}$$

Take $M = \max_{i,j \in \{1,\ldots,n\}} \frac{-m_i}{m_j}$ to obtain a bound that works for all viable combinations of i and j. This is enough to say that any efficient solution is actually substantially properly efficient. ■

This leads one to wonder if substantial efficiency is equivalent to proper efficiency in general LMOPs. However, despite every efficient solution being properly efficient in LMOPs, it is not always the case that every efficient solution is automatically a substantially efficient solution. A counter-example is provided.

Example 12.1. Consider the LMOP

$$\begin{aligned}\text{Minimize: } f(x) &= f(x_1, x_2, x_3)\\ &= (f_1(x), f_2(x), f_3(x))\\ &= (x_1 + x_2 - x_3, x_1 - x_2, -x_1 + x_2 + x_3),\\ \text{subject to: } x &= (x_1, x_2, x_3) \geqslant 0.\end{aligned} \tag{12.2}$$

The solution $x^* = (0, 0, 0)$ is an efficient solution with $f(x^*) = 0$. Indeed, consider any other feasible solution $x = (x_1, x_2, x_3) \geqslant 0$ with $f(x) \leqslant 0$. If $f_1(x) \leqslant 0$ then $x_3 \geqslant x_1 + x_2$. And if $f_2(x) \leqslant 0$ then $x_2 \geqslant x_1$. Using those, if $f_3(x) \leqslant 0$, then $0 \geqslant f_3(x) = -x_1 + x_2 + x_3 \geqslant 2x_2 \geqslant 2x_1 \geqslant 0$. That implies that $x_1, x_2, x_3 = 0$. This is enough to say $x^* = (0, 0, 0)$ is an efficient solution because there is no other solution for which $f(x) \leqslant 0$. Proposition 12.1 implies this solution is properly efficient.

Proper efficiency can also be demonstrated using a method from [10] which states that x^* is properly efficient if and only if there exists $\lambda \in \mathbb{R}_+^m$ such that x^* minimizes

$$\sum_{k=1}^{m} \lambda_k f_k(x).$$

Indeed if $\lambda = (1, 1, 1)$ then

$$\sum_{k=1}^{3} \lambda_k f_k(x) = x_1 + x_2 \tag{12.3}$$

which is minimized when $x_1, x_2 = 0$. The solution $x^* = (0, 0, 0)$ has that property and so it is properly efficient.

Now it will be shown that x^* is not substantially efficient. Let $M > 1$ be any potential bound on the trade-off quotients. Consider the solution $x = (1, 1 - \frac{1}{M}, 5)$, which gives $f(x) = (-3 - \frac{1}{M}, \frac{1}{M}, 5 - \frac{1}{M})$. Since $f_1(x) < f_1(x^*)$ and $f_2(x) > f_2(x^*)$, if x^* is SE then it should be the case that the trade-off quotient between f_1 and f_2 is bounded by M. However that is not the case,

$$\frac{f_1(x^*) - f_1(x)}{f_2(x) - f_2(x^*)} = \frac{3 + \dfrac{1}{M}}{\dfrac{1}{M}} = 3M + 1 > M. \tag{12.4}$$

So given any $M > 1$, an x can be found so that the trade-off quotient is greater than M. So x^*, while efficient and properly efficient, cannot be SE. □

So it is not the case that every efficient solution is SE in LMOPs, but it would be nice if a substantially efficient solution were present in any given problem. That would give problem analysts a target to aim for when solving problems. However, it is not always the case that a substantially efficient solution exists.

Example 12.2. Consider the feasible solution set

$$X = \{(x, y, z) \in \mathbb{R}^3 : 0 \leqslant x, y, z \leqslant 1, y \geqslant x, z \geqslant y - x\}$$

for the problem of minimizing $f(x, y, z)$ defined by the objective functions

$$\begin{aligned} f_1(x, y, z) &= -x + y \\ f_2(x, y, z) &= z + x - y \\ f_3(x, y, z) &= z - x + y \\ f_4(x, y, z) &= x \\ f_5(x, y, z) &= -x \end{aligned}$$

and a feasible solution (x_1, y_1, z_1), where $x_1 < y_1$ and $z_1 > y_1 - x_1 > 0$. Whether or not $x_1 = y_1$, f_2 and f_3 can be made smaller without influencing any other objective by taking $z_2 = y_1 - x_1$. So an efficient solution must have its third component equal to the second minus the first. Observe that $f_1(x_1, y_1, z_2) = -x_1 + y_1 > 0$ and $f_3(x_1, y_1, z_2) = z_2 - x_1 + y_1 = 2y_1 - 2x_1 > 0$ since $y_1 > x_1$. Both can be lessened by taking $y_2 = x_1$. This will not affect f_2 since f_2 will stay at 0. Also f_4 and f_5 will remain unchanged since x_1 has not changed. So a new solution that will lower the first three components has the form $(x_1, y_2, z_2) = (x_1, x_1, 0)$. So if a solution is to be efficient it must have the form $(x, x, 0)$ and $f(x, x, 0) = (0, 0, 0, x, -x)$. From this it is seen that every solution of the form $(x, x, 0)$ is efficient as a change in x will result in a fair exchange in the fourth and fifth component of f.

However, none of these efficient solutions turn out to be substantially efficient. Two cases will be examined for the efficient solution $(x, x, 0)$. First when $0 < x \leqslant 1$ and second when $x = 0$.

Case 1. Assume that $0 < x \leqslant 1$ and let $M > 0$. Take ϵ so that $(x - \epsilon, x - \epsilon, \epsilon^2) \in X$ and $\epsilon < \frac{1}{M}$. Then

$$f_4(x, x, 0) = x > x - \epsilon = f_4(x - \epsilon, x - \epsilon, \epsilon^2)$$

and

$$f_3(x - \epsilon, x - \epsilon, \epsilon^2) = \epsilon^2 > 0 = f_3(x, x, 0)$$

so the trade-off quotient takes the form

$$\frac{f_4(x, x, 0) - f_4(x - \epsilon, x - \epsilon, \epsilon^2)}{f_3(x - \epsilon, x - \epsilon, \epsilon^2) - f_3(x, x, 0)} = \frac{x - (x - \epsilon)}{\epsilon^2 - 0} = \frac{\epsilon}{\epsilon^2} = \frac{1}{\epsilon} > \frac{1}{\frac{1}{M}} = M. \quad (12.5)$$

This means that $(x, x, 0)$ is not SE when $x > 0$.

Case 2. Assume that $x = 0$ and $M > 1$ is given. Then take $\epsilon < \frac{1}{M}$ and note that $(\epsilon, \epsilon, \epsilon^2) \in X$. Then

$$f_5(0, 0, 0) = 0 > -\epsilon = f_5(\epsilon, \epsilon, \epsilon^2)$$

and

$$f_3(\epsilon, \epsilon, \epsilon^2) = \epsilon^2 > 0 = f_3(0, 0, 0)$$

so the trade-off quotient takes the form

$$\frac{f_5(0, 0, 0) - f_5(\epsilon, \epsilon, \epsilon^2)}{f_3(\epsilon, \epsilon, \epsilon^2) - f_3(0, 0, 0)} = \frac{\epsilon}{\epsilon^2} = \frac{1}{\epsilon} > \frac{1}{\frac{1}{M}} = M \quad (12.6)$$

so in this case $(0, 0, 0)$ is also not SE. □

A simple criterion for the existence of SE solutions has eluded the authors. The difficulty of not having SE solutions abundantly available leads to investigating what may be responsible for their irregularity. That could reveal what types of problems do have SE solutions. The following theorem helps one to understand why SE solutions are not ubiquitous.

Theorem 12.1 *If (L) is an LMOP such that $S_{(L)} = \varnothing$ then either X is unbound in at least one component or for each $x^* \in E_{(L)}$ there is an m_j perpendicular to a non-zero vector in $T_X(x^*)$.*

Proof. Note that the definition of (L) ensures X is closed and convex. Let $x^* \in E_{(L)}$ but $x^* \notin S_{(L)}$, then for all $M > 0$ there exists $x_M \in X$ and indices $i_M, j_M \in \{1, \ldots, n\}$,

$$\begin{aligned} m_{i_M} \cdot x^* &> m_{i_M} \cdot x_M, \\ m_{j_M} \cdot x^* &< m_{j_M} \cdot x_M, \\ \frac{m_{i_M} \cdot (x^* - x_M)}{m_{j_M} \cdot (x_M - x^*)} &> M. \end{aligned} \tag{12.7}$$

This defines a sequence $\{x_k\}_{k \in \mathbb{N}}$. Moreover by the pigeon hole principle it is possible to pass to a subsequence so that the objective indices i and j are fixed because there is only a finite number of combinations of two elements from a collection of size n but infinitely many x_k. Let M_k be the corresponding bounds of that subsequence.

There are then two cases since

$$\lim_{k \to \infty} \frac{m_i \cdot (x^* - x_k)}{m_j \cdot (x_k - x^*)} > \lim_{k \to \infty} M_k = \infty. \tag{12.8}$$

Case 1. $m_i \cdot (x^* - x_k) \to \infty$.

In this case $m_i \cdot x^* - m_i \cdot x_k > N_k$ for some scalar sequence $\{N_k\}_{k \to \infty}$. Since $m_i \cdot x^*$ is fixed then $m_i \cdot x_k$ must go to $-\infty$ to stay ahead of N_k as $N_k \to \infty$. Since m_i is fixed this is only possible if at least one of the components of x_k goes to $\pm\infty$.

Case 2. $\lim_{k \to \infty} m_j \cdot (x_k - x^*) = 0$.

In this case either $x_k \to \bar{x} \neq x^*$, $x_k \to x^*$, or $\|x_k\| \to \infty$ with $m_j \perp \lim_{k \to \infty} \frac{x_k - x^*}{\| x_k - x^* \|}$. If $\|x_k\| \to \infty$ then X is unbounded in some component. If $x_k \to \bar{x}$ then since X is convex, every point between x_k and x^* will be in X. In particular $\frac{x_k}{k} - \frac{(k-1)x^*}{k} \in X$. This forms a sequence in X with

$$\frac{\frac{x_k}{k} + \frac{(k-1)x^*}{k} - x^*}{\frac{1}{k}} = x_k - x^* \to \bar{x} - x^*. \tag{12.9}$$

The definition of a tangent cone states that $\bar{x} - x^* \in T_X(x^*)$. So m_j is perpendicular to $\bar{x} - x^*$ since $m_j \cdot (\bar{x} - x^*) = 0$.

If $x_k \to x^*$ then $\|x_k - x^*\| \to 0$ so

$$\lim_{k\to\infty} \frac{x_k - x^*}{\|x_k - x^*\|} \in T_X(x^*) \tag{12.10}$$

by definition. So

$$\frac{m_i \cdot (x^*-x_k)/\|x_k - x^*\|}{m_j \cdot (x_k - x^*)/\|x_k - x^*\|} = \frac{m_i \cdot (x^*-x_k)}{m_j \cdot (x_k - x^*)} > M_k. \tag{12.11}$$

Observe that

$$\begin{aligned} m_i \cdot \frac{(x^*-x_k)}{\|x_k - x^*\|} &= \cos(\theta_i)\|m_i\|\frac{\|x^*-x_k\|}{\|x_k - x^*\|} \\ &= \cos(\theta_i)\|m_i\| \leqslant \|m_i\|, \end{aligned} \tag{12.12}$$

where θ_i is the angle between m_i and (x^*-x_k). So the numerator of the left most quotient from (12.11) is bounded above. This means that in order for (12.11) to hold,

$$\lim_{k\to\infty} m_j \cdot \frac{(x_k - x^*)}{\|x_k - x^*\|} = m_j \cdot \lim_{k\to\infty} \frac{x_k - x^*}{\|x_k - x^*\|} = 0. \tag{12.13}$$

This implies that m_j is perpendicular to a non-zero vector in $T_X(x^*)$. ■

Comparing an efficient solution with any other solution in problems in a one-dimensional feasible space ensures one solution will be greater than the other. So there are only two directions of trade-off when the feasible space is one-dimensional, greater than or less than. However, the concepts of greater than or less than are not enough to determine a solution's direction away from another when the feasible space has two or more dimensions. A set of angles can express the direction of a solution from another in a feasible space of dimension two or more. This means that for a given efficient solution x, infinitely many solutions with a unique direction from x may have to be considered when checking the collection of trade-offs for boundedness. When an objective is constant in a direction α away from x, other feasible solutions which have a direction arbitrarily close to α can be considered for trade-off. These are problematic if one objective is increasing in the directions arbitrarily close to α and another objective is decreasing but not constant in the direction α, resulting in an unbounded trade-off quotient. Having infinite directions away from any efficient point and more than two objectives creates a difficulty in determining what types of problems have SE solutions. While possible, testing every efficient solution's tangent cone's perpendicularity to the directions of the objective functions (akin to looking for directions away from the efficient solution that leave an objective constant) is not as useful as knowing *a priori* if a substantially efficient solution exists in the problem. A corollary to theorem 12.1 for checking the substantial efficiency of solutions is now provided.

Corollary 12.1. *Let (L) be an LMOP as in definition 12.2, and let x^* be an efficient solution. If X is bounded and there is no m_j perpendicular to any non-zero vector in $T_X(x^*)$ then $x^* \in X$ is substantially efficient.*

Proof. In theorem 12.1 if it was assumed that x^* is not SE then at then end of the proof a contradiction would be arrived at, implying that x^* must have been SE. ■

Remark. Extensions can be made on this corollary by assuming that each m_j is entirely positive, then X only needs to be bounded below.

The following is an example illustrating the usefulness of the previous corollary in identifying if a solution is substantially efficient.

Example 12.3. Consider the following LMOP,

$$\begin{aligned} \text{minimize:} f(x) = \{f_1(x), f_2(x), f_3(x)\} := \{m_1 \cdot x,\ m_2 \cdot x,\ m_3 \cdot x\} \\ \text{where } m_1 = (1, -1, -1),\ m_2 = (2, -1, -2),\ m_3 = (-1, 2, 1) \end{aligned} \tag{12.14}$$

$$\text{subject to: } x \in X := \{(x_1, x_2, x_3): 0 \leqslant x_1, x_2, x_3 \leqslant 2,\ x_1, +x_2 + x_3 \geqslant 4\}. \tag{12.15}$$

Observe that X is a closed bounded irregular tetrahedron. Let $x^* = (0, 2, 2)$, and note that x^* is the unique maximizer of f_3 and unique minimizer of f_1 in X, showing that x^* is efficient. Since x^* is efficient, the corollary gives a sufficient condition to check for substantial efficiency. The first condition is that X be bounded, which it is.

The second condition to check is if m_1, m_2 and m_3 are not perpendicular to any non-zero vector in $T_X(x^*)$. Since

$$T_X(x^*) = \{(x_1, x_2, x_3): x_1 + x_2 + x_3 \geqslant 0,\ x_2, x_3 \leqslant 0,\ x_1 \geqslant 0\}$$

and $m_i(1)$ has the opposite sign of both $m_i(2)$ and $m_i(3)$ for all i, it would be impossible for there to be $x \in T_X(x^*)$ with $x \cdot m_i = 0$ for any i unless $x = (0, 0, 0)$. That is to say that there is no non-zero x in the tangent cone perpendicular to any of the m_i. So the corollary then implies that $x^* = (0, 2, 2)$ is substantially efficient.

To demonstrate how the corollary compares to a direct method, the direct method for showing substantial will also be given. Again, note that x^* is the unique minimizer of f_1 and f_2 and the unique maximizer of f_3. So it only needs to be shown that there exists $M > 0$ such that for every $x \in X/\{x^*\}$,

$$\frac{f_3(x^*) - f_3(x)}{f_1(x) - f_1(x^*)} = \frac{6 - (-x_1 + 2x_2 + x_3)}{x_1 - x_2 - x_3 - (-4)} \leqslant M \tag{12.16}$$

and

$$\frac{f_3(x^*) - f_3(x)}{f_2(x) - f_2(x^*)} = \frac{6 - (-x_1 + 2x_2 + x_3)}{2x_1 - x_2 - 2x_3 - (-6)} \leqslant M. \tag{12.17}$$

To show this, first observe

$$\begin{aligned} x_1 + x_2 + x_3 &\geqslant 4 \\ x_1 + x_2 &\geqslant 4 - x_3 \\ &\geqslant 4 - 2 = 2 \geqslant x_3 \end{aligned}$$

because x_3 is between 0 and 2. From this,

$$\begin{aligned} x_3 &\leqslant x_1 + x_2 \\ -x_3 &\leqslant x_1 + x_2 - 2x_3 \\ -2x_2 - x_3 &\leqslant x_1 - x_2 - 2x_3 \\ 6 + x_1 - 2x_2 - x_3 &\leqslant 2x_1 - x_2 - 2x_3 + 6 \\ \frac{6 - (-x_1 + 2x_2 + x_3)}{2x_1 - x_2 - 2x_3 - (-6)} &\leqslant 1. \end{aligned} \tag{12.18}$$

Also,

$$\begin{aligned} x_3 &\leqslant 2 \\ -x_1 + x_3 &\leqslant 2 \\ x_1 - x_3 &\leqslant 2x_1 - 2x_3 + 2 \\ 6 + x_1 - 2x_2 - x_3 &\leqslant 2x_1 - 2x_2 - 2x_3 + 8 \\ 6 + x_1 - 2x_2 - x_3 &\leqslant 2(x_1 - x_2 - x_3 + 4) \\ \frac{6 - (-x_1 + 2x_2 + x_3)}{x_1 - x_2 - x_3 - (-4)} &\leqslant 2. \end{aligned} \tag{12.19}$$

So taking $M = 2$ it has been shown that x^* is substantially efficient. □

The comparison of these two methods shows that there are circumstances when looking at the tangent cone to check substantial efficiency is easier than a direct method. The same problem gives a non-substantially efficient solution. This example is given in hopes of explaining why no vector can be perpendicular to m_j as mentioned in corollary 12.1.

Example 12.4. Within the same problem framework as example 12.3 consider the solution $y^* = (2, 0, 2)$ which is also efficient with $f(2, 0, 2) = (0, 0, 0)$. This is seen because

$$\begin{aligned} f_1(x) = x_1 - x_2 - x_3 < 0 &\Rightarrow x_1 < x_2 + x_3 \\ &\Rightarrow f_3(x) > 0 \end{aligned} \tag{12.20}$$

$$\begin{aligned} f_2(x) = 2x_1 - x_2 - 2x_3 < 0 &\Rightarrow 2x_1 < x_2 + 2x_3 \leqslant 4x_2 + 2x_3 \\ &\Rightarrow f_3(x) > 0 \end{aligned} \tag{12.21}$$

$$\begin{aligned} f_3(x) = -x_1 + 2x_2 + x_3 < 0 &\Rightarrow x_1 > 2x_2 + x_3 \geqslant x_2 + x_3 \\ &\Rightarrow f_1(x) > 0. \end{aligned} \tag{12.22}$$

Given $M > 0$, consider the point $y_M = (2, \frac{2+M}{1+M}, 1) \in X$ and note that

$$f_1(y_M) = 2 - \frac{2+M}{1+M} - 1 < 0 = f_1(y^*)$$

and

$$f_2(y_M) = 4 - \frac{2+M}{1+M} - 2 > 0 = f_2(y^*).$$

So

$$\frac{f_2(y^*) - f_2(y_M)}{f_1(y_M) - f_1(y^*)} = \frac{-4 + \dfrac{2+M}{1+M} + 2}{2 - \dfrac{2+M}{1+M} - 1} = \frac{-2(1+M) + 2 + M}{(1+M) - 2 - M} = M. \quad (12.23)$$

This is true for every $M > 0$ so y^* cannot be substantially efficient.

This relates to the corollary. Note that $\lim_{M\to\infty} y_M - y^* = (0, 1, -1)$, and

$$(0, 1, -1) \in T_X(y^*) = \{(x_1, x_2, x_3) \subset \mathbb{R}^3: x_1 + x_2 + x_3 \geqslant 0, \text{ and } x_1, x_3 \leqslant 0 \text{ and } x_2 \geqslant 0\}.$$

This is the type of vector perpendicular to the m_j mentioned in the corollary. □

This looks like a promising characterization of all SE solutions, however, the converse of corollary 12.1 is not true. If X is bounded, x^* is efficient and there exists m_i perpendicular to some vector in the tangent cone of x^*, then x^* can still be SE. Consider the following counter-example.

Example 12.5. Let $X := \{(x_1, x_2, x_3) \subset \mathbb{R}^3: 0 \leqslant x_1, x_2, x_3 \leqslant 1\}$. The identity mapping $f(x) = x$ has only one efficient solution at (0, 0, 0), which is substantially efficient by proposition 12.1. However, the vector $m_1 = (1, 0, 0)$ from $f_1(x) = (1, 0, 0) \cdot x$ is perpendicular to $(0, 1, 0) \in T_X(0, 0, 0)$. □

Checking every efficient point's tangent cones in the pursuit of SE solutions may be too difficult. So seeing the limitations of that approach, the authors turn to another way to characterize SE solutions. Since substantial efficiency gives boundaries for the trade-offs of pairs of objectives, it is reasonable to look at all the restrictions of the problem to two objectives. The next proposition is for a general problem connecting substantial efficiency to proper efficiency when a problem is restricted.

Proposition 12.5. *Consider a problem (P) (X not necessarily closed). If $x^* \in X$ is such that the restriction of (P) to any two objective functions has x^* as a properly efficient solution then x^* is SE for (P).*

Proof. If x^* is properly efficient when (P) is restricted to any two objective functions then x^* must be efficient to (P) on its own. Indeed if there exists an $x \in X$ with

$f_i(x) < f_i(x^*)$ then for any other index j, $f_j(x) > f_j(x^*)$ otherwise x^* would not be efficient with regards to the restriction of (P) to the ith and jth components.

Let $M_{a,b}$ be the bound of the trade-off quotient from the properly efficient criterion when (P) is restricted to the ath and bth objective components. Let $M = \max_{a,b\in\{1,2,\dots,m\};a\neq b}(M_{a,b})$. Then for each objective component f_i and $x \in X$ satisfying $f_i(x) < f_i(x^*)$, $f_j(x^*) < f_j(x)$ implies

$$\frac{f_i(x^*) - f_i(x)}{f_j(x) - f_j(x^*)} \leqslant M.$$

So x^* is substantially efficient. ■

The converse does not hold in general, however, with a slight alteration proposition 12.5 can be made necessary and sufficient. First a counter-example is given.

Example 12.6. Consider the minimization problem (P) where $f(x) = (-x, -x + 1, x)$ and $X = [0, 1]$. Note that every feasible solution is efficient. Also note that $x^* = 0$ is a substantially efficient solution. Indeed, if x is such that $f_i(x^*) > f_i(x)$ and $f_j(x^*) < f_j(x)$ then either $i = 1$ or 2 and $j = 3$. Also

$$\frac{f_1(x^*) - f_1(x)}{f_3(x) - f_3(x^*)} = \frac{f_2(x^*) - f_2(x)}{f_3(x) - f_3(x^*)} = \frac{0 - x}{x - 0} = 1 \leqslant 2, \tag{12.24}$$

so if $M = 2$ the criterion for substantial efficiency is satisfied. However, when (P) is restricted to the output functions f_1 and f_2, $x^* = 0$ turns out not to be efficient and thus cannot be properly efficient. □

This shows what is keeping the converse of the previous proposition from being true. When the feasible solution is not efficient in all the restrictions, substantial efficiency is not equivalent to every restriction being properly efficient. So a slight change is made to arrive at a bijective statement.

Theorem 12.2. *Consider the general problem (P). If $x^* \in X$ is a solution such that the restriction of (P) to any two objective functions is efficient, then x^* is a properly efficient solution for the restriction of (P) to any two objective functions if and only if x^* is a substantially efficient solution of (P).*

Proof. (⇒) Proposition 12.5.

(⇐) Assume that x^* is SE with the trade-off quotient bound M. Take $a, b \in \{11\dots, n\}$ and let ($P^{a,b}$) be problem (P) restricted to the objective functions f_a and f_b. By assumption, x^* is efficient for ($P^{a,b}$). Without loss of generality for the choice of a or b, assume x is such that $f_a(x^*) > f_a(x)$ and $f_b(x^*) < f_b(x)$. Then because x^* is substantially efficient, $\frac{f_a(x^*)-f_a(x)}{f_b(x)-f_b(x^*)} \leqslant M$. So the criterion for proper efficiency of (P) restricted to f_a and f_b is also trivially satisfied. ■

Looking at the intersection of properly efficient solutions for all the pairwise problem restrictions allows one a method for finding an SE solution, but it is still a lot to ask. In LMOPs, only the intersection of efficient solutions is necessary as efficiency implies proper efficiency. An empty intersection does not necessarily mean that no SE solution exists either. However, this still leads to a corollary of the theorem 12.2 that finally elicits a sufficient condition for the guaranteed presence of an SE solution in a special case of (L). This is proven directly and using theorem 12.2.

Corollary 12.2. *Let* $X = \{(x_1, \ldots, x_n): a_1 \leqslant x_1 \leqslant b_1, \ldots, a_n \leqslant x_n \leqslant b_n\}$ *(a closed bounded box) for a problem (L). If the results of all the pairwise sums of* m_i *have non-zero components with the same signs (i.e. for any possible combination of i, j, k, l, p, that* $\mathbf{Sgn}\{(m_i + m_j)(p)\} = \mathbf{Sgn}\{(m_k + m_l)(p)\} \neq 0$) *then an SE solution exists.*

Proof. (Without using theorem 12.2.) Let y be the vector in X that minimizes $(m_i + m_j) \cdot x$ for all i, j and $x \in X$. So for all $k \in \{1, \ldots, n\}$ it will be that $y(k) = a_k$ when $Sgn\{(m_i + m_j)(k)\} = 1$ and $y(k) = b_k$ when $Sgn\{(m_i + m_j)(k)\} = -1$. The solution y is efficient, for if it were not there would exist $z \in X$ and i for which $f_i(z) < f_i(y)$, and for all other j it would be that. But then summing these inequalities would give $(m_i + m_j) \cdot z < (m_i + m_j) \cdot y$ which contradicts how y is defined.

Now, it will be shown that y is substantially efficient. Assuming it is not, let i, j and $x \in X$ be such that $f_i(y) > f_i(x)$ and $f_j(y) < f_j(x)$ and

$$\frac{f_i(y) - f_i(x)}{f_j(x) - f_j(y)} > 1. \tag{12.25}$$

But then

$$\begin{gathered} \frac{m_i \cdot y - m_i \cdot x}{m_j \cdot x - m_j \cdot y} > 1 \\ m_i \cdot y - m_i \cdot x > m_j \cdot x - m_j \cdot y \\ m_i \cdot y + m_j \cdot y > m_i \cdot x + m_j \cdot x, \end{gathered} \tag{12.26}$$

but then this again contradicts how y is defined. So y must be substantially efficient.

(Using theorem 12.2.) whenever (L) is restricted to two objective functions, y will be an efficient solution and thus properly efficient as well. Using theorem 12.2 it must be the case that y is an SE solution. ■

This is a very particular set of problems for which SE solutions can be guaranteed to exist. The rarity of the assurance of existence leads one to wonder what can be said of the set of SE solutions when they do exist. The next theorem ties the topology of the collection of SE solutions to the trade-off quotient's bounds.

Theorem 12.3. *Consider the problem (P). If the collection of trade-off quotient bounds* $\{M_x\}_{x \in S_{(P)}}$ *is bounded (i.e. there is a uniform trade-off quotient bound* $M > 0$ *that can be applied for any* $x \in S_{(P)}$) *then* $S_{(P)}$ *is closed.*

Proof. Assume $S_{(P)}$ is not closed, thus there is a limit point of $S_{(P)}$ that $S_{(P)}$ does not contain. Let $\{x_n\}_{n \in \mathbb{N}} \subseteq S_{(P)}$ be a sequence that converges to $x \notin S_{(P)}$. Since X is closed, $x \in X / S_{(P)}$. Let $M \in \mathbb{N}$ with $M > 2$ be a common bound for the trade-off quotients of every element in $\{x_n\}_{n \in \mathbb{N}}$.

Let $\{z_n\}_{n\in\mathbb{N}} \subseteq X$ be a sequence for which

$$f_{i_n}(x) - f_{i_n}(z_n) > 0$$

$$\text{and} \quad f_{j_n}(z_n) - f_{j_n}(x) > 0$$

$$\text{but} \quad \frac{f_{i_n}(x) - f_{i_n}(z_n)}{f_{j_n}(z_n) - f_{j_n}(x)} > n.$$

Since there are infinitely many $n \in \mathbb{N}$ but only finitely many combinations of i_n and j_n then by the pigeon hole principle, a subsequence can be passed to in order to keep the i_n and j_n fixed. Do so to obtain a fixed i and j for which

$$f_i(x) - f_i(z_n) > 0$$

$$\text{and} \quad f_j(z_n) - f_j(x) > 0,$$

$$\text{but} \quad \frac{f_i(x) - f_i(z_n)}{f_j(z_n) - f_j(x)} > n.$$

Then pass to another subsequence so that $f_i(x)$ is always greater or lesser than $f_i(x_n)$ and so that $f_j(x)$ is also always greater or lesser than $f_j(x_n)$. Now there are some cases to consider.

Case 1. $f_i(x_n) > f_i(x)$ and $f_j(x_n) > f_j(x)$.

Choose N large enough that $f_j(z_M) > f_j(x_N)$. Then

$$M < \frac{f_i(x) - f_i(z_M)}{f_j(z_M) - f_j(x)} < \frac{f_i(x_N) - f_i(z_M)}{f_j(z_M) - f_j(x_N)} \leqslant M, \tag{12.27}$$

which is a contradiction.

Case 2. $f_i(x_n) > f_i(x)$ and $f_j(x_n) < f_j(x)$.

Choose N large enough so $\frac{f_j(z_{2M}) - f_j(x)}{f_j(z_{2M}) - f_j(x_N)} < 2$, $f_j(z_{2M}) - f_j(x_N) > 0$, $f_i(x_N) - f_i(z_{2M}) > 0$ and $f_j(x) < f_j(x_N) + \epsilon$, where $\epsilon < \frac{f_j(z_{2M}) - f_j(x)}{2M}$. Then

$$\begin{aligned} f_i(x_N) - f_i(z_{2M}) &> f_i(x) - f_i(z_{2M}) \\ &> 2M(f_j(z_{2M}) - f_j(x)) \\ &> 2M(f_j(z_{2M}) - f_j(x_N) - \epsilon) \end{aligned}$$

$$\Rightarrow M \geqslant \frac{f_i(x_N) - f_i(z_{2M})}{f_j(z_{2M}) - f_j(x_N)} > 2M - \frac{2M\epsilon}{f_j(z_{2M}) - f_j(x_N)} \tag{12.28}$$

$$> 2M - \frac{2M\dfrac{f_j(z_{2M}) - f_j(x)}{2M}}{f_j(z_{2M}) - f_j(x_N)} > 2M - 2.$$

But that is a contradiction when $M \geqslant 2$, which it is.

Case 3. $f_i(x_n) < f_i(x)$ and $f_j(x_n) > f_j(x)$.

Take N large enough that $\frac{f_j(z_{2M}) - f_j(x)}{f_j(z_{2M}) - f_j(x_N)} < 2$, $f_j(z_{2M}) - f_j(x_N) > 0$, $f_i(x_N) - f_i(z_{2M}) > 0$ and $f_i(x) < f_i(x_N) + \epsilon$, where $\epsilon < f_j(z_{2M}) - f_j(x)$. Then

$$\begin{aligned} f_i(x_N) + \epsilon - f_i(z_{2M}) &> f_i(x) - f_i(z_{2M}) \\ &> 2M(f_j(z_{2M}) - f_j(x)) \\ &> 2M(f_j(z_{2M}) - f_j(x_N)) \\ \Rightarrow M \geqslant \frac{f_i(x_N) - f_i(z_{2M})}{f_j(z_{2M}) - f_j(x_N)} &> 2M - \frac{\epsilon}{f_j(z_{2M}) - f_j(x_N)} \\ &> 2M - \frac{f_j(z_{2M}) - f_j(x)}{f_j(z_{2M}) - f_j(x_N)} > 2M - 2, \end{aligned} \tag{12.29}$$

which is again a contradiction.

Case 4. $f_i(x_n) < f_i(x)$ and $f_j(x_n) < f_j(x)$.

Take N large enough so $\frac{f_j(z_{2M}) - f_j(x)}{f_j(z_{2M}) - f_j(x_N)} < 2$, $f_j(z_{2M}) - f_j(x_N) > 0$, $f_i(x_N) - f_i(z_{2M}) > 0$, and $f_i(x) < f_i(x_N) + \epsilon$ and $f_j(x) < f_j(x_N) + \epsilon$, where $\epsilon < \frac{f_j(z_{2M}) - f_j(x)}{2M+1}$. Then

$$\begin{aligned} f_i(x_N) + \epsilon - f_i(z_{2M}) &> f_i(x) - f_i(z_{2M}) \\ &> 2M(f_j(z_{2M}) - f_j(x)) \\ &> 2M(f_j(z_{2M}) - f_j(x_N) - \epsilon) \\ \Rightarrow M \geqslant \frac{f_i(x_N) - f_i(z_{2M})}{f_j(z_{2M}) - f_j(x_N)} &> 2M - \frac{(2M+1)\epsilon}{f_j(z_{2M}) - f_j(x_N)} \\ &> 2M - \frac{(2M+1)\frac{f_j(z_{2M}) - f_j(x)}{2M+1}}{f_j(z_{2M}) - f_j(x_N)} > 2M - 2, \end{aligned} \tag{12.30}$$

which is again a contradiction.

Since these are all the possible cases and each ends in a contradiction then the original assumption that $S_{(P)}$ is not closed is not true. ■

This also shows that when the set of SE solutions is not closed the collection of trade-off bounds is unbounded. The next thing to be considered is when the set of SE solutions is a collection of isolated points. There is a common situation when this can occur. First a lemma involving the existence of a particular vector is needed.

Lemma 12.1. *Let $a, b \in \mathbb{R}^n$ with $a \not\parallel b$ and $1 > \epsilon > 0$ be given. Then there is a $\delta > 0$ such that there exists $c \in \partial N_\epsilon(0)$ with $a \cdot (-c) > \delta\|a\|\epsilon$ while $b \cdot (c) = \frac{1}{m}\|b\|\epsilon$ for a choice of m that is large.*

Proof. For this proof let $\angle xy$ denote the positive angle between x and y; so $\angle xy = \angle yx$. Let Q be the plane containing both a and b. If $a \perp b$ then take $c \in \partial N_\epsilon(0) \cap Q$ so $\angle a(-c) < 45°$ and $\angle bc < 90°$ as close to $90°$ as necessary to obtain $\cos(\angle bc) = \frac{1}{m}$ for some choice of m that is large. Then $\delta = \frac{\sqrt{2}}{2}$ will suffice because

$$a \cdot (-c) = \cos(\angle a(-c))\|a\|\|c\| > \frac{\sqrt{2}}{2}\,\|a\|\,\epsilon = \delta\,\|a\|\,\epsilon, \tag{12.31}$$

and

$$b \cdot c = \cos(\angle bc)\|b\|\|c\| = \frac{1}{m}\,\|b\|\,\epsilon.$$

On the other hand, if $a \not\perp b$ take z to be the unit vector perpendicular to b in Q with $a \cdot z > 0$. So z and a are on the same 'side' of b, so to speak. Let $\theta = \min\{\angle ab, \angle za\}$. Then take c to be the vector in $\partial N_\epsilon(0) \cap Q$ such that $\angle c(-z) < \theta/2$ and $b \cdot c = \frac{1}{m}\|b\|\|c\| = \frac{1}{m}\|b\|\epsilon$ with m being any value for which $\frac{1}{m} < \cos(\theta/2)$. So c is the vector in Q perpendicular to b, with magnitude ϵ, and whose angle with $-z$ measures between 0 and $\theta/2$. So $-c$ is within an angle of $\theta/2$ of z and is also within the same plane as a and b. Since $\angle za = |90° - \angle ab| \neq 0°$ then

$$\angle a(-c) \leqslant \begin{cases} \angle za + \frac{\theta}{2} & \text{if } 0° < \angle ab \leqslant 45° \\ \angle za + \frac{\theta}{2} & \text{if } 45° < \angle ab < 90° \\ \angle za - \frac{\theta}{2} & \text{if } 90° < \angle ab < 180° \end{cases}$$
$$= \begin{cases} 90° - \angle ab + \frac{1}{2}\angle ab & \text{if } 0° < \angle ab \leqslant 45° \\ 90° - \angle ab + \frac{1}{2}(90° - \angle ab) & \text{if } 45° < \angle ab < 90° \\ \angle ab - 90° - \frac{1}{2}(\angle ab - 90°) & \text{if } 90° < \angle ab < 180°. \end{cases} \tag{12.32}$$

In every case $\angle a(-c)$ is a fixed angle less than $90°$. This means that one may set $\cos(\angle a(-c)) =: \delta > 0$ so $a \cdot (-c) = \delta\|a\|\ \|-c\| = \delta\|a\|\epsilon$. ■

A shifted variant of lemma 12.1 is used in the following proposition.

Proposition 12.6. *Consider the LMOP (L) with all m_i non-parallel and spanning a subspace of dimension at least n, and an open feasible set $X \subseteq \mathbb{R}^n$ with $n \geqslant 2$. Then $S_{(L)} \cap X^\circ$ consists of isolated points. That is, $S_{(L)} \cap X^\circ$ contains no limit points.*

Proof. Let $(x_n)_{n\in\mathbb{N}} \subseteq S_{(L)} \cap X^\circ$ be some sequence converging to a point $x \in \mathbb{R}^n$. The goal is to show that $x \notin S_{(L)} \cap X^\circ$. It may be assumed that $x \in X^\circ$ and $x \in E_{(L)}$ or else $x \notin S_{(L)} \cap X^\circ$ automatically. For every x_n there is a corresponding M_n such that for all i and $y \in X$ with $f_i(x_n) > f_i(y)$,

$$\frac{f_i(x_n) - f_i(y)}{f_j(y) - f_j(x_n)} \leqslant M_n$$

for all j with $f_j(y) > f_j(x_n)$. Since $x \in X^\circ$ and X° is open, there exists $\epsilon > 0$ and N such that for all $n > N$, the neighborhoods $\overline{N_{\epsilon/2}(x_n)} \subset N_\epsilon(x) \subseteq X^\circ$.

The fact that the m_i span a subspace of at least dimension n shows that there is no vector $v \in \mathbb{R}^n$ such that v is perpendicular to each m_i. Therefore, for each x_n it must be the case that for some j_n, $f_{j_n}(x_n) - f_{j_n}(x) = m_{j_n} \cdot (x_n - x) > 0$; but since each x_n and x itself are efficient there is some other i_n for which $f_{i_n}(x) - f_{i_n}(x_n) > 0$. Since there are infinitely many x_n and only finitely many combinations of i_n and j_n, then by the pigeon hole principle there must a subsequence where the i_n match and the j_n match. Pass to that subsequence and fix i and j to be those values so that $f_i(x) > f_i(x_n)$ and $f_j(x) < f_j(x_n)$ for all n.

For each $n > N$, using lemma 12.1 take $y_n \in \partial N_{\epsilon/2}(x_n)$ so that $m_i \cdot (x_n - y_n) > \delta\|m_i\|\ \|x_n - y_n\| > 0$ for some fixed δ and $m_j \cdot (y_n - x_n) = \frac{1}{n}\|m_j\|\ \|y_n - x_n\|$. The freedom of choice of y_n and the fact that the m_i and m_j cannot be scalar multiples of one another ensure that such a choice of y_n is possible. Note that $\|x_n - y_n\| = \frac{\epsilon}{2}$.

Take $\delta\frac{\epsilon}{3} > \gamma > 0$ and n large enough so that $\|x - x_n\| < \gamma$. This means that $f_i(x) > f_i(x_n) - \gamma\|m_i\|$ because f_i is continuous. Similarly true is $f_j(x) > f_j(x_n) - \delta\|m_j\|$. Now observe that $f_i(x) > f_i(x_n) > f_i(y_n)$ and $f_j(x) < f_j(x_n) < f_j(y_n)$. Now consider the trade-off quotient

$$\frac{f_i(x) - f_i(y_n)}{f_j(y_n) - f_j(x)} \geqslant \frac{f_i(x_n) - \gamma - f_i(y_n)}{f_j(y_n) - f_j(x_n) + \gamma} > \frac{\delta\|m_i\|\frac{\epsilon}{2} - \gamma\|m_i\|}{\frac{1}{n}\|m_j\|\frac{\epsilon}{2} + \gamma\|m_j\|} > 0. \tag{12.33}$$

Since the choice of γ was arbitrary, and $\gamma \to 0$ implies $n \to \infty$, then the whole trade-off quotient $\frac{f_i(x) - f_i(y_n)}{f_j(y_n) - f_j(x)} \to \infty$ as $\gamma \to 0$. This means that the trade-off quotient cannot be bounded and thus the sequence x_n does not converge to an SE solution.

This quickly leads to a corollary. ■

Corollary 12.3. *Consider the LMOP (L) with all m_i non-parallel and spanning a subspace of dimension at least n, and with a closed feasible set $X \subseteq \mathbb{R}^n$ with $n \geqslant 2$. If $x \in S_{(L)}$ is not isolated then x must lie on the boundary of the feasible collection.*

One may wonder under what circumstances would an SE solution be found on the interior. The following shows what an SE point on the interior would entail.

Proposition 12.7. *For a problem (L) with X closed and bounded, if there exists an $x \in X^\circ$ that is SE, then all $y \in X$ are SE.*

Proof. Let $x \in X^\circ$ be SE and assume that $y \in X$ is not SE. First note that y must be efficient, for if it is not then $f(y) \geqslant f(x)$ implies that for all i that $m_i \cdot y \geqslant m_i \cdot x$ and for some j that $m_j \cdot y > m_j \cdot x$. So for some very small $\epsilon > 0$ it must be that $x + \epsilon(x - y) \in X$ since x is on the interior of X. But that implies that $f(x) \geqslant f(x + \epsilon(x - y))$ and in the jth component there is a strict inequality. That implies x is not efficient which is a contradiction so y must be efficient.

Now if y is not substantially efficient, then there exist sequences

$$\{z_k\}_{k \in \mathbb{N}} \subseteq X$$
$$\{i_k\}_{k \in \mathbb{N}}, \{j_k\}_{k \in \mathbb{N}} \subseteq \{1, \dots, n\}$$

for which

$$\frac{f_{i_n}(y) - f_{i_n}(z_n)}{f_{j_n}(z_n) - f_{j_n}(y)} > n.$$

As earlier, by the pigeon hole principle a subsequence can be passed to hold i_n and j_n fixed at i and j.

So given any $N > 0$ there exists an $\epsilon > 0$ for which $x + \epsilon(y - z_N) \in X$. Note that since the objectives are linear $f_i(x) - f_i(x + \epsilon(y - z_N)) = \epsilon(f_i(y) - f_i(z_N)) > 0$. Similarly it must be that $f_j(x + \epsilon(y - z_N)) - f_j(x) = \epsilon(f_j(y) - f_j(z_N)) > 0$. However,

$$\frac{f_i(x) - f_i(x + \epsilon(y - z_N))}{f_j(x + \epsilon(y - z_N)) - f_j(x)} = \frac{\epsilon(f_i(y) - f_i(z_N))}{\epsilon(f_j(z_N) - f_j(y))} > N, \tag{12.34}$$

which is a contradiction to $x \in S_{(L)}$.

Corollary 12.3 and proposition 12.7 combined show that in a common situation substantially efficient points will only be found on the boundary.

Theorem 12.4. *Consider the LMOP (L) with all m_i non-parallel and spanning a subspace of dimension at least $n \geqslant 2$, where the feasible set $X \subseteq \mathbb{R}^n$ is closed. If $x \in S_{(L)}$ then it must lie on the boundary of X.*

Proof. If x is not on the boundary then it is on the interior. By proposition 12.7 all the points must be substantially efficient, so x is not an isolated substantially efficient point. But then by corollary 12.3 if x is not isolated then it must lie on the boundary. That is a contradiction, so x must lie on the boundary to begin with. ■

This is good because for the most part, analysts will only need to look at the boundary of X to find SE solutions. For clarification, SE solutions on the boundary of the feasible collection need not be isolated as the last example makes evident.

Example 12.7. Let $X = \{(x_1, x_2, x_3) \in \mathbb{R}^3: x_2, x_3 \geqslant 0\}$ be the collection of feasible solutions. Let

$$F(x) = (f_1(x), f_2(x), f_3(x)) = (m_1 \cdot x, m_2 \cdot x, m_3 \cdot x),$$

$$\text{where } m_1 = (0, 1, 1),$$

$$m_2 = (0, 1, 2),$$

$$m_3 = (0, 2, -1)$$

be the multi-valued function to be minimized over X. It is evident that the set $S = \{(x_1, 0, 0) \in \mathbb{R}^3: x_1 \in \mathbb{R}\}$ is within the collection of efficient solutions. In order to check the substantial efficiency of any given $s \in S$ only points $x \in X$ where $f_3(x) < f_3(s)$ need to be considered because there is no $x \in X$ where $f_1(x) < f_1(s)$ or $f_2(x) < f_2(s)$. If x has $f_3(x) < f_3(s)$ then $f_1(x) > f_1(s) = 0$ and $f_2(x) > f_2(s) = 0$. So $f_1(s), f_2(s), f_3(s) = 0$, $x_3 > 2x_2$ and $x_2, x_3 \geqslant 0$ give

$$\frac{f_3(s) - f_3(x)}{f_1(x) - f_1(s)} = \frac{-2x_2 + x_3}{x_2 + x_3} \leqslant 1 \tag{12.35}$$

$$\text{and } \frac{f_3(s) - f_3(x)}{f_2(x) - f_2(s)} = \frac{-2x_2 + x_3}{x_2 + 2x_3} \leqslant 1. \tag{12.36}$$

This is enough to say that any $s \in S$ is substantially efficient.

12.5 Conclusion

To conclude, a brief history of SE solutions was provided and then followed by some new information regarding substantial efficiency in general and in the context of linear problems. It was shown that SE solutions do not always exist in linear problems, even when the feasible space is bounded. Substantial efficiency was also shown to connect the perpendicularity of the tangent cone to the direction vectors of objective functions in linear problems. Various restrictive existence criteria were provided to ensure a substantially efficient point in very specific cases. Finally, information about the topology of the set of SE solutions is presented. The theme of the new results was that finding SE solutions is difficult but not impossible. The insight gained was that SE solutions are worth pursing, but when they cannot be found, analysts and decision makers need to be aware of the potential for disproportional trade-offs to occur between some objectives which can have negative consequences.

Engineers can find uses in SE solutions as they may find the potential anomaly of an unbound trade-off between any of their objectives disconcerting. The implications for market analysts are many if their problems can be made into LMOPs. First, the knowledge that SE solutions are rare and difficult to detect should keep the analyst forever on guard against market manipulators and could call for perpetual regulation through things such as taxation, fines, premiums, time constraints on transactions and so on. Second, if the feasible collection of solutions is bounded, they have a method of checking if their solution is SE by comparing the tangent cone to the direction vectors m_j in the problem. Third, they have a method for trying to locate an SE solution by restricting the problem to pairs of objectives and finding the collections of properly efficient solutions then intersecting those collections. Lastly, they know to only consider the boundary of the feasible solutions for most LMOPs. Engineers may also find use in taking advantage of a non-market system in the same way when they have a non-substantially efficient solution to their problem.

All things considered, it seems to the authors that substantial efficiency would be desired above proper efficiency when available in a solution set. However the extremely restrictive nature of substantial efficiency makes their existence, if not rare, hard to predict. This is a good reason to continue pursuing solutions similar to substantially efficient solutions but with some relaxed conditions. In [23], Kaliszewski himself introduced Δ-substantial efficiency which excludes infinitesimal changes around the solutions in consideration. Conditions such as that provided in [25], where Pourkarimi and Karimi introduced quasi-substantially efficient solutions, or solutions that are substantially efficient only on a subset of objectives. Another option would be the extension of the concept of ϵ-proper efficiency in [17]. These may make the existence of solutions that continue to avoid the anomalies produced by improperly efficient solutions in many pairs of objectives more probable.

References

[1] Chaboud A P, Chernenko S V, Howorka E, Iyer R S K, Liu D and Wright J 2004 The high-frequency effects of US macroeconomic data releases on prices and trading activity in the global interdealer foreign exchange market *International Finance Discussion Papers* 823, Board of Governors of the Federal Reserve System (US) https://EconPapers.repec.org/RePEc:fip:fedgif:823

[2] CBOE 2014 2014 CBOE market statistics *Technical report, CBOE Global Markets*

[3] CBOE 2016 2016 CBOE market statistics *Technical report, CBOE Global Markets*

[4] Brunnermeier M K 2009 Deciphering the liquidity and credit crunch 2007–2008 *J. Econ. Perspect.* **23** 77–100

[5] Healy P M and Palepu K G 2003 The fall of Enron *J. Econ. Perspect.* **17** 3–26

[6] Wah E and Wellman M P 2013 Latency arbitrage, market fragmentation, and efficiency: a two-market model *Proceedings of the Fourteenth ACM Conference on Electronic Commerce, EC '13* (New York: ACM), pp 855–72

[7] Pareto V 1971 *Manual of Political Economy* (New York: Kelley)

[8] Koopmans T C 1951 Efficient allocation of resources *Econometrica: J. Economet. Soc.* **19** 455–65

[9] Kuhn H W and Tucker A W 1950 Nonlinear programming *Proc. Second Berkeley Symp. on Mathematical Statistics and Probability* pp 481–92

[10] Geoffrion A M 1968 Proper efficiency and the theory of vector maximization *J. Math. Anal. Appl.* **22** 618–30

[11] Seinfeld J and McBride W 1970 Optimization with multiple performance criteria, application to minimization of parameter sensitivities in a refinery model *Ind. Eng. Chem. Process Des. Develop.* **9** 53–7
[12] Belenson S and Kapur K 1973 An algorithm for solving multicriterion linear programming problems with examples *Oper. Res. Q.* **24** 65–77
[13] Borwein J 1977 Proper efficient points for maximizations with respect to cones *SIAM J. Control Optimization* **15** 57–63
[14] Benson H and Morin T 1977 The vector maximization problem: proper efficiency and stability *SIAM J. Appl. Math.* **32** 64–72
[15] Wendell R and Lee D 1977 Efficiency in multiple objective optimization problems *Math. Program.* **12** 406–14
[16] Benson H 1979 An improved definition of proper efficiency for vector maximization with respect to cones *J. Math. Anal. Appl.* **71** 232–41
[17] Liu J 1999 ϵ-properly efficient solutions to nondifferentiable multiobjective programming problems *Appl. Math. Lett.* **12** 109–13
[18] Jiang Y and Deng S 2014 Enhanced efficiency in multi-objective optimization *J. Optim. Theory Appl.* **162** 577–88
[19] Rockafellar R T and Wets R J B 1998 *Variational Analysis* (Berlin: Springer)
[20] Chen G, Huang X and Yang X 2005 *Vector Optimization, Set-Valued and Variational Analysis* (Berlin: Springer)
[21] Ehrgott M 2005 *Multicriteria Optimization* 2nd edn (Berlin: Springer)
[22] Miettinen K M 1999 *Nonlinear Multiobjective Optimization* (Norwell, MA: Kluwer Academic)
[23] Kaliszewski I 1994 *Quantitative Pareto Analysis by Cone Separation Technique* (Norwell, MA: Kluwer Academic)
[24] Khaledian K, Khorram E and Soleimani-damaneh M 2016 Strongly proper efficient solutions: efficient solutions with bounded trade-offs *J. Optim. Theory Appl.* **168** 864–83
[25] Pourkarimi L and Karimi M 2016 Characterization of substantially and quasi-substantially efficient solutions in multiobjective optimizations problems *Turk J. Math.* **41** 293–304

Chapter 13

A machine learning approach for engineering optimization tasks

Arpana Rawal, Mamta Singh and Jyothi Pillai

The evolving world is continuously tuning itself with the famous underlying Darwinian principle 'survival of the fittest species'. This naturally existing successful optimization system continues to inspire philosophers, analysts and practitioners to address real-world optimization tasks in our daily lives as well. While dealing with the combinations of possible worlds that satisfy combinations of constraints, researchers often attempt to build preference relations for possible worlds, and try to find a best possible world according to the preferences. The preference for choosing a set of available alternatives is often based on minimizing some negative performance parameters, say an error, or maximize some favorable parameters, say accuracy. Both constraints and alternatives act as functioning world parameters in optimization terminology.

The conventional taxonomy of optimization methods has seen a multitude of mathematical programming techniques formulated based on either the types of objectives or constraints [1]. The principles of mathematical programming/optimization are borrowed from the field of operations research (OR) and are found to be suitable for analyzing a set of model functioning parameters under a precise set of constraints so as to minimize the defined objective function of the model.

Optimization tasks are inherently combinatorial in nature and can be described with a finite set of constraints and more importantly have a cost function attached to them. The main goal of carrying out optimization tasks is to find the problem's solution or the parameters of the problem in order to arrive at a 'good' or 'acceptable' solution, if not the 'best' solution. Some degree of intelligence ought to be applied to arrive at optimal, good or acceptable solutions, because of the infeasibility of applying brute-force to determine all possible solutions in problem search spaces. One of the most significant trends in the realm of artificial intelligence

doi:10.1088/978-0-7503-2404-5ch13

(AI) during the past fifteen years has been the integration of optimization methods throughout. Virtually all machine learning (ML) algorithms are reduced to optimal solution results by mixing models and optimization methods.

13.1 Optimization: classification hierarchy

The initial classification used the distinction criterion of the domain type of the problem settings, namely discrete valued domains over real-valued domains. Some models only make sense if the variables take on values from a discrete set, often a subset of integers, whereas other models contain variables that can take on any real value.

Models with discrete variables are discrete (combinatorial) optimization problems, while models with continuous variables are continuous optimization problems. Continuous optimization problems tend to be easier to solve than discrete optimization problems; the smoothness of the functions means that the objective function and constraint function values at a point x can be used to deduce information about points in a neighborhood of x. Continuous optimization algorithms are important in discrete optimization because many discrete optimization algorithms generate a sequence of continuous sub-problems (table 13.1).

Conventional methods of addressing combinatorial optimization problems range from diversified application domains of resource scheduling, element-routing in transportation networks and minimization of cost, to performance in real-modeling problems of Netflix recommendations, driving navigation by GPS, etc. These methods encompass exact algorithms over shortest-path or minimum-cost spanning trees or branch-and-bound spaces using integer linear programming formulations and 'polynomial-time' algorithms of linear programming for maximum-speed flow and minimum-cost flow in networks, multi-commodity flow, and coloring and matching problems over graph structures, while approximate algorithms are used for the vertex cover and set cover Steiner-tree and traveling salesman problems. Owing to the disadvantages observed in these algorithms due to the sub-optimal solutions arrived at, with weak worst-case bounds and poor empirical performance in achieving real-world domain constraints, combinatorial approaches have been used to seek faster and more effective but tailored heuristic algorithms to obtain nearly optimal solutions using real-time problem-specific constraints. The reader is referred to [2] and [3] to obtain sound concepts of convex optimization methods by mathematical programming techniques, exploiting structure using graphs, stochastic processes and statistical methods [2, 3].

The motivation behind deploying learning heuristics in combinatorial optimization tasks is to replace fast approximation computations against heavy ones using domain knowledge and make these computations generic for similar problem families. In the case that partially acceptable solutions are obtained, ML heuristics aim to explore sub-optimal solution spaces, out of which the best solution can be selected for final real-time problem modeling. Some appropriate exemplary domains in ML that can be captured to highlight the use of meta-heuristics for combinatorial optimization are the minimum vertex cover social network routing of messages

Table 13.1. Successful optimization implementations in machine learning domains.

Machine learning domain	Application domain	Case study (instance)	Optimization technique
Combinatorial methods (exact + approximate methods)	Network routing problems	Traveling salesman problems	Continuous approximation methods
Deep neural networks		Traveling salesman problems	Continuous approximation methods
Constraint satisfaction		Traveling salesman problems	Heuristic search methods
Multi-attribute utility theory (MAUT)	Decision-making problems	Product line consolidation and selection (family of staplers)	Statistical methods
Bayesian classifiers		Candidate recruitment	Heuristic search methods
Statistical ML	Classification problems	Text classification	Continuous approximation (convex optimization) methods
Bayesian networks + non-negative matrix factorization (NMF-FE)		fMRI brain imaging	Hidden Markov models
SVM classifiers		Classification (imbalanced datasets)	Mathematical programming
Multi-objective learning		Multiple-digit classification, scene understanding (citiscape identification)	Perceptual task learning, gradient descent search
Probabilistic inference (ranking) models	Classification problems	Search cost optimization for early prediction of good protein torsion angles for protein structure reconstruction	Polynomial (linear) programming
Reinforcement learning	Planning and control	Job-shop scheduling problems	Mathematical programming
Deep neural networks	Perceptual task learning	Speech and image recognition	Gradient descent search

owing to changing influence patterns of neighbor-contacts over the social network, the formulation of maximum-cut user clusters owing to changing user behaviors thus resulting in changing maximum-cut weights, and the classical traveling salesman problem for the minimum-cost and optimal resource utilization solutions for large-scale marine transportation network routing problems for the shipment of goods.

In contrast to mono-objective optimization problems, multi-objective optimization problems, if modeled with one equation, can introduce a bias in the modeling phase, hence multiple tasks need to be solved jointly, sharing an inductive bias. Instead of arriving at a consistently unique solution in the case of a mono-objective optimization problem, we obtain a set of solutions obtained in multi-objective optimization problems, also called Pareto solutions or the trade-off surface. The latter spectrum of optimization problems has been addressed to date by following two theoretical principles, as follows.

Definition A multi-objective optimization problem (MOOP) can be defined as the search for the best set of solutions $X = (x_1, x_2, \ldots\ldots, x_n)^T$ (where n is the number of optimization parameters) that minimizes a number of M conflicting (or not) objective functions:

$$\frac{\text{Minimize}}{\text{Maximize}} = f_m(\bar{X})\,\bar{m} = 1, \ldots, M. \tag{13.1}$$

In the multi-attribute utility theory (MAUT) approach, optimization in complex engineering decision making can be brought about by integrating the risks and uncertainties involved with two or more relatively important multiple attributes, assuming that these multiple attributes are considered in choosing the decision alternatives and, second, the notion that one attribute is 'relatively more important' than another. This optimization step is executed by maximizing the utility function defined on these attributes or by making tradeoffs depending on the current attribute levels. Consider a decision problem such as selection of a job, choice of an automobile or resource allocation, where the choice of the decision alternatives shares a common feature; decision alternatives are affected by the relative importance of multiple attributes of interest arising from the input feature vector of the problem. Applications of MAUT in the public sector are numerous since most public sector problems involve multiple conflicting objectives, such as in public healthcare systems, environmental policy, regulatory issues, site selection, and energy and public policy.

The multi-criteria decision-making theory (MCDM) approach refers to simulating optimized decision making in the presence of multiple, usually conflicting, situations and constraints. MCDM problems are present in almost all decision-making situations ranging from common household decisions to complex strategic and policy level decisions in corporations and government undertakings. An ML optimization has been successfully implemented on Bayesian learning using utility functions by Thomas [4], who mapped a subjective-answer evaluation framework to multi-objective decision making incorporating the concept of utility functions as one of the candidate answer evaluation methodologies. The evaluation method using utility functions was extended to fuzzy-Bayesian decision making of answer evaluation categories using fuzzy states for no information, probabilistic imperfect, probabilistic perfect, fuzzy information and fuzzy perfect information. The reader is referred to fuzzy-Bayesian learning experiments on the subjective question–answer evaluation framework [4].

Despite the existence of such nice academic principles of optimization, real complex industrial domains still face difficulties in performing real-time optimization. One such application case study is highlighted—industrial test-cases of material processing numerical simulation which allow a strategy of trial and error to improve virtual processes without incurring material costs or interrupting production and therefore save a lot of money. In the case study, in a French founded ANR project, ten industrial partners have been selected to cover the different areas of the mechanical forging industry and provide different examples of the simulation tools. It aims to demonstrate that it is possible to obtain industrially relevant results on a very large range of applications within a few tens of simulations and without any specific automatic optimization technique knowledge, although the simulations required large computational times and also user time to analyze the results, adjust the operating conditions and restart the simulation. Therefore, the simulations often have to be interrupted before the optimum is reached due to a lack of time.

This called for handling the large computational time with a meta-modeling approach. An evolutionary algorithm coupled with meta-modeling allowed interpolating the objective function on the entire parameter space by only knowing the exact function values at a reduced number of 'master points'. Two algorithms are used: an evolution strategy combined with a Kriging metamodel and a genetic algorithm combined with a meshless finite difference method. The latter approach is extended to multi-objective optimization. The population-based approach, aiming to find the set of solutions that allowed the best possible compromises between the different conflicting objectives, was found to be highly computationally efficient with the use of the parallel capabilities of the new multi-core hardware in the Forge 2009 IHM computer being trained over all the defined examples and computing several simulations at reduced time complexities. The reader is referred to [5] for reading on the presented examples of mono-objective optimization of forging the billet shapes of a common rail as well as the cogging of a bar, and multi-objective optimization on the two design parameters in a wire drawing industry process.

13.2 Optimization problems in machine learning

The ML community has produced a tremendous number of successful applications in recent years, ranging from image recognition to autonomous driving and outplaying humans in complex games. This spree of newsworthy results renewed the interest in combining the perspectives of AI and operations research. The state-of-the-art ML algorithms are gradually being augmented by better optimal solutions, as advancing businesses and technology are being impacted by the new paradigm shift of data science and analytics. This is due to the fact that the fundamental natural and human resources still remain finite along with the legal and ethical bounds on real-world complex problem-solving frontiers. Many decades of early works saw conventional AI and ML algorithms solve real-world optimization problems using both OR and ML analysis. These works were the consequence of the need to arrive at the most suitable option from a family of ML models that could perform fairly well according to some minimum estimate of the generalization error based on the given training data. This search typically involves some combination of data pre-processing, optimization and heuristics; these are still constrained by three popular sources of error: error distribution introduced by learning bias, noise in datasets and the difficulty of the search problem that underlies the given modeling problem. These standard optimization packages suffered from efficiency and scalability issues, owing to the large-scale ML model designs. This in turn led to the need to design suitable optimization methods using ML heuristics. This is rightly described as 'the interplay of optimization and machine learning' [6].

Optimization lies at the heart of ML. Most ML problems reduce to optimization problems. The readers of this book must have sound knowledge of operations research, mathematical programming and baseline ML algorithms. The chapter does not intend to deal with the mathematical foundations of optimization types, statistical methods, meta-heuristics nor does it drill down into optimization classification hierarchy but attempts to explore the depth to which optimization has invaded different ML tasks based on assimilated exhaustive works performed by ML analysts. The aim of this chapter is to explore the increasing degree of coupling observed between available optimization techniques and ML models with the help of case studies in diverse domains. The authors seek to examine the cross-boundaries of the following types of works in the literature. First, how small changes to existing ML models which use methods such as multi-kernel, ranking, clustering, structured learning and similar have adapted the optimization tasks into newer dimensions. The second category of works explores how extended sets of optimization methods have enhanced the scalability and efficiency of ML models, which have also promoted the use of newer exploratory domains. We have already introduced chronologically the classification hierarchy of optimization methods as applied in diversified domains of ML tasks. The rest of the chapter is organized as follows. Section 13.3 deals with popular optimization methods used for ML in supervised learning. Section 13.4 explains some case studies revealing how optimal solutions can be used to address the most challenging task of feature selection in ML.

13.3 Optimization in supervised learning

Supervised learning is the process of data mining for deducing rules from training datasets. A broad array of supervised learning algorithms exists, every one of them with its own advantages and drawbacks. There are some basic issues that affect the accuracy of the classifier while solving a supervised learning problem, such as bias–variance trade-off, dimensionality of the input space and noise in the input data space. All these problems affect the accuracy of the classifier and are the reason that there is no global optimal method for classification. The classification results are also altered by the noise in the data, that is, redundant records, incorrect records, missing records, outliers and so forth. All these problems affect the accuracy of a classifier.

Definition The phenomenon of the dimensionality curse arises due to a large number of attributes acting as feature vectors in a dataset, wherein, even if the decision depends on a subset of this high dimensional feature vector, the performance of the classifier will be clouded by high variance due to the high dimensionality of the dataset.

In contrast to single-task learning, multi-task learning can be perceived as a learning paradigm in which data from multiple tasks are used with the hope of obtaining superior performance over learning each task independently. In multi-task learning, multiple tasks are solved jointly, sharing the inductive bias between them. Multi-task learning is inherently a multi-objective problem because different tasks may conflict, necessitating a trade-off. The potential advantages of MTL can be realized over even seemingly unrelated real-world tasks, albeit they share strong dependencies due to the shared processes, for example autonomous driving and object manipulation are seemingly unrelated, but the underlying data are governed by the same laws of optics, material properties and dynamics. This motivates the use of multiple tasks as an inductive bias in learning systems. Other multi-task learning problems that have already been addressed are digit classification, scene understanding (joint semantic segmentation, instance segmentation and depth estimation), and multi-label classification [7].

A typical MTL system is given a collection of input points and sets of targets for various tasks per point. A common way to set up the inductive bias across tasks is to design a parameterized hypothesis class that shares some parameters across tasks. Typically, these parameters are learned by solving an optimization problem that minimizes a weighted sum of the empirical risk for each task. However, the linear-combination formulation is only sensible when there is a parameter set that is effective across all tasks. In other words, minimization of a weighted sum of empirical risk is only valid if tasks are not competing, which is rarely the case. MTL with conflicting objectives requires modeling of the trade-off between tasks, which is beyond what a linear combination achieves, hence this forms the most suitable case study in illustrating how optimization works in such a learning paradigm.

13.3.1 Bayesian optimization

The use of the naïve Bayes' theorem (also called the maximum *a posteriori* hypothesis) as part of probability statistics provides a quantitative approach to learning algorithms by manipulating over probabilities in order to weigh the evidence supporting the hypothesis. In a simple sense, it can accommodate the classifying of new data instances (prior knowledge is given for observed data instances) by combining the predictions of multiple hypotheses, weighted by their probabilities. Even if the Bayesian learning algorithm does not explicitly manipulate probabilities, it can help in identifying the model characteristics under which the learning behaves optimally [8]. Such Bayesian optimality can be observed in the learning of problems with continuous-valued domains, where Bayesian analysis is used for minimizing the squared error between the output hypothesis predictions exhibiting maximum likelihood estimates. The reader is referred to the basic concepts of probability densities over normal distributions for understanding Bayesian classifiers in [8].

13.3.2 Bayesian optimization for weight computation: a case study

We examine a case study that describes a problem setting where the maximum likelihood hypothesis is used for predicting probabilities that are learnt as weights for training a fuzzy weighted associative classifier on a collection of patient heart-disease data taken from the heart.D53.N303.C5.num database [9]. Initially, a criterion is defined upon which the optimization step is performed using the maximum likelihood hypothesis upon any learnt target function, say f' in the problem setting; f': $X \rightarrow [0, 1]$, such that $f' = P(f(x) = 1)$. Training data D is extracted in the form $D = \{\langle x_1, d_1\rangle, \langle x_2, d_2\rangle, \ldots, \langle x_m, d_m\rangle\}$, where d_i are the observed occurrences of the disease, with a 0 or 1 value for each patient instance, x_i. With the assumption that the probability of encountering a patient x_i is independent of the hypothesis (h) about his/her chances of being a victim of disease occurrence, we attempt to compute the maximal probability $P(d_i|h, x_1)$ of observing $d_i = 1$ for a single instance x_i, given a world in which hypothesis h holds:

$$P(d_i/h, x_i) = \begin{cases} h(x_i) & \text{if } d_i = 1 \\ (1 - h(x_i)) & \text{if } d_i = 0 \end{cases} \tag{13.2}$$

$$P(d_i/h, x_i) = h(x_i)^{d_i}(1 - h(x_i))^{1-d_i}. \tag{13.3}$$

Thus, treating the x_i and d_i variables of a binomial distribution of the given training data samples, we compute the likelihood ratios by iterating over all probability values around the initially assigned probability value of disease occurrence as observed from the training samples and obtain the maximum of all $h(.)$ values as a maximum likelihood estimate. The expression for MLE is:

$$h_{\mathrm{ML}} = \underset{h \in H}{\operatorname{argmax}} \prod_{i=1}^{m} h(x_i)^{d_i}(1 - h(x_i))^{1-d_i}. \tag{13.4}$$

Table 13.2 provides the binomial distribution of 11 patient instances in age group 1 and 17 patient instances under age group 2, whose weights need to be computed using Bayesian optimization over the maximum likelihood hypothesis. The likelihood estimate computations, $P(d_i|\ h, x_i)$ can be seen in table 13.3 for a set of different iterating probabilities, $h(.)$. The maximum of all $h(.)$ values is declared as the MLE and is assigned as the weight for learning the fuzzy weighted associative classifier.

Table 13.2. Patient dataset (training sample).

Patient_ID	age_group	$P(.)$	Patient_ID	age_group	$P(.)$
1	I	0	4	II	0
2	I	0	5	II	0
3	I	0	6	II	0
4	I	0	7	II	0
5	I	0	8	II	0
6	I	0	9	II	0
7	I	0	10	II	1
8	I	0	11	II	1
9	I	1	12	II	1
10	I	1	13	II	1
11	I	1	14	II	0
1	II	0	15	II	1
2	II	0	16	II	1
3	II	0	17	II	1

Table 13.3. Bayesian optimization over maximum likelihood hypothesis.

Patient_age_group_type	Prior probability ($h(x_i)$)	$(1 - h(x_i))$	Training sample size (m)	$h(.)$ (iterating probabilities)	hML (MLE)
I	3/11	8/11	11	{1/11, 2/11, 3/11, 4/11}	0.2727
II	7/17	10/17	17	(5/17, 6/17, 7/17, 8/17, 9/17}	0.412

13.3.3 Bayesian optimal classification: a case study

Organizational dynamics falling in line with employee career dynamics, the selection of candidates to be placed in a certain position by any organization, requires careful planning. This case study deals with managerial research conducted on a quantitative scale that aims to formulate a decision-making system that assigns a suitable designation alternative to the participating candidates on the basis of quantitative assessments [10]. The human resources (HR) section of company X is responsible for taking the initiative in selecting candidates in accordance with the assessment of their superiors by collecting data on a five-point rating using a questionnaire method applied over the candidate's behavioral data and interviews. Three managers were nominated to assess the 11 candidates for subordinates and fill in the questionnaire comprising twelve indicators, as can be seen in table 13.4.

Table 13.4. Bayesian optimization over the maximum likelihood hypothesis. Adapted as illustration from [10] © 2018 IOP Publishing. Reproduced with permission.

Indicator	5	4	3	2	1
Work quality	Always exceeds the set standards.	Sometimes exceeds the set standards.	Always meets the set standards.	Sometimes does not meet the set standards.	Always achieves less than the set standards.
Work quantity	The quantity, volume, frequency or speed of completion of work always exceeds the set standards.	The quantity, volume, frequency or speed of completion of work sometimes exceeds the set standards.	The quantity, volume, frequency or speed of completion of work always meets the set standards.	The quantity, volume, frequency or speed of completion of work is sometimes insufficient to meet the set standards.	The quantity, volume, frequency or speed of completion of work is always less than the set standards.
Knowledge and work skills	Very good at mastering and understanding the knowledge and skills for the range of duties and responsibilities.	Good at mastering and understanding the knowledge and skills for the range of duties and responsibilities.	Masters and understands the knowledge and skills for the range of duties and responsibilities.	A lack of mastering and understanding the knowledge and skills for the range of duties and responsibilities.	Very lacking in mastering and understanding the knowledge and skills for the range of duties and responsibilities.
Initiative	Very fast, precise and correct in acting to carry out and complete duties without waiting for orders and instructions.	Very fast, precise and correct in acting to carry out and complete duties, and sometimes waits for instructions that are general in nature.	Very fast, precise and correct in acting to carry out and complete duties, and sometimes waits for relatively detailed instructions.	Slow in acting to carry out and complete duties, and always waits for detailed instructions.	Slow in acting to carry out and complete duties, and always waits for very detailed instructions.
Cooperation	Active in helping and supporting all co-workers, gives very serious attention to the achievement of group/company goals.	Active in helping and supporting co-workers in certain sections/groups only, gives serious attention to the achievement of group/company goals.	Needs to be reminded to help and support co-workers, needs to pay serious attention to the achievement of group/company goals.	Not enthusiastic in helping or supporting colleagues despite being reminded/reprimanded, little attention to the achievement of group/company goals.	Provides no help or support for colleagues and/or always becomes an obstacle in achieving group/company goals.

Reliability	Always willing to be assigned outside duties and responsibilities and results are as expected.	Always willing to be assigned outside duties and responsibilities, results are relatively close to what is expected.	Always willing to be assigned outside duties and responsibilities, results are less than expected.	Always willing to be assigned outside duties and responsibilities, results are rarely as expected.	Always unwilling to be assigned outside duties and responsibilities, results are rarely as expected.
Learning ability and willingness	A great passion to learn for self-development and improvement of work ability, earnestly applies what has been learned in duties.	A good spirit to learn for self-development and improvement of work ability, seriously wants to apply what has been learned in duties.	Sufficient interest to learn for self-development and improvement of work ability, but needs to request support from superiors to apply what has been learned in duties.	Less enthusiasm for learning for self-improvement and improvement in work ability, and/or needs to be controlled /reminded to always apply what has been learned in duties.	Little interest in learning for self-development and improvement of work ability, although has been reminded.
Attendance	Never late, never leaves early, no work lost.	Performance appraisal is affected by early logout and/or 1 day lost work.	Performance appraisal is affected by 3–5 late arrivals and early departures, and/or 2–3 days lost work.	Performance appraisal is affected by late logins and/or 4–5 days lost work.	Performance appraisal is affected by habitual late logins and/or more than 5 days lost work.
Planning and organizing	Excellent ability in setting task priorities and placing and managing existing resources.	Good ability in setting task priorities and placing and managing existing resources.	Fairly good ability in setting task priorities and placing and managing existing resources.	Less able to set task priorities and less effective in placing and managing existing resources.	Not able to assign task priorities and manage existing resources.
Controlling	Good at monitoring and controlling all resources very optimally.	Good at monitoring and controlling all resources optimally enough.	Sometimes needs guidance to monitor and control all resources, but less than optimally.	Always needs guidance to monitor and control all resources, but less than optimally.	Always needs a reminder to simply monitor and control resources.

(*Continued*)

Table 13.4. (*Continued*)

Indicator	5	4	3	2	1
Decision making	Makes a decision on a problem very quickly and accurately, and the results can be justified.	Makes a decision on a problem quickly and accurately, and the results can be justified.	Capable of making a decision on a problem quickly and accurately, although often needs support and direction.	Sometimes needs a great deal of support and direction to make decisions quickly and accurately, and/or decisions are sometimes less justified.	Always slow in making decisions, even if they have been given guidance, and their directives or decisions are not justifiable.
Development of subordinates	Very attentive in the improvement and development of subordinates and/or has a planned program for staff.	Attentive in the improvement and development of subordinates, but lacks a planned program for staff.	Sufficient attention on the improvement and development of subordinates, but does not have a planned program for staff.	Lack of attention on the improvement and development of subordinates, and does not have a planned program for staff.	No attention on the improvement and development of subordinates, and does not have a planned program for staff.

Table 13.5. The ranked criteria for candidate assessment. Adapted as illustration from [10] © 2018 IOP Publishing. Reproduced with permission. All rights reserved.

Criteria of scale	Description
Criterion 1	Has a good ability to understand, a high morale and a satisfactory quality of work.
Criterion 2	The results obtained are very effective with a near perfect level of accuracy.
Criterion 3	Self-organization and subordinates are optimal, and is responsible for all the work done.
Criterion 4	Has a high level of creativity and initiative in working with the task of applying the knowledge mastered.
Criterion 5	Has the courage to make decisions and dares to be accountable for all the risks that may occur.

These indicators are: (a) work quality, (b) work quantity, (c) knowledge and work skills, (d) initiative, (e) cooperation, (f) reliability, (g) learning ability and willingness, (h) attendance, (i) planning and organizing, (j) controlling, (k) decision making and (l) development of subordinates, each expressed with five-point rated descriptive options that can closely measure the participating candidate. The indicators and their rating scale of responses is pre-defined by the HR department of company X after analyzing several factors that are sure to influence the decision-making process, including the past experience of the candidate, cognitive differences, age and individual differences, belief in personal relevance and lastly their escalation of commitment. Another constraint imposed by the HR department is the fulfillment of the criteria in selecting candidates according to the needs of the position.

The criteria are defined in ascending order of preference for recruiting candidates for top managerial positions of the company, as can be seen in table 13.5, as a consequence of previous review; this study is internal and has been recognized by the management of the company.

The understanding of Bayes' optimal classification in the case study refers to the fulfillment of the recommendation criteria by any contesting candidate (K_j) for whom the probability $P(K_j|D)$ is maximum. This most probable classification of instance k_i can be computed from an expression of Bayesian optimal classification:

$$\underset{v_j \in V}{\text{argmax}} \sum_{h_{ij} \in H} P(v_j|h_i)P(h_i|D). \tag{13.5}$$

It should be noted that in the case study experiments the set of five criteria is defined as a set of five hypotheses h_i in the whole hypothesis space H. While the candidates were being personally judged, the managers felt that some of the decision-making indicators were relatively more important and occupied the top priority in lieu of the fulfillment of the specific criteria that were defined in table 13.5. Hence, distinct sets of preferred indicators needed to be screened while interviewing candidates under distinct criteria, in order to decide whom to recommend for the related vacant positions. This causes the formulation of a criteria-specific feature-vector mapping table that could be drawn from the response value patterns that were captured by the managers during the candidates' interview procedures, as can be seen in table 13.6.

Eventually, each of the candidates were assessed with above average values of responses in a set of not more than 6–7 indicators, as listed in table 13.7. The rest of the responses are assumed to be '0' or '—' (do not care) for obtaining the resulting table of naïve Bayesian probability computations for the contesting candidates under each criterion of assessment.

The posterior probabilities of each candidate being recommended into each criterion needs the computation of separate hypothesis combinations, governed by the distinct set of feature vectors (acting as evidential support) mentioned above. For example, the posterior probability of candidate K_j, $P(K_j \mid D)$ can be computed by the expression that follows from equation (13.5) over all five hypothesis (criterion) combinations.

Each $P(K_j \mid h_i)$ is the conditional probability of candidate K_j being recommended in criterion (hypothesis combination) h_i in our case study; $P(h_i \mid D)$ is the frequency of occurrence of the relative sample of criterion h_i assessed over each of the 12 indicators. The latter component can be visualized as the consequence of training

Table 13.6. The criteria-specific feature-vector mapping matrix.

Criterion	a	b	c	d	e	f	g	h	i	j	k	l
h_1	X	X	X				X	X	X			
h_2	X	X	X			X		X	X		X	
h_3	X		X	X	X			X	X		X	X
h_4	X		X	X	X		X	X		X	X	X
h_5		X		X	X	X	X	X		X	X	

Table 13.7. Naïve Bayesian computations for candidates K1–K11.

Candidate	Criterion	X1	X2	X3	X4	X5	X6	X7	X8	X9	X10	X11	X12	p(X|C)
K1	1	4	3	3	—	—	—	5	5	3	—	—	—	0.7138
K2	4	4	—	5	5	—	—	3	5	—	—	3	3	0.7188
K3	2	5	5	—	—	—	4	—	4	3	—	3	—	0.7119
K4	3	3	—	—	5	—	—	—	5	5	—	4	5	0.6853
K5	5	—	3	—	5	3	—	3	4	—	3	5	—	0.7875
K6	4	3	—	5	5	3	—	5	5	—	3	—	—	0.7478
K7	1	4	4	3	—	—	—	4	5	3	—	—	—	0.7069
K8	3	3	—	3	4	5	—	—	4	5	—	4	5	0.816
K9	5	—	3	—	4	—	3	3	4	—	—	5	—	0.6958
K10	2	5	4	3	—	—	4	—	4	—	—	3	—	0.678

the recruitment domain with past employee promotion histories and is available in table 13.8;

$$P_{h_5}(K_5 \mid D) = \sum P(K_5 \mid h_5).\ P(h_5 \mid D) \tag{13.6}$$

$$P^*(K_5 \mid D) = \max[P(K_5 \mid h_5).\ P(h_5 \mid D),\ P(K_5 \mid h_4).\ P(h_4 \mid D),\ P(K_5 \mid h_3).\ P(h_3 \mid D),\ P(K_5 \mid h_2).\ P(h_2 \mid D),\ P(K_5 \mid h_1).\ P(h_1 \mid D)]. \tag{13.7}$$

It can be observed that the candidates were assessed under five different criteria using the relevant set of decision-making indicators, and their posterior probabilities of falling in those criteria were computed using the conditional probabilities of table 13.8 relevant to those criteria. These observations are listed in table 13.9 along with the recommended criterion fulfilled by each candidate K_j.

The candidate K5, who is ranked one under criterion 5, was recommended for the top managerial position for decision making in the company's management, while candidate K6, who satisfies criterion 4 with maximum marginal probability (0.7478)

Table 13.8. Naïve Bayesian conditional probability distribution over 12 indicators, five criteria. Adapted as illustration from [10] © 2018 IOP Publishing. Reproduced with permission. All rights reserved.

Criterion		Conditional probability distribution											
		a	b	c	d	e	f	g	h	i	j	k	l
1	1/5	9/58	8/58	7/58	1/58	1/58	1/58	10/58	11/58	7/58	1/58	1/58	1/58
2	1/5	11/59	10/59	4/59	1/59	1/59	9/59	1/59	9/59	4/59	1/59	7/59	1/59
3	1/5	7/75	1/75	4/75	8/75	11/75	1/75	1/75	10/75	11/75	1/75	9/75	11/75
4	1/5	8/69	1/69	11/69	11/69	4/69	1/69	9/69	11/69	1/69	4/69	4/69	4/69
5	1/5	7/48	1/48	1/48	10/48	4/48	4/348	7/48	9/48	1/48	4/48	11/48	1/48

Table 13.9. Naïve Bayesian posterior probability computations for candidates K1–K11.

Candidate	Criterion	$P_{h5}(k_i \mid D)$	$P_{h4}(k_i \mid D)$	$P_{h3}(k_i \mid D)$	$P_{h2}(k_i \mid D)$	$P_{h1}(k_i \mid D)$	$P^*(k_i \mid D) = \max(.)$
K1	1	1.73	2.39	1.64	1.71	3.57	3.57
K2	4	3.10	3.59	2.64	2.31	2.69	3.59
K3	2	1.88	1.39	1.80	2.67	2.59	2.67
K4	3	2.90	2.46	3.43	2.40	2.02	3.43
K5	5	3.94	2.46	2.11	1.96	1.69	3.94
K6	4	3.21	3.74	2.19	1.48	2.88	3.74
K7	1	1.60	2.26	1.64	1.73	3.53	3.53
K8	3	2.92	2.91	4.08	2.27	2.19	4.08
K9	5	3.48	1.96	1.56	2.21	1.69	3.48
K10	2	1.85	1.87	1.52	2.65	2.45	2.65
K11	5	3.56	2.23	1.89	1.96	2.03	3.56

was recommended as the assistant of the elected head of office. The candidates K8, K3 and K1 satisfying criteria 3, 2 and 1 with maximum *a posteriori* probabilities 0.816, 0.712 and 0.714, respectively, will be placed around the factory with the responsibility of holding machines and necessary job-scheduling tasks. The reader is referred to [10] for the details of the other objectives of the mentioned case study.

13.3.4 Bayesian optimization via binary classification: a case study

When solving an optimization problem, we usually obtain a set of a large number of solutions. What if we learn about the optimum using these solutions? Shylo and Shams [11] worked on an ML approach to formulate a statistical learning problem as an optimal binary classification problem by implementing a logistic regression model on a collection of solutions of empirical datasets. Job scheduling problems and exploring their predictive accuracy computationally in this application domain happens to be the earliest of all case-study domains [11].

A binary optimization problem can be formally expressed as

$$\begin{aligned}&\text{Minimize}\, f(x)\\ &\text{s.t. } x \in X \subset (1, 0),\end{aligned}$$

where the objective function f is an arbitrary function, and $\{0, 1\} \cap x \neq \varnothing$.

A feasible vector x is defined as one possible optimal solution defined in solution space χ with its feature dimensionality denoted by index parameter j iterating on index set $\overline{1, n} \equiv \{1, 2, \ldots\ldots, n\}$. Also, the learning classifier is assumed to partition each solution x using an iterative method; the learning transits from one solution to another after each iteration i, resulting in a set of m such binary optimal solutions, $\{x^1, x^2, \ldots, x^m\} \subseteq \chi$, discovered after m iterations.

Mathematically, each solution x^i can be partitioned into two subsets:

$$x^1 : C_1^1 = \{j|x_j^* = 1\} \text{ and } C_1^0 = \{j|x_j^* = 0\},\quad j \in \overline{1, n} \tag{13.8}$$

$$\vdots$$

$$x^* : C_*^1 = \{j|x_j^* = 1\} \text{ and } C_*^0 = \{j|x_j^* = 0\},\quad j \in \overline{1, n} \tag{13.9}$$

$$\vdots$$

$$x^{\mathrm{m}} : C_m^1 = \{j|x_j^* = 1\} \text{ and } C_m^0 = \{j|x_j^* = 0\},\quad j \in \overline{1, n}. \tag{13.10}$$

For such a sequence of binary optimal solutions, we evaluate the corresponding objective function values. For $j \in \overline{1, n}$, let $I_j(.)$ denote a list of m pairs formed by the jth components of these solutions and the corresponding objective values:

$$I_j(.) = \left[\left(xj_j^1, f(x^1)\right), \left(x_j^2, f(x^2)\right), \ldots\ldots\left(x_j^m, f(x^m)\right)\right], \tag{13.11}$$

where each jth component falls either as an element of binary class 1 or class 0 for any mth solution instance x_m. Hence the two partitions constructed out of $I_j(t)$ can be defined by the two sequences:

$$I_j^1(.) = \{(x_j^k,\ f(x^k)) : (x_j^k,\ f(x^k)) \in I_j(t),\ x_j^k = 1\}, \tag{13.12}$$

$$I_j^0(.) = \{(x_j^k,\ f(x^k)) : (x_j^k,\ f(x^k)) \in I_j(t),\ x_j^k = 0\}. \tag{13.13}$$

In other words, in order to build a prediction model, we need to map n such vectors, $I_j(t), j \in \overline{1, n}$, into the interval [0, 1], estimating the larger of the two conditional probabilities $P(j \in C^1 \mid I_j(t))$ and $P(j \in C^0 \mid I_j(t))$, and classifying the class label with that class, whose probability value is greater. Now, this binary classification problem can be solved with a variety of statistical methods. The case study aims to classify such a problem using the logistic regression method, a widely used statistical technique and, hence, often included as a combinatorial optimization method, whose sigmoidal functional representation is defined as

$$h_\theta(X) = \frac{1}{1 + e^{-(\theta_0 + \theta X)}}. \tag{13.14}$$

Usually, the maximum likelihood estimate (MLE) metric Θ is used to fit the parameters of the above-mentioned dataset $\{x^k, f(x^k)\}$ and can be obtained as

$$h_{\theta(t)}(I_j(t)) = \frac{1}{1 + \exp\left[\sum_{(1, f(x_k)) \in I_j^1(t)} \theta(t) f(x_k) - \sum_{(0, f(x_k)) \in I_j^0(t)} \theta(t) f(x_k)\right]}. \tag{13.15}$$

Now, substituting the sequences of optimal objectives corresponding to the two binary class labels C^1 and C^0 in the denominator component of the above expression, we obtain the revised expression as

$$h_{\theta(t)}\left(D_j^1, (t), , , D_j^0, (t)\right) = \frac{1}{1 + \exp\left(\theta(t)\left[D_j^1(t) - D_j^0(t)\right]\right)}, \tag{13.16}$$

where

$$D_j^0(i) = \min\left(\left\{f\left(x_j^k\right) \mid \left(x_j^k,\ f(x^k)\right) \in I_j(t),\ x_j^k = 1\right\}\right) \tag{13.17}$$

$$D_j^1(i) = \min\left(\left\{f\left(x_j^k\right) \mid \left(x_j^k,\ f(x^k)\right) \in I_j(t),\ x_j^k = 1\right\}\right). \tag{13.18}$$

An example of a training dataset is considered for implementing logistic regression, as shown in table 13.10, where two independent regressor variables $D^1j(i)$ and $D^0j(i)$ are listed from the execution of a baseline search algorithm (the TABU and Guided TABU search techniques are used in the case study) that reports the best objective function $f(x)$ for each solution component. The regression modeling for the component has been

Table 13.10. Training samples (logistic regression model).

$D_j^0(i)$	$D_j^1(i)$	$D^1(t) - D^0(t)$	O(.)	Θ * Δ	$h_{\Theta=0.5}$	Θ * Δ	$h_{\Theta=0.2}$	Θ * Δ	$h_{\Theta=0.15}$	Θ * Δ	$h_{\Theta=0.1}$	Θ * Δ	$h_{\Theta=0.05}$
1395	1366	29	0	14.5	0.00	5.80	0.00	4.35	0.01	2.9	0.05	1.45	0.19
1368	1400	−32	1	−16	1.00	−6.40	1.00	−4.8	0.99	−3.2	0.96	−1.6	0.83
1366	1438	−72	1	−36	1.00	−14.40	1.00	−10.8	1.00	−7.2	1.00	−3.6	0.97
1373	1366	7	0	3.5	0.03	1.40	0.20	1.05	0.26	0.7	0.33	0.35	0.41
1379	1365	14	1	7	0.00	2.80	0.06	2.1	0.11	1.4	0.20	0.7	0.33
1365	1389	−24	1	−12	1.00	−4.80	0.99	−3.6	0.97	−2.4	0.92	−1.2	0.77

computed for various probabilities $P(.) = \{0.5, 0.2, 0.15, 0.1, 0.05\}$ and the model is found to fit the best for the predicted probability (0.5), as can be seen in the calculations performed in the same table.

13.4 Optimization for feature selection

As ML aims to address larger, more complex tasks, such as personalization of filtering systems on the world wide web or electronic mailboxes, Netnews and the like, the problem of screening out the most relevant information from the huge volumes of both necessary and low-quality information is becoming an increasing concern. Revisiting the curse of dimensionality, which increases the computational time exponentially with an increase in the number of features to tens or hundreds in an induction task, a set of either irrelevant or redundant features plays no role in prediction except for serving as noise and increasing the induction time. A very popular exemplary domain that can be used to illustrate the above-mentioned phenomenon is automatic image recognition in more realistic contexts involving noise, changing lighting conditions and shifting viewpoints. Since each feature used as part of a classification procedure can increase the cost and running time of a recognition system, there is strong motivation within the image processing community to design and implement systems with small feature sets. Similarly, it is not uncommon in a text classification task to represent examples using 10^4 to 10^7 attributes, with the expectation that only a small fraction of these will contribute in learning the patterns. In contrast, in an attempt to include a sufficient set of features to achieve high recognition rates by generating effective classification rules, the image learning community seeks an 'optimal' subset of features from a larger set of possible features for various recognition tasks.

In order to remove irrelevant or redundant features, the feature selection relevancy criterion can be defined in one of the following patterns. We adhere to the below mentioned standard notations throughout this section for representing our data and variables:

- The relevance of a feature with an output class label: Given an N-dimensional feature vector, described by an instance space of feature domains, F_1 X F_2 X … X F_n, we can model the sample S as having been extracted from this

instance space and associate them with classification labels 1 ... C (whether Boolean, multiple valued or continuous) according to the target function, c. A feature x_j can be considered as relevant to target concept c, if there exists a pair of samples s and s' ($s, s' \in S$) such that $c(s) \neq c(s')$ and $x_j(s) \neq x_j(s')$ only apply in their feature vectors.
- Extending the concept of relevance with respect to the distribution of samples in instance space, a feature x_j can be considered as strongly relevant to sample set S, if there exists a pair of samples s and s' ($s, s' \in S$) such that $c(s) \neq c(s')$ and $x_j(s) \neq x_j(s')$ only applies in their feature vectors. In other words, a feature x_j is said to be weakly relevant to sample S, where $S = f(c, D)$, and if it is possible to remove half a subset of features so that x_j becomes strongly relevant. Several other works have also concluded that a feature can be regarded as irrelevant, if it is conditionally independent of class labels which means that the feature having no influence on class labels can be discarded [19–22]. Even if the feature is independent of the input data, it cannot be independent of the class label for constructing a suitable learning model.

Discussing the vast spectrum of feature selection algorithms for dealing with datasets that contain large numbers of irrelevant attributes, these have been characterized as either 'wrapper', 'filter' or 'embedded' approaches, based on the relation between the selection scheme and the basic induction algorithm. Irrespective of any of the methods adopted, a convenient paradigm for viewing many of these algorithms is to find a sub-optimal procedure to compute a subset of possible features out of a heuristic search, with each state in the search space specifying a subset of the possible features. Instead of directly evaluating all subsets of (2^N) features in N-dimensional instance space, we adopt one of the following approaches to implement our heuristics:

1. One might begin with nothing or a finite set of attributes representing a starting point in the search space and successively add attributes; this approach is called forward selection.
2. One might start with all attributes and successively remove them on the basis of some heuristics; this approach is known as backward elimination.

This is nicely explained by Blum and Langley through a step-wise constructed state-transition diagram of partially ordered feature search space [12].

Filter methods use variable ranking techniques as the principle criteria for variable selection by ordering; hence, they are used for the pre-processing step, wherein highly ranked features are selected above a threshold using a suitable ranking criterion. They can be applied before classification to filter out the less relevant variables. Two of the well-known filter methods are RELIEF and its variants [13] and FOCUS [14]. While RELIEF computations provide the relevance weighting of each feature with reference to its class label; the FOCUS algorithm iterates on all feature subsets to arrive at a minimal set of features that can provide consistent labeling of training data. The reader is advised to go through these methods before working on optimal feature-extraction methods, i.e. see [13, 14].

The major difference between the two filter methods is that RELIEF computations need an extra heuristic to compute a threshold on selecting the final feature subset. Although effective in removing redundant features, all combinations of highly correlated features in the given feature vector will obtain high relevance weightings in RELIEF, while the FOCUS algorithm is very sensitive to noise or inconsistencies residing in the training dataset. Another less popular category of wrapper methods performs a search through the space of feature subsets using the estimated accuracy obtained after 'wrapping around' the particular selection from that induction (learning) model. These methods are computationally expensive. Moreover, both feature-extraction methods lose their practicality with an exponential rise in the dimensionality of the total feature vector given at the input.

13.4.1 Feature extraction using precedence relations: a case study

Normally, educational data mining (EDM) predictor attributes include students' personal files, demographic data, academic data, learning patterns, behavioral attributes, institutional infrastructure and facilities. Usually, the student's performance evaluation tools do not declare the kind of effort they still need to apply in order to pass their ongoing course of study. Very few prediction models are able to optimize to what extent these academic efforts must be enforced both at the student level and at the teacher level so that they can be predicted into the category of pass students in their forthcoming end-of-semester examinations. One such case study deals with a sequence of experiments carried out by Singh [16], where in the first experiment one takes in four independent experimental parameters, namely attendance (x_1), assignment credit (x_2), internal score (x_3) and subject count (x_4).

In subsequent experiments, the attribute schema was increased by two more features, namely laboratory credit (x_5) and previous year or semester marks (x_6) of the students to formulate a six-attribute feature-extraction (FE)-cum-ranking model. For formulation of a nine-attribute model, three additional parameters were included, that is marks in higher secondary exam (x_7), medium of study (x_8) and student's location of residence (x_9).

The implemented feature-extraction-cum-ranking model was developed with two-phase functionality: in the first phase, a naïve Bayesian (NB) mining model framework was implemented to predict 'at-risk' and 'above-risk' levels of students belonging to ongoing courses. The posterior probability computations on students' fitness and unfitness that determine their 'risk' class labels are shown in the expressions below:

$$P(\text{fit} \mid \{x_1, x_2, x_3, x_4\}) = \frac{\sum_{i=1}^{4} p(\text{fit}).\ p\left(\frac{x_i}{\text{fit}}\right)}{\sum_{i=1}^{4} p(\text{fit}).\ p\left(\frac{x_i}{\text{fit}}\right) + \sum_{i=1}^{4} p(\text{unfit}).\ p\left(\frac{x_i}{\text{unfit}}\right)} \quad (13.19)$$

$$P(\text{unfit} \mid \{x_1, x_2, x_3, x_4\}) = \frac{\sum_{i=1}^{4} p(\text{unfit}).\, p\left(\frac{x_i}{\text{unfit}}\right)}{\sum_{i=1}^{4} p(\text{fit}).\, p\left(\frac{x_i}{\text{fit}}\right) + \sum_{i=1}^{4} p(\text{unfit}).\, p\left(\frac{x_i}{\text{unfit}}\right)}. \quad (13.20)$$

These NB *a posteriori* computations on class labels were further used to arrive at the relative relevance of attributes contributable to success/failure grades in the second phase. Here, the individual portions of the summative numerator components of expressions (4.2) and (4.3) that pertain to the single attribute, reflected the posterior effect of that attribute in evaluating relative attribute fitness/unfitness. The comparisons among such individual portions helped in ranking the attributes of the experimental feature vector. In this way, *attribute precedence relations* were obtained by revisiting these individual numerator components, i.e. the average fitness (average_fit(x_i, t_j)) and average unfitness (average_unfit(x_i, t_j)) of the students owing to the degree of involvement from each attribute.

Kira and Rendell [13] designed the RELIEF algorithm that assigns a relevant weight to each attribute of the feature vector by computing the difference between the selected test instance with reference to the nearest-hit and the nearest-miss training instances. Assuming the training instances are denoted by a p-dimensional feature vector X, the RELIEF algorithm makes use of a p-dimensional Euclidean distance to select 'near-hit' and 'near-miss' instances from the training dataset. If the test instance x_j is predicted as a positive instance then the near-hit instance (x_j) is assigned as $Z+$ and the near-miss instance (x_j) is assigned a Z-value. The reverse happens if the test instance is predicted as a negative instance. In such a situation, the nearest negative training neighbor is assigned $Z+$ and the nearest positive training neighbor, possessing the opposite class value, is assigned a Z-component. These components are computed as part of the preparation for computing weight updates, as described in expression (13.21). This weight update operation was performed on each of the participating attributes in the experimental feature vector. These updated attribute weights act as rank values of the attributes when sorted in increasing order of relevance. The author also appreciates the nearest-neighbor approach to finding the 'nearest-hit' and 'nearest-miss' training instances to compute the weight updates as defined above:

$$\begin{aligned} w_i' = w_i &- \text{diff}(x_i, \text{near-hit-instance}_i)^2 \\ &+ \text{diff}(x_i, \text{near-miss-instance}_i)^2 \end{aligned}. \quad (13.21)$$

Sun and Wu [15], during their in-depth study of feature selection methods, proved that RELIEF is the most successful algorithm that solves a convex optimization problem with a margin based objective function [15]. As it was observed that the RELIEF model could not filter out redundant attributes as well as weakly relevant ones, this case study provided variant logistics to the approach. Attribute precedence relations were thus introduced as an innovative mining metric for personalized counseling of students and the performance of such a hybrid feature-extraction-cum-ranking model is evaluated by extending the experiments with the RELIEF method as the benchmark, i.e. by comparing equivalently generated attribute precedence relations from the RELIEF feature-extraction model. We will not go into the details of the heuristics, or on how the residual weights are initialized to arrive at the weighted feature vector by performing weight update computations using RELIEF logistics. The reader is referred to the feature-vector ranking computations performed by Singh as a part of doctoral work in the educational data mining realm [16].

The performance of the hybrid model is shown in table 13.11 in terms of percentage accuracy for three types of experimental set-ups as already discussed in the beginning of this case study. The feature-extraction model accuracy was found to be the highest in the four-attribute FE model and the lowest in the nine-attribute FE model. The reasons for the above results revealed that when some academic parameters (static attributes) were appended as add-on attributes (making the total number of parameters $p = 6$), the accuracy of the model is significantly decreased. This suggested that the features should be directly co-related with the students' academic features.

Table 13.11. Performance comparisons of attribute precedence relations of fitness.

Experiment_ID	Dataset		Modeling attributes	Accuracy (%)	Accuracy (%)
Model type	Training tuple count	Test tuple count	Proposed FE model	Comparison with RELIEF weights (normalized)	Comparison with RELIEF weights (prior-probabilities)
Four-attribute model	87	**20**	Attendance, assignment credit, internal score, subject count	82%	83%
Six-attribute model	87	20	Attendance, assignment credit, internal score, subject count, laboratory credit, previous year percentage	60%	58.33%
Nine-attribute model	87	20	Attendance, assignment credit, internal score, subject count, laboratory credit, previous year percentage, score in higher secondary, living location, medium of study	39.44%	45%

The above processing steps on feature-vector ranking were still deviated further by computing an optimal set of fitness precedence relations from both the precedence relation sets obtained due to average probabilities of fitness and unfitness. Initially the attribute precedence relation of unfitness is converted into equivalent attribute precedence relations of fitness by simply reversing the increasing order of attributes so that the least unfit attributes becomes the most fit in equivalent fitness precedence and vice versa. These two sets of fitness precedence relations were compared in order to identify a consistent position j, defined either as exact corresponding position j occupied by attribute x_i in both relations or, at the most, occupying either of the adjoining position combinations such as $(j, j+1)$ or $(j-1, j)$. In this way, the attributes find their final positions of optimized fitness precedence owing to the heuristics applied over their consistent valid and conflicting positions.

Having observed increased model accuracies in the latter set of experiments, it could be concluded that the feature-vector ranking becomes more robust for the Bayesian driven hybrid FE approach, if optimized attribute precedence relations of fitness are used instead of conventional attribute precedence relations of fitness (table 13.12).

13.4.2 Feature extraction via ensemble pruning: a case study

Ensemble learning methods are gaining popularity among the ML and data mining communities. Ensemble learning models are widely accepted to make better predictions than individual classifiers, given the same amount of training information, for example the bagging, boosting, arcing and random forest algorithms. The effectiveness of the ensemble methods relies on creating a collection of diverse, yet accurate learning models. Two costs, namely the huge requirements of memory and computational time, are the great disadvantages when implementing a large-scale ensemble learning on real-world datasets that are prone to easily generate an ensemble with thousands of learning models. In addition, the larger the size of an ensemble, the more prone it is to overfitting problems in an attempt to minimize the classification error to zero. Hence, the complexity of such ensemble learning models is eventually corrected by incorporating pruning over ensembles, such as KL-divergence pruning and kappa pruning over an Adaboost model, kappa-error convex hull pruning, back-fitting pruning and many more. All these pruning methods suffered from greedy search criteria, which they adopted in order to optimize the selected subset of dataset feature vectors and thus improve the performance on the lower end.

Unlike previous heuristic approaches, it was found that ensemble pruning looks for a subset of classifiers that has the optimal accuracy–diversity trade-off [17]. Ensemble pruning can be viewed more or less similarly to weight-based ensemble optimization that aims to improve the learning accuracy of the ensemble by tuning the weight on each ensemble member, but without compromising its performance. The individual accuracy and pairwise independence of classifiers in an ensemble is often referred to as the strength and divergence of the ensemble. The generalization error rule of an ensemble states that the error is loosely bounded by $\frac{\bar{\rho}}{s^2}$, where $\bar{\rho}$ is

Table 13.12. Performance comparisons of optimal attribute precedence relations of fitness.

Experiment_ID	Data set		Modeling attributes	Accuracy (%)	Accuracy (%)
Model type	Training tuple count	Test tuple count	Proposed FE model	Comparison with RELIEF weights (normalized)	Comparison with RELIEF weights (prior-probabilities)
Four-attribute model	87	**20**	Attendance, assignment credit, internal score, subject count	83%	84%
Six-attribute model	87	20	Attendance, assignment credit, internal score, subject count, laboratory credit, previous year percentage	63.33%	64.17%
Nine-attribute model	87	20	Attendance, assignment credit, internal score, subject count, laboratory credit, previous year percentage, score in higher secondary, living location, medium of study	39.44%	44.4%

the average correlation between classifiers and s is the overall strength of the classifiers. Thus, the ensemble pruning dealt with in the case study initially formulates the ensemble error function by computing strength and diversity measurements for a classification ensemble followed by minimizing this approximate ensemble error function through a quadratic integer programming formulation.

At first, this ensemble pruning method was tested on twenty-four UCI repository datasets with Adaboost as the ensemble generation technique and was reported to perform favorably over two other metric-based ensemble pruning algorithms: diversity-based pruning and Kappa pruning, picked as the benchmarks. This was followed by testing the same subset selection procedure in the form of a cross-domain classifier-sharing strategy on a publicly available marketing dataset—here it was a catalog marketing dataset from the direct marketing association (DMEF academic dataset three, specialty catalog company, Code 03DMEF)—to select a good subset of classifiers from the entire ensemble for each problem domain. The essence of sharing classifiers is sharing common knowledge among different but closely related problem domains. In that study, classifiers trained from different but closely related problem domains are pooled together and then a subset of them is selected and assigned to each problem domain. The computational results show that the selected subset performs as well as, and sometimes better than, including all elements of the ensemble. The reader is referred to Zhang *et al* [17] for the details of the methodology in both benchmarking and empirical experiments.

13.4.3 Feature-vector ranking metrics

Assimilating the works on feature extraction as perceived by Wu *et al* [18], a few more ranking methods were found that help in the computation of feature relevance levels, and these are described in the set of expressions 13.22–13.28. One of the simplest metrics is Pearson's correlation coefficient, meant for detecting linear correlation ranking between variables and target class labels:

$$R(i) = \frac{\mathrm{cov}(x_{i,} Y)}{\sqrt{\mathrm{var}(X_i) * \mathrm{var}(Y)}}, \tag{13.22}$$

where x_i is the ith variable, Y is the output (class labels), cov() is the covariance and var() is the variance. Correlation ranking can only detect linear dependencies between the variable and target. Another feature selection metric measures the dependence of two variables and is based on information entropy of class label c, which is defined as

$$H(Y) = -\sum\nolimits_y p(y)\log(p(y)), \tag{13.23}$$

while the conditional entropy of class label c, given attribute x_j, can be defined as

$$H(Y|X) = -\sum\nolimits_x \sum\nolimits_y p(x, y)\log(p(y|x)), \tag{13.24}$$

so that the mutual information (MI) can be computed as

$$I(Y, X) = H(Y) - H(Y|X). \tag{13.25}$$

It can be observed that the uncertainty in output class c is reduced, constrained by the inclusion of the attribute x_j. The simple interpretation of the MI metric is that x_j and c will not be correlated if $\mathrm{MI}(x_j, c) = 0$ otherwise there exists some correlation between them. In other words, the greater the MI ($I(.)$) value, the more the feature x_j contributes to classifying the output class label c.

Mutual information is an important concept to be used in embedded methods of feature extraction. We can incorporate both backward elimination and forward selection approaches to resolve the feature ranking based on conditional mutual information computation in successive iterations, which needs to be maximized for obtaining optimal feature selection that takes care of a trade-off between interdependence and discrimination.

Having invaded through insights of feature selection methods based on relevance and feature interaction of samples, another important aspect of the feature-extraction method is to evaluate how well that method helps in resolving the induction (learning) model. For this, we need to understand the baseline objective of any learning model in terms of characterizing the relationship between F and C. For a feature-extraction model defined as a function $f(S, F, C)$, where $S = \{s_1, s_2, \ldots, s_n\}$, $F = \{ F_1 \text{ X } F_2 \text{ X} \ldots \text{ X } F_n\}$, $C = \{ C_1, C_2, \ldots, C_k\}$. Assuming that F_{s1} is the subset of already-selected features, F_{s2} is the subset of unselected features, and $F = F_{s1} \cup F_{s2}$, $F_{s1} \cap F_{s2} = \phi$, any optimal feature subset obtained by selection algorithms should preserve the existing relationship between F and C hidden in the dataset. The most appropriate evaluation metrics that extract the best of the feature subset are classification accuracy and F-score, that also tend to preserve the above relationship between F and C [18].

Definition The classification accuracy can be formulated as the function 'classify (s)', which returns the classification accuracy rate of $s_i \mid s_i \in S$, S is the set of data items to be classified and s_c is the class of the item, s_i. Mathematically, this can be expressed as

$$\text{acc}(s) = \frac{\sum_{i=0}^{|s|} \text{ass}(s_i)}{|s|}, \qquad s_i \in S, \tag{13.26}$$

$$\text{ass}(s) = \begin{cases} 1, & \text{classify}(s) = sc \\ 0, & \text{otherwise} \end{cases}, \tag{13.27}$$

where $|S|$ represents the number of the elements in the collection S, $s_i \in S$.

Definition Another metric of feature discrimination is the F-score. The larger the F-score is, the more this feature is discriminative. Given training vectors X_k if the number of the jth dataset is n_j, then the F-score of the ith feature is defined as

$$F(s_i) = \frac{\sum_{j=1}^{m} (\overline{x_{i,j}} - \bar{x}_i)^2}{\sum_{j=1}^{m} (1/(n_j + 1)) \sum_{k=1}^{n_j} \left(x_{i,j}^{k} - x_{i,j}\right)^2}, \tag{13.28}$$

where $\bar{x}_i$, $\overline{x_{i,j}}$ are the average of the ith feature of the whole dataset and the jth dataset, respectively; $x_{i,j}^{k}$ is the ith feature of the kth instance in the jth dataset and m is the number of datasets, $k = 1, 2, \ldots, m$ and $j = 1, 2, \ldots l$.

References

[1] Rao S S 2009 *Engineering Optimization: Theory and Practice* (New York: Wiley)

[2] Trevisan L 2011 Combinatorial optimization: exact and approximate algorithms *CS261: Optimization and Algorithmic Paradigms Lecture Notes* (Stanford, CA: Stanford University)

[3] Collette Y and Siarry P *Multi-Objective Optimization: Principles and Case Studies* (New York: Springer)

[4] Thomas A, Sharma H R and Sharma S 2014 *A Tool Design of Subjective Question Answering Using Text Mining* (Saarbrücken: Lambert Academic)

[5] Fourment L, Duclocix R, Marie S, Ejday M, Monnereau D, Masse T and Montmitonnet P 2010 Mono and multiobjective optimization techniques applied to a large range of industrial test cases using metamodel assisted evolutionary algorithms *AIP Int. Conf. Proc.* **1252** 833–40

[6] Bennet K P and P-Hernandz E 2006 The interplay of optimization and machine learning research *J. Machine Learn. Res.* **7** 1265–81

[7] Sener O and Koltum V 2018 Multi-task learning as multi objective optimization *Proc. of the 32nd Conf. on Neural Information Processing Systems (NeurIPS) (Montreal, Canada)* pp 1–15

[8] Niculescu R S, Mitchell T M and Rao R B 2006 Bayesian network learning with parameter constraints *J. Machine Learn. Res.* **7** 1357–83

[9] Soni S and Vyas O P 2011 Performance evaluation of weighted associative classifier in health care data mining and building fuzzy weighted associative classifier *International Conference on Parallel Distributed Computing Technologies and Applications Communications in Computer and Information Science* vol 203 (Berlin: Springer), pp 224–37

[10] Kadar J A, Agustono D and Napitupala D 2018 Optimization of candidate selection using naïve Bayes: case study in company X *J. Phys: Conf. Ser.* **954** 012028

[11] Shylo O and Shams H 2018 Boosting Binary Optimization Via Binary Classification: A Case Study of Job Shop Scheduling arXiv: 1808.10813v1

[12] Avrim L and Langley P 1997 *Selection of Relevant Features and Examples in Machine Learning* (Amsterdam: Elsevier), pp 245–71

[13] Kira K and Rendell L 1992 The feature selection problem: traditional methods and a new algorithm *10th National Conference on Artificial Intelligence* (San Francisco: Morgan Kaufmann), pp 128–34

[14] Almuallim H and Dietterich T G 1991 Learning with many irrelevant features *9th National Conference on Artificial Intelligence* (Cambridge, MA: MIT Press), pp 547–52

[15] Sun Y and Wu D 2008 A RELIEF based feature extraction algorithm *Conf. Proc. of the SIAM Int. Conf. on Data Mining (Atlanta, GA)* pp 188–95

[16] Singh M 2017 Prediction of academic performance of students using machine learning techniques *Doctoral thesis* Dr C V Raman University, Kota Bilaspur, Chhattisgarh, India

[17] Zhang Y, Burer S and Street W N 2006 Ensemble pruning via semi-definite programming *J. Mach Learn. Res.* **7** 1315–38

[18] Wu S, Hu Y, Wang W, Feng X and Shu W 2013 Application of global optimization methods for feature selection and machine learning *Math. Prob. Eng.* **2013** 241517

[19] Chandrashekar G and Sahin F 2014 A survey on feature selection methods *Comput. Electr. Eng.* **40** 16–28

[20] Dietterich T G 2000 Ensemble methods in machine learning ed J Kittler and F Roli *MCS 2000, LNCS 1857* (Berlin: Springer), pp 1–15

[21] Koller D and Sahani M 1996 Toward optimal feature selection *Conf. Proc. of Machine Learning* pp 1–14

[22] Guyan I 2003 An Introduction to variable and feature selection *J. Mach. Learn. Res.* **3** 1157–82

IOP Publishing

Modern Optimization Methods for Science, Engineering and Technology

G R Sinha

Chapter 14

Simulation of the formation process of spatial fine structures in environmental safety management systems and optimization of the parameters of dispersive devices

Sergij Vambol, Viola Vambol, Nadeem Ahmad Khan, Kostiantyn Tkachuk, Oksana Tverda and Sirajuddin Ahmed

The topical scientific applied issue of the creation of control systems for ecological safety through the use of dispersive devices is considered. For suppression of the formation processes of toxic substances and limitation of their distribution in the atmosphere during extraction, processing and transportation of bulk materials (which produce dust) and during fire suppression and thermal waste treatment, an analysis of the systems that use spatial fine structures is offered. The results of numerical simulation of these ecologically hazardous technological processes are described. The physical model of controlling such processes is based on injecting a cooling liquid using dispersive devices. Mathematical models of the gas and dispersed phases of spatial fine structures are developed. The mathematical formulation of the conservation laws for viscous gas (steam) is achieved through the Navier–Stokes equations; for drops, it is given as an equation of the balance of forces that affect the drop and equalize the inertia force and the resultant forces of gravity and aerodynamic resistance. For dispersion of the fluid, irrigation systems of the nozzle type, atomizers and centrifugal atomizer have been suggested. The dependence of the ability to create effective fine structures on the characteristics of the technical devices in the context of natural and man-made hazards of different origins is presented. Using the numerical simulation of the formation processes of spatial fine structures, the most efficient modes of supplying liquid to various hazardous factors are defined.

14.1 The use of spatial finely dispersed multiphase structures in ensuring ecological and technogenic safety

By spatial finely dispersed multiphase structures, we understand to mean structures consisting of finely dispersed drops of fluid, air and particles (such as dust particles, chemical components and biodestructors), which are present in working areas and can exist there for a given time. The simplest structure is a finely dispersed water suspension, that is, a stream of sprayed water, which consists of small droplets and is able to partially or completely separate one part of space from another (figure 14.1).

Dispersed water suspensions are widely used in modern stationary systems for ensuring technogenic safety, in fire protection systems, and are also one of the means of protecting people and equipment during emergency response situations. Creating a system of water suspension from neutralizing solutions is the main way to limit the spread and neutralization of the clouds formed when chemically hazardous substances are released into the atmosphere (such as ammonia, chlorine, nitrogen oxides, sulfur dioxide, hydrogen chloride and fluoride, ethylene oxide, phosgene, etc).

To create a finely dispersed water suspension, sprayers of various designs are used: slotted sprayers, spray guns (such as fire-hoses, figure 14.2(a)), bag sprayers, combined installations (mobile dispersed structures) and drencher sprinklers (stationary dispersed structures, figure 14.2(b)).

14.1.1 Analysis of recent research and publications

In the process of loading and unloading loose dusty materials, in the destruction of rocks, etc, a significant amount of finely dispersed (soaring) particles (including nanoparticles) enter the atmospheric air and adversely affect natural components and human health [1–5]. This applies to different countries where mining facilities exist, and the importance of the mining sector is steadily increasing [6, 7]. During the storage and transportation of coal after the grinding operation, the release of coal dust also takes place. To reduce such emissions, various emulsions are used based on synthetic polymeric materials, pulp and paper waste, and refined products and waste [8]. However, this method complicates the further use of the coal.

Emergencies in waste storage facilities [9, 10], at industrial enterprises [11, 12] and other technogenic objects are accompanied by high-intensity energy release and the formation of environmentally hazardous molecular compounds [13–15]. At the same

Figure 14.1. Creation of a finely dispersed water suspension by an atomizer.

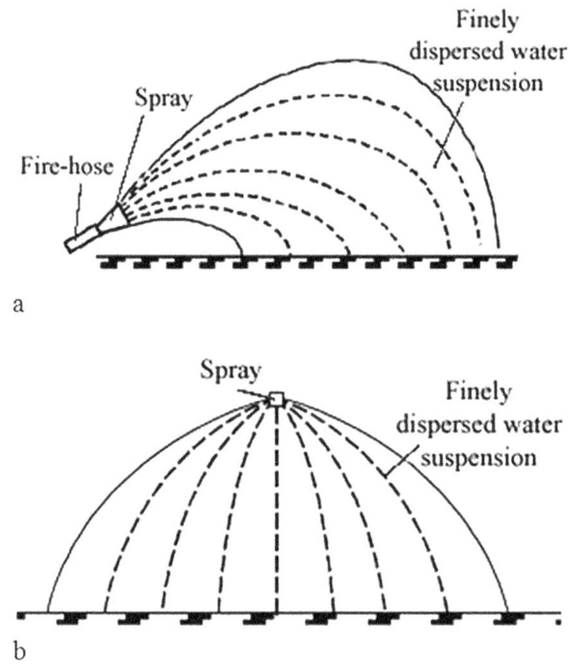

Figure 14.2. Schemes of dispersed structures (finely dispersed water suspension).

time, such emergency situations are characterized by the entry into the atmosphere of a significant amount of carbon monoxide, carbon dioxide, soot, etc, but no special measures are taken to localize (prevent the spread of) these substances [16–18].

The use of thermal methods for the disposal of solid waste products improves the environment by reducing the number of landfills. However, the process of thermal utilization itself is accompanied by harmful emissions into the environment [19, 20]. This process can be ecologically effective in the case of preventing the formation of highly toxic substances (such as dioxins and furans) at the stage of thermo-chemical treatment of waste. The study of the mechanism of the formation of dioxins during heat treatment of waste was given much attention in [21, 22], and in [23] the authors proposed a purification system, scientifically substantiated its effectiveness and confirmed this experimentally. It should be emphasized that it is difficult to implement such a system for economic reasons.

The most modern and promising methods of waste disposal are based on the use of high-temperature treatment [19, 24]. This is due to the fact that under conditions of high temperature (1200 °C and above), dioxins and other highly toxic substances decompose into simple fragments [20]. However, the mechanisms of dioxin re-formation have also been investigated. Re-formation of dioxins during cooling of the high-temperature multicomponent gas is observed if the gas temperature is from 450 °C to 300 °C. The process of their formation is affected by the presence of chlorine and oxygen as well as the rate of gas cooling [20]. This fact gives us reason to believe that the waste recycling process can be more environmentally friendly.

This is possible if we exclude the conditions that contribute to the formation of dioxins after the gas leaves the plasma reactor. Thus, the search for technological solutions that increase the level of ecological safety of solid waste thermal utilization is a very important scientific task. Scientists from different countries have already proposed various methods for cleaning the hot flue gases of various industrial processes. However, as a result of this cleaning, a new problem arises, for example, toxic filtration materials, other hazardous chemical compounds, etc. Since dioxins are not formed at temperatures below 300 °C, the most logical solution in this situation is the instantaneous cooling of the flue gases to a temperature below 300 °C. In [25, 26], the authors considered mathematical models and performed research on the creation of dispersed structures. It should be noted that in solving the problems listed above, specific technical conditions and requirements were considered. Based on the classical equations of fluid dynamics, there currently already exists a general approach for the physical and mathematical formulation of this class of problems.

14.1.2 Statement of the problem and its solution

Creating a finely dispersed water suspension from neutralizing solutions is a good approach to suppress the formation processes and limit the spread of toxic substances and (or) neutralize them. For this purpose sprayers of various designs are used.

At present, theoretically justified methods for calculating the regimes of formation of dispersed water structures are needed. This will allow determining their main parameters through calculations: geometrical dimensions in various conditions, spatial distribution of droplet concentration, interaction with air and heat flows, influence of technical parameters and methods of optimization.

The experience of creating a finely dispersed water suspension shows that there is a need for in-depth study of the delivery processes of dispersed fluids and their further sedimentation to design the necessary spatial structure of the water aerosol and to develop effective design solutions on this basis. The possibility of using sprayers in the development and creation of an ecological friendly process of thermal utilization is also promising.

By virtue of economy and convenience in the modern scientific world, numerical experiments are more often used for a detailed analysis of any complex process. To study the process of creating a dispersed water suspension in order to eliminate sources of environmental hazards, a numerical experiment is also the most convenient tool. This chapter presents the methodology and results of a numerical experiment of the possibility of delivering dispersed water for a required distance and height for dust suppression. At the same time, the effectiveness of dust suppression is investigated depending on the modes of water supply by the atomizer. It also presents the results of numerical modeling to create a dispersed water suspension for cooling the gas to prevent the formation of dioxins, to allow thermal utilization to be an ecologically friendly technological process.

The object of this research is finely dispersed multiphase structures. The subject of the research is the dependence of the parameters of ecologically effective finely dispersed water structures on the technical features of the devices used for their creation.

14.2 Physical and mathematical simulation of the creation process of spatial finely dispersed structures

The most universal theoretical description of the creation process of spatial finely dispersed structures is based on the laws of conservation of mass and momentum in a medium that is not uniform in phase composition, including atmospheric air and water droplets. These equations with given boundary conditions can be jointly solved by modern numerical methods, which are separated into an independent branch of knowledge—computational aero-hydrodynamics. In the mathematical description of the dispersed structure, the following assumptions were made:

- an incompressible, isothermal and turbulent carrier gas phase;
- isotropic turbulence;
- spherical, non-evaporating drops;
- the volume occupied by the drops is neglected.

The interaction of the phases was taken into account as a 'drop—source in a cell' model. In accordance with this model, the presence of particles in the flow manifests itself through an additional source of momentum in Reynolds-averaged Navier–Stokes equations, which are closed by a semi-empirical model of k-ε–type turbulence.

Thus, to fully describe the creation process of spatial fine-dispersed structures, it is necessary to consider models of the gas phase, the dispersed phase and the transition process—the model of interfacial interaction.

14.2.1 Gas phase study and mathematical model description

The mathematical model of a three-dimensional quasistationary turbulent gas flow in a working area is based on the system of Navier–Stokes equations averaged according to Reynolds. The equations of conservation of mass and momentum in the vector record form are [27]

$$\nabla \cdot (\rho\vec{u}) = 0, \tag{14.1}$$

$$\nabla \cdot (\rho\vec{u}\vec{u}) = -\nabla p + \nabla \cdot (\tau) + S_f, \tag{14.2}$$

where ρ is density, $\vec{u}$ is the speed vector, p is static pressure, S_f is the source of momentum due to interfacial interaction, and τ is the tension tensor, determined by the expression

$$\tau = (\mu + \mu_t)(\nabla\vec{u} + \nabla\vec{u}^T),$$

where μ is molecular viscosity and μ_t is turbulent viscosity.

For the closure of the system of equations averaged according to Reynolds (14.1), (14.2), the Lauder–Spalding turbulence k-ε–model was used [28]. The transport equations for the kinetic energy of turbulence k and its dissipation rate ε are

$$\frac{\partial}{\partial x_i}(\rho k u_i) = \frac{\partial}{\partial x_j}\left[\left(\mu + \frac{\mu_t}{\sigma_k}\right)\frac{\partial k}{\partial x_j}\right] + G - \rho\varepsilon, \tag{14.3}$$

$$\frac{\partial}{\partial x_i}(\rho\varepsilon u_i) = \frac{\partial}{\partial x_j}\left[\left(\mu + \frac{\mu_t}{\sigma_\varepsilon}\right)\frac{\partial \varepsilon}{\partial x_j}\right] + C_{1\varepsilon}\frac{\varepsilon}{k}G - C_{2\varepsilon}\rho\frac{\varepsilon^2}{k}, \tag{14.4}$$

where ρ is density, k is the kinetic energy of turbulence, u_i is the the projection of the averaged gas velocity on the axis of a three-dimensional rectangular Cartesian coordinate system, x_j are the coordinates of a three-dimensional rectangular Cartesian coordinate system, ε is the turbulence kinetic energy dissipation rate and σ_k, σ_ε, $C_{1\varepsilon}$, $C_{2\varepsilon}$ are empirical coefficients. G is the term characterizing the emergence of the kinetic energy of gustiness due to shear stresses and is defined by the formula

$$G = -\rho\overline{u_i'u_j'}\frac{\partial u_j}{\partial x_i}.$$

Turbulent viscosity is determined by the Kolmogorov–Prandtl formula [29]

$$\mu_t = C_\mu\rho\frac{k^2}{\varepsilon}, \tag{14.5}$$

where C_μ is the empirical coefficient.

To determine S_f in equation (14.2) an interfacial interaction model was used.

The border conditions

The system of partial differential equations (PDEs) (14.1)–(14.4) was complemented by the boundary conditions for independent variables.

At the boundaries of the computational domain, the following boundary conditions for the continuous phase were set:

- At the outlet of the atomizer nozzles: air speed and wind speed corresponding to the atmosphere.
- At the remaining boundaries: wind speed corresponding to the atmosphere, and static pressure.
- On solid surfaces: adhesion condition, approximated by the wall function.

Direct application of the adhesion condition is necessary—modification of the turbulence model in the near-wall region, where the turbulent viscosity is close to molecular and significant refinement of the computational grid near the wall. The experience of numerical simulation of three-dimensional flows shows that the complexity of the geometric form of the computational domain often leads to the fact that it is the required dimension of the computational grid, which becomes the critical parameter, which determines the possibility of conducting numerical experiments. Therefore, instead of the sticking condition, the wall functions were used to describe the turbulent boundary layer—a system of semi-empirical functions for independent variables in the center of the near-wall computational cell (point P) and variables on the wall that correspond to these values. Moreover, these functions are based on the assumption of Lunder and Spalding [30, 31]. The law of the wall for the averaged velocity is

$$U^* = \begin{cases} y^* & \text{when } y^* \leqslant 11.225, \\ \dfrac{1}{K}\ln(Ey^*) & \text{when } y^* > 11.225, \end{cases} \tag{14.6}$$

where K is Karman's constant, E is the empirical constant, U^* and y^* are dimensionless parameters, which are determined by the formulas

$$U^* = \frac{U_P C_\mu^{1/4} k_P^{1/2}}{\tau_w / \rho}, \tag{14.7}$$

$$y^* = \frac{\rho C_\mu^{1/4} k_P^{1/2} y_P}{\mu}, \tag{14.8}$$

where U_P is the average velocity of gas at a point P, k_P is the kinetic energy of gustiness at point P, τ_w is friction tension on the wall, ρ is gas density, y_P is the distance between point P and the wall, and μ is dynamic viscosity.

Next, it is necessary to solve equation (14.4) in the near-wall cells and the entire computational domain. At the same time the boundary condition on the wall was set for k, which is written as

$$\frac{\partial k}{\partial n} = 0, \tag{14.9}$$

where n is the local coordinate normal to the wall.

The turbulence kinetic energy generation G and its dissipation rate ε, which are included in the source term of equation (14.4), are calculated in the near-wall cells based on the local equipoise hypothesis. With this assumption we believe that the emergence of turbulence kinetic energy and its dissipation rate in the near-wall control volume are the same. Then equation (14.5) for ε in the near-wall cells is not solved; instead, the turbulence kinetic energy dissipation rate is determined by the formula

$$\varepsilon_P = \frac{C_\mu^{3/4} k_P^{3/2}}{\kappa y_P}, \tag{14.10}$$

where κ is the empirical constant.

14.2.2 Dispersed phase study and mathematical model description

Consider the movement of a two-phase flow. One phase is air and the second phase is polydisperse water aerosol. Since the volume concentration of water droplets in the flow is small, we can consider the motion of noninteracting droplets of various sizes separately. Regarding the dispersed phase, we accept the following basic assumptions:

- The dispersed phase is a completely atomized fluid (water) spray, consisting of a finite set of evaporating spherical droplets of various diameters.
- All forces acting on the drop, except for the forces of gravity and aerodynamic drag, are neglected.
- The processes of secondary crushing and coagulation of droplets are neglected.

Under the above assumptions, the behavior of the dispersed phase (water droplets) is conveniently considered in the Lagrangian description. For sprayed fluids, the Rosin–Rammler expression is the generally accepted droplet size distribution [32]. The entire range of initial sizes of droplets was divided into detached intervals; each of them is represented by the average initial diameter for which the trajectory calculation is performed. In addition, each simulated drop is actually a 'package' of drops with identical trajectories. The Rosin–Rammler equation describes the distribution of droplets in size, and the droplets' mass fraction with a diameter greater than d is described by the formula

$$Y_d = e^{-(d/\bar{d})^n}, \tag{14.11}$$

where $\bar{d}$ is the average median diameter of droplets in the spray and n is a distribution parameter.

The motion of a particle of the dispersed phase (droplets) is described by Newton's second law. Under the above assumptions, the equation of motion for a single drop in vector form is

$$m_p \frac{d\vec{u}_p}{dt} = \vec{G} + \vec{F}_R, \tag{14.12}$$

where m_p is drop mass, $\vec{u}_p$ is drop rate, t is time and $\vec{G}$ is gravity, determined by

$$\vec{G} = m_p \vec{g}, \tag{14.13}$$

where $\vec{g}$ is the acceleration of gravity and $\vec{F}_R$ is the aerodynamic drag force determined by the expression

$$\vec{F}_R = 0,5 C_R A_p \rho |\vec{u} - \vec{u}_p| (\vec{u} - \vec{u}_p), \tag{14.14}$$

where C_R is the coefficient aerodynamic drag resistance, A_p is the middle sectional area of the drop, ρ is gas density and $\vec{u}$ is gas velocity.

Substituting the expressions (14.13) and (14.14) into equation (14.12) and taking into account that the mass of a spherical drop and the area of its average cross section are determined by the formulas

$$m_p = \rho_p \frac{\pi d_p^3}{6}, \tag{14.15}$$

$$A_p = 0.25\pi d_p^2, \tag{14.16}$$

and having projected the vectors of both parts of equation (14.12) on the axis of the fixed Cartesian coordinate system, we will have the motion equation system of the drop in the form

$$\frac{du_{pj}}{dt} = \rho_p \frac{\pi d_p^3}{6} g_j - \frac{3\rho C_R}{4\rho_p d_p}(u_{pj} - u_j)\left[\sum_j (u_{pj} - u_j)^2\right]^{\frac{1}{2}}, \tag{14.17}$$

where $j = 1, 2, 3$.

To calculate the trajectory of a drop, we supplement the system (14.17) with the following equation

$$\frac{dx_{pj}}{dt} = u_{pj}, \tag{14.18}$$

where $j = 1, 2, 3$, x_{pj} are the Cartesian coordinates of a drop.

At values of droplet speeds, when compressibility can be neglected, the coefficient of aerodynamic resistance C_R of a spherical droplet is an unambiguous function of the relative Reynolds number and can be calculated from the formula

$$\text{Re}_p = \frac{\rho d_p \left| \vec{u} - \vec{u}_p \right|}{\mu}, \tag{14.19}$$

where d_p is the droplet diameter and μ is the dynamic viscosity of gas.

For approximating the dependency, the empirical Zhen–Trizek formula is used [33]

$$C_R = \frac{24}{\text{Re}_p} + \frac{6}{1 + \sqrt{\text{Re}_p}} + 0.27. \tag{14.20}$$

Thus, in a known gas-dynamic field the changing of the parameters of a single non-evaporating drop are described by ordinary differential equation (ODE) systems (14.17) and (14.18). In this system, the desired functions are the projections of the absolute velocity of the drop u_{pj} and its coordinates x_{pj}. The remaining values serve as parameters. Some of these parameters are functionally related to independent variables. Therefore, to close the system, it must be supplemented with algebraic equations (14.11), (14.19) and (14.20).

The initial conditions

For the numerical solution of the system of ODEs (14.17) and (14.18), it is necessary to specify the initial conditions, that is, the values of independent variables at the initial time $t = 0$: u_{pj0} and x_{pj0}.

When simulating irrigation by a fire-hose with a sprayer, it is assumed that the drops start from the point of collapse of a compact jet located at a distance from the nozzle

$$r = \psi h_{\text{drop}}, \tag{14.21}$$

where ψ is the ratio of the height of raising individual drops h_{drop} to the height of raising a compact jet h_{jet}, determined by the table 20.10 from [34]; h_{drop} is the height of raising individual drops, determined by the formula [34]

$$h_{\text{drop}} = \frac{h}{1 + \psi_1 h}, \tag{14.22}$$

where h is the full head at the start of the sprayer and ψ_1 was the empirical coefficient depending on the shape of the sprayer [34].

The initial velocity of the droplets was set equal to the local velocity of the fluid in a compact jet

$$u_{p0} = v = \phi\sqrt{2gh}, \tag{14.23}$$

where φ is the speed coefficient ($\varphi = 0.95$).

The direction of the drops' initial velocity vector was determined by the value of the first derivative of the equation of the trajectory of the compact jet flowing from the nozzle at the point of its decay,

$$y = xtg\Theta - \frac{gx^2}{2v^2 \cos^2 \Theta}, \tag{14.24}$$

where Θ is the angle of inclination of the nozzle to the horizon, and x and y are the horizontal and vertical Cartesian coordinates, respectively.

When simulating atomizer irrigation, the coordinates of the points of emission of drops x_{pj0} were chosen at the centers of the faces of the computational cells belonging to the output section of the atomizer. It has been assumed that the drops' initial velocity and the local gas velocity in the outlet section of the atomizer are equal: $u_{pj0} = u_j$.

In all cases, the diameters of droplets d_p were specified by a histogram of the initial distribution of droplets by size, constructed using equation (14.11).

14.2.3 Mathematical model of interfacial interaction

There is interaction between the phases. This interaction is taken into account using the discrete model 'particle—source in a cell'. In accordance with this model, the drop availability in the flow was taken into account as an additional component in the equation for the conservation of the momentum of a continuous phase [35].

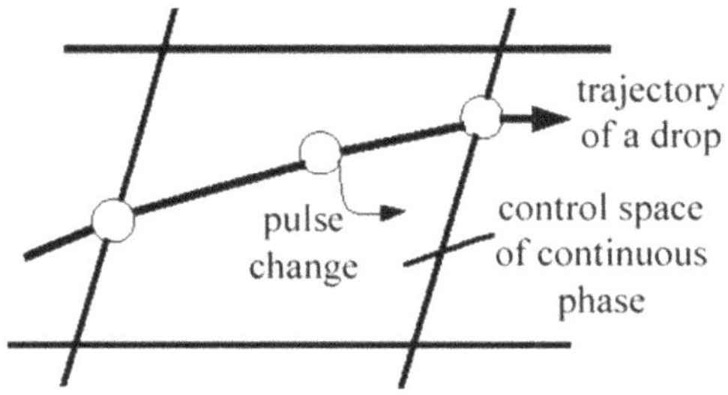

Figure 14.3. Interfacial interaction scheme.

During the calculation of the trajectories of droplets, the amount of movement acquired or lost by the 'package' of droplets following along a given trajectory is tracked. The resulting values should then be taken into account as the source term for calculation of the gas phase S_{fi} in equation (14.2). Thus, since the gas phase already affects the dispersed phase through equations (14.17) and (14.18), the response impact of the dispersed phase into continuum is also counted. That is, by alternately solving the equations of the dispersed phase and the continuous phase until the solution of both phases is established, this two-way interaction will be taken into consideration. A diagram of such a model of interfacial interaction is shown in figure 14.3.

The transfer of the amount of movement from the continuous phase to the dispersed phase is calculated by estimating the change in the droplet movement amount as it passes through each control volume of a study area geometric model.

This change in the amount of movement is calculated as

$$\Delta S_{fi} = \sum\left(\frac{2\mu C_R \operatorname{Re}_p}{3\rho_p d_p^2}(u_{pi} - u_i)\dot{m}_p \Delta t\right), \tag{14.25}$$

where $\dot{m}_p$ is the drops' mass flow and Δt is the time step.

14.3 Numerical simulation of the formation of spatial dispersed structures and the determination of the most effective ways of supplying fluid to eliminate various hazards

Initially, it is necessary to ensure the numerical solution's stability, convergence and accuracy.

14.3.1 Ensuring numerical solution stability, convergence and accuracy

The numerical solution stability can be ensured by using the method of lower relaxation of independent variables.

The number of iterations required for solving stationary problems of aero-hydrodynamics is determined by both the algorithm of the difference scheme and the criterion for evaluating the convergence of the solution. The criteria used by researchers for convergence of the solution can be classified as follows:

1. *Definition of the discrepancy for each solved differential equation.* In this case, as a rule, to achieve the convergence of the entire solution, it is necessary for each difference equation to provide the specified level of the residual.
2. *Integral discrepancy.* In this case, a uniform criterion is determined for all equations, which allows analyzing the convergence of the solution.

To assess convergence, applying an integral criterion with respect to the vector of conservative variables is proposed.

The condition of convergence of the solution can be represented as

$$\sqrt{\frac{\sum R_i^2}{V^2}} \leqslant \varepsilon, \tag{14.26}$$

where R_i is the error for difference equations modeling the transfer of independent variables, V is th volume of control space and ε is the convergence criterion.

A numerical solution can be considered converged if one of the following conditions is met:

- Condition (14.26) is satisfied.
- The solution no longer changes with the continuation of the iterations.

As an alternative criterion for the convergence of problem solving, the pulsation of the mass flow ratio in the input section to the output section can be taken.

The accuracy of the numerical solution can be assessed by the degree to which the convergence criteria are met, the degree of independence of the decision on the size of the computational grid, and the degree of qualitative correspondence of the calculation results to known physical concepts. Solution independence can be estimated by numerical results comparing obtained on several computational grids differing in the number of computational cells. Thus, using this computational method, at any point of the computational domain one can determine the gas parameters.

14.3.2 Description of the numerical integration method of the dispersed phase equations

The numerical integration of the ODE system (14.17) and (14.18) with the given initial conditions was performed by the fourth order Runge–Kutta method with a variable step chosen by the formula

$$\Delta t = \frac{\tau_a}{\chi}, \tag{14.27}$$

where χ is the the coefficient of 'grinding', given depending on the desired accuracy of integration, and τ_a is the aerodynamic relaxation time determined by the expression

$$\tau_a = \frac{\rho_p d_p^2}{18\mu}, \tag{14.28}$$

where d_p is the droplet diameter, μ is the dynamic viscosity of gas and ρ is the drop density.

For the numerical simulation of the specific situation of eliminating the dust cloud, we assume:

1. Bulk material taken from the center of the pile.
2. When material is collected, a spherical dust cloud with a radius of 2 m is formed.
3. Water is delivered to the dust cloud by an atomizer, which can be installed in three positions:
   - lower position (at ground level, that is, 6 m below the level of backfill).
   - average position (6 m above ground level, that is, at the level of backfill).
   - top position (12 m above ground level, that is, 6 m above the level of backfill).

When simulating irrigation of zones of occurrence of natural or man-made hazards of up to several tens of meters (for example, when a dust cloud is formed during loading or unloading of dusty bulk materials) in conditions of calm, headwind and tailwind, the calculated area covered a fragment of space with, located in the middle of it, a stack of bulk material that on the x-axis had a length of 18 m and on the y-axis 6 m in opposite directions from the dust cloud center, with an adjacent area of the atmosphere with a height of 24 m.

When simulating the hazard described above, the computational domain was covered with a non-uniform computational grid that included 199 656 tetrahedral cells. The boundaries of the dust cloud were modeled by a permeable spherical surface with a center located at the top of the stack.

When simulating irrigation under cross-wind conditions, the length of the computational domain was increased to 35 m (for the full length of the stack) in the leeward direction from the center of the dust cloud. In this case, the computational domain included 1 187 702 tetrahedral cells.

Under different conditions of wind action, the boundaries of the dust cloud were simulated by a permeable spherical surface with a center located at the top of the coal pile.

In the following, the discretization algorithm for the governing equations (14.1)–(14.4) was considered on the example of the generalized transfer equation of an arbitrary scalar quantity written in integral form for an arbitrary control volume V using the appropriate dependences.

The task of transporting finely dispersed drops of the fluid to the dust-release zone using single-phase jet-centrifugal atomizers was solved in [36] with the participation of the authors of this work. The same software product was used to solve the problem of transporting process fluid droplets using two-phase nebulizers and atomizers [37]. Using this approach, it is possible to present options for transporting fine droplets to the zone of occurrence of natural or man-made hazards using the example of suppressing dust using various sprayers: single-phase jet-centrifugal, two-phase sprayers and atomizers.

14.3.3 Results of numerical simulation of a spatial finely dispersed structure creation process which suppresses dust

As an applied study of the problem of transporting processed fluid droplets to the dust emission zone, it was proposed to consider and numerically simulate creating a water dispersed structure with two different devices. The first chosen technology is the fire-hose and the second device is the atomizer. As a criterion assessment for comparing the effectiveness of these devices, the following parameters were chosen: the total residence time of all the fluid droplets in the dust cloud and the mass fraction of fluid droplets that did not fall into the dust cloud.

14.3.3.1 Numerically simulating creating a water dispersed structure above a dust cloud using a fire-hose

Several variants of irrigation of a dust cloud with a fire-hose, which differed in the delivery angle, full head, speed and wind direction, were investigated. In all the variants, the following parameters of droplet size distribution were used:

- The minimum diameter of droplets in the spray is 100 μm.
- The maximum diameter of the droplets in the spray is 1500 μm.
- The average median diameter of the droplets in the spray is 1000 μm.
- The exponent in the Rosin–Rammler formula is 2.

The histogram of droplet size distribution corresponding to the above parameters is shown in figure 14.4.

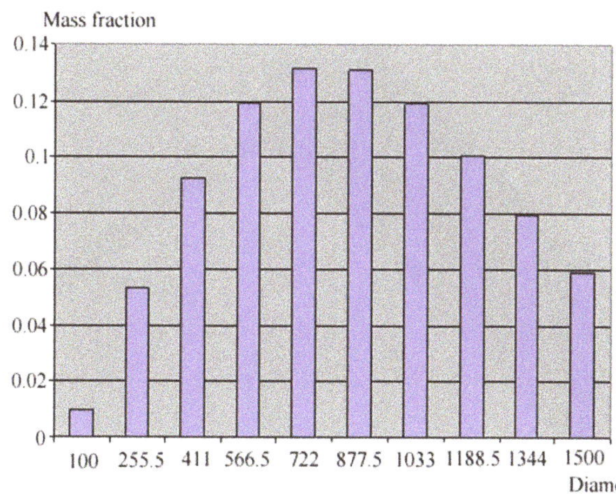

Figure 14.4. The histogram of the distribution of droplets by size with irrigation using a fire-hose.

As a result of the calculations, it was found that at a feed angle of 45°, the compact jet reaches the surface of the coal pile without disintegrating. Therefore, for further analysis, four variants with a feed angle of 45° were selected (table 14.1, figures 14.5–14.10).

From figure 14.5 it can be seen that in variant no. 1 most of the drops, with the exception of the smallest and largest, fall into the dust cloud. In this case, all the drops are precipitated within the storage area.

From figure 14.6 it can be seen that in variant no. 2 all the drops fall into the dust cloud. However, the smallest droplets are carried by the wind (see figure 14.7) outside the storage area.

From figure 14.8 it can be seen that in variant no. 3, none of the droplets reach the dust cloud—they all are blown back by the headwind (see figure 14.9). At the same time, most drops are deposited within the storage area. The smallest drops are carried by headwinds outside the storage area.

Table 14.1. Irrigation variants of the dust cloud using a fire-hose.

Variant	Angle	Pressure, m	Wind speed, m s^{-1}	Wind direction
1	45°	40	0	—
2	45°	25	3	tailwind
3	45°	40	3	headwind
4	45°	40	3	cross wind

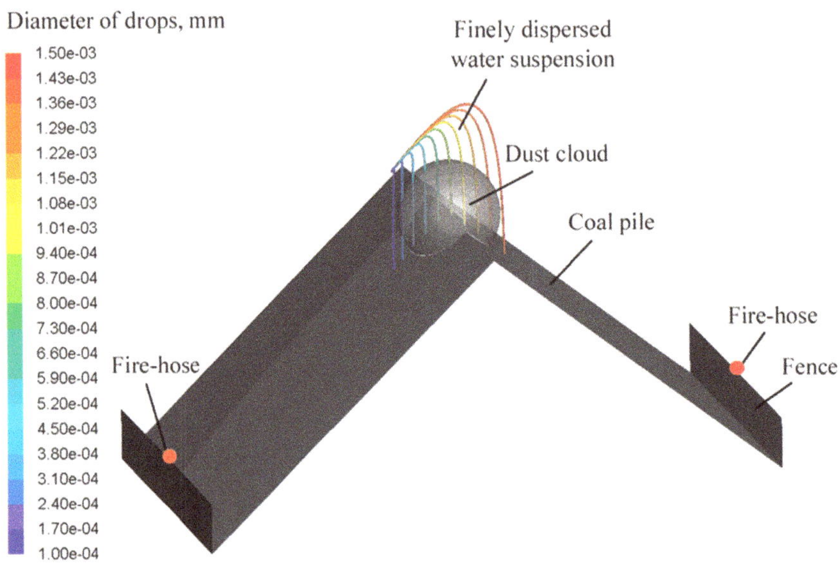

Figure 14.5. Trajectories of 'packages' of drops, painted in accordance with their diameter, in meters (variant no. 1).

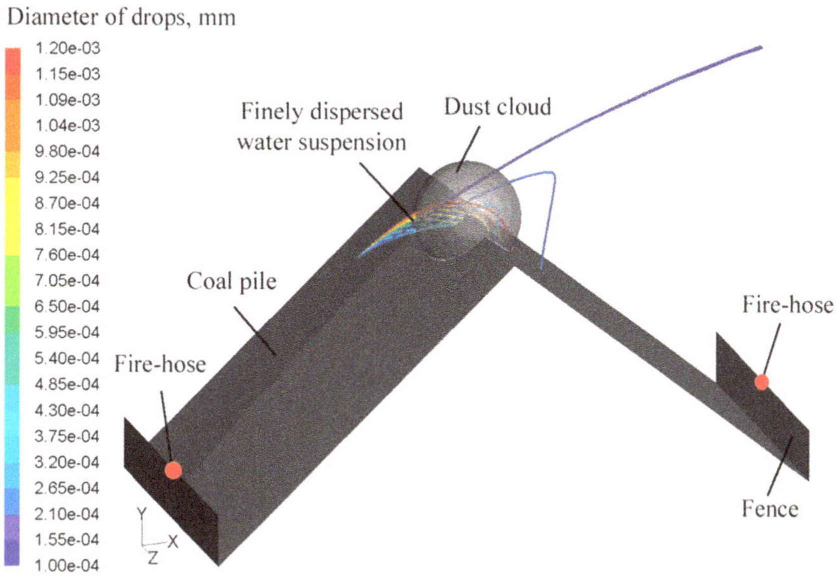

Figure 14.6. Trajectories of 'packages' of drops, painted in accordance with their diameter, in meters (variant no. 2).

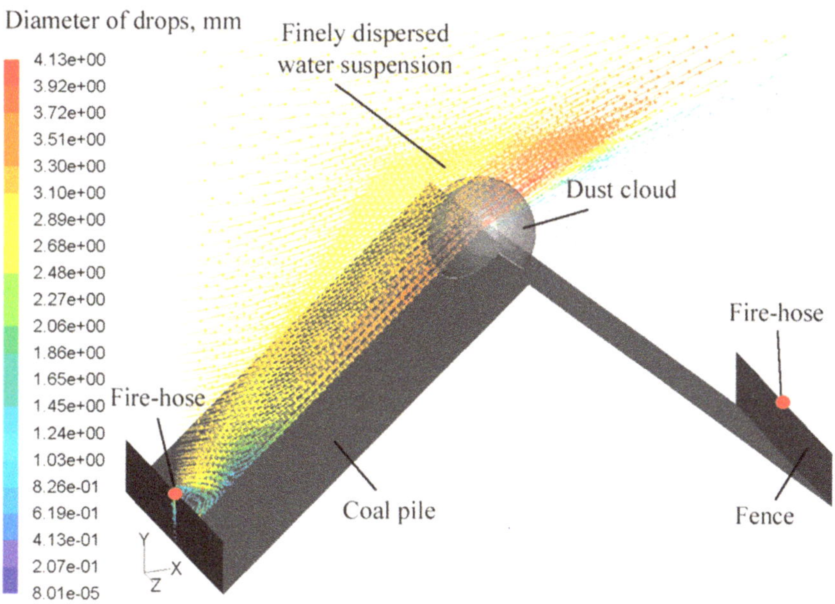

Figure 14.7. Air velocity vectors in the plane of symmetry of the computational domain, painted in accordance with the absolute value of the velocity, in m s^{-1} (variant no. 2).

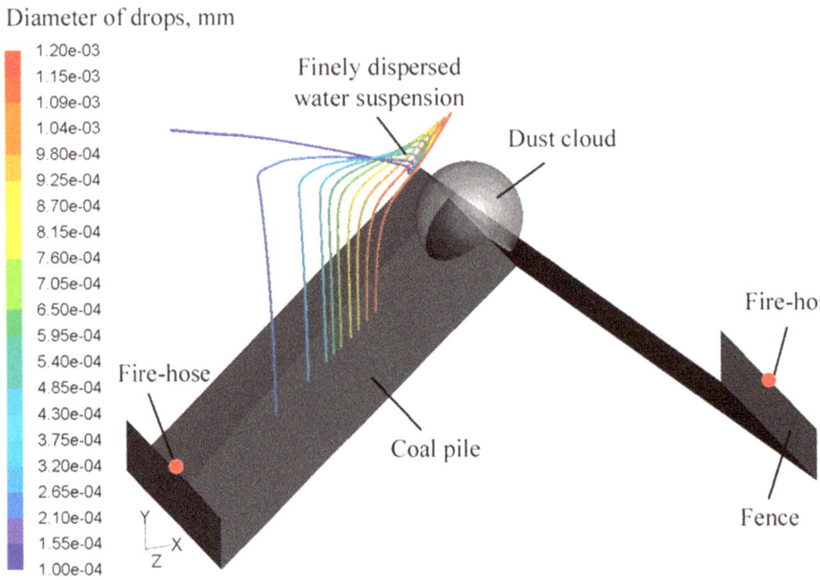

Figure 14.8. Trajectories of 'packages' of drops, painted in accordance with their diameter, in meters (variant no. 3).

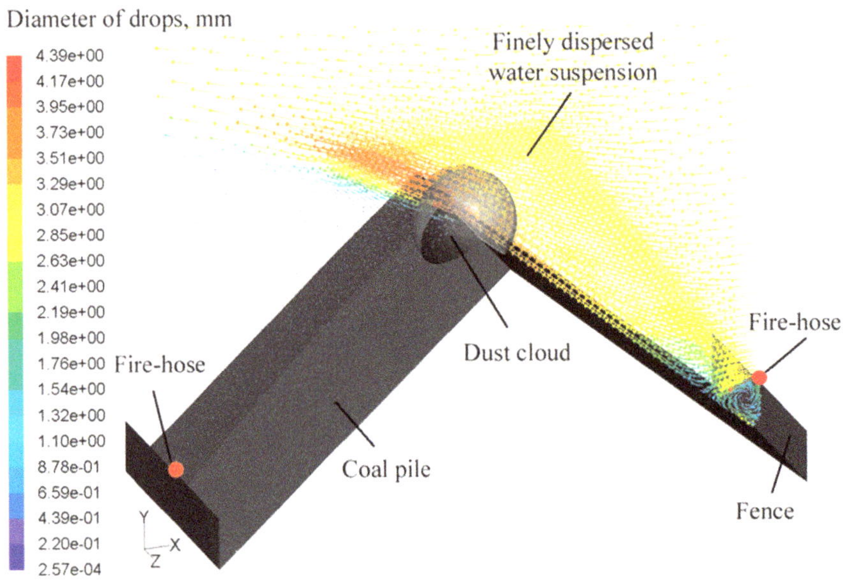

Figure 14.9. Air velocity vectors in the plane of symmetry of the computational domain, painted in accordance with the absolute value of the velocity, m s^{-1} (variant no. 3).

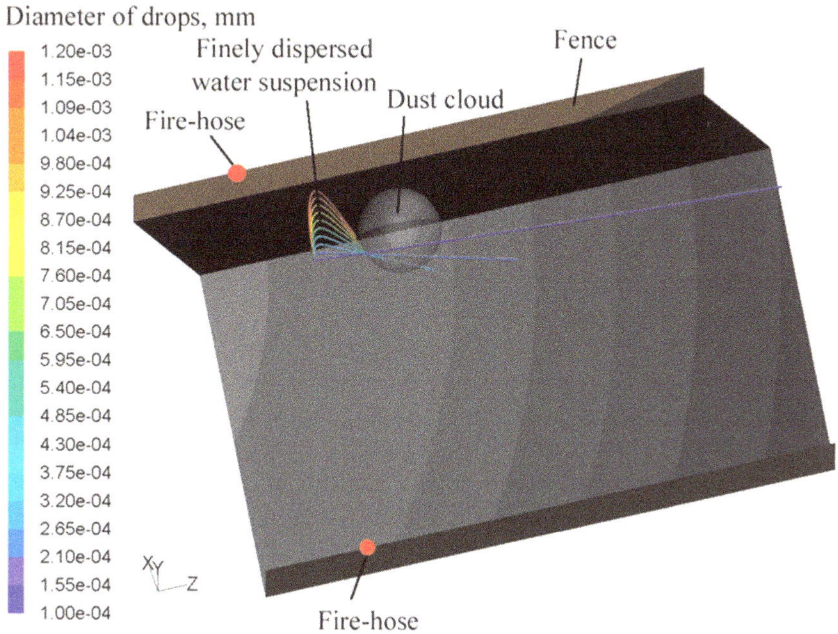

Figure 14.10. Trajectories of 'packages' of drops, painted in accordance with their diameter, in meters (variant no. 4).

From figure 14.10 it can be seen that in variant no. 4 part of the representative droplets reaches the dust cloud. At the same time, most drops are deposited within the storage area. The smallest drops are carried by the side wind outside the storage area.

According to the results of numerical simulations, it was found that in variant no. 1, the total residence time of all the drops in the dust cloud was 21.6 s. At the same time, the mass fraction of droplets that did not fall into the dust cloud is 7.7%. In variant no. 2, the total residence time of all the drops in the dust cloud was 23 s. At the same time, the mass fraction of droplets that did not fall into the dust cloud is 1.1%. For variant no. 4, the total residence time of all drops in a dust cloud was 9.2 s. At the same time, the mass fraction of droplets that did not fall into the dust cloud is 22.4%. The obtained data allow us to further consider the issue of optimizing the irrigation of a dust cloud with a fire-hose.

14.3.3.2 Numerically simulating creating a water dispersed structure above the dust cloud using an atomizer

As in the previous case, several options for the dust cloud irrigation by an atomizer were investigated. The variable parameters selected are different angles of flow, the speed of the blowing aerosol, and the speed and direction of the wind.

In all the variants, the following parameters of droplet size distribution were used:

- The minimum diameter of droplets in the spray is 60 μm.
- The maximum diameter of the droplets in the spray is 130 μm.

- The average median diameter of the droplets in the spray is 100 μm.
- The exponent in the Rosin–Rammler formula is 3.

The histogram of droplet size distribution, corresponding to the above parameters, is shown in figure 14.11.

For further analysis, 15 variants with feed angles of 45° and 60° were selected (table 14.2, figures 14.12–14.18).

In variants no. 1 to no. 4, most of the droplets, with the exception of the largest ones, fall into the dust cloud. At the same time most of the drops, with the exception of the smallest, are deposited within the storage area.

For variant no. 5 (figure 14.12), a large area of irrigation of the storage area is characteristic, with only a fraction of the droplets falling into the dust cloud.

From figure 14.13, it can be seen that in variant no. 6 only part of the droplets fall into the dust cloud. In this case, all the drops are deposited within the storage area.

From figure 14.14, it can be seen that in variant no. 7 most of the droplets are deposited on the windward slope of the pile, not reaching the dust cloud. Only the smallest droplets fall into the dust cloud and are carried away by a tailwind outside the storage area.

From figure 14.15, it can be seen that in variants no. 8 and no. 9 larger droplets are deposited on the windward slope of the pile, not reaching the dust cloud. Smaller droplets fall into the dust cloud, which are carried away by a tailwind outside the storage area.

From figure 14.16, it can be seen that in variant no. 10 only the largest drops are deposited on the windward slope of the pile, not reaching the dust cloud. All other droplets fall into the dust cloud and are carried away by a tailwind outside the storage area. An exception is made for medium-sized droplets which, being involved

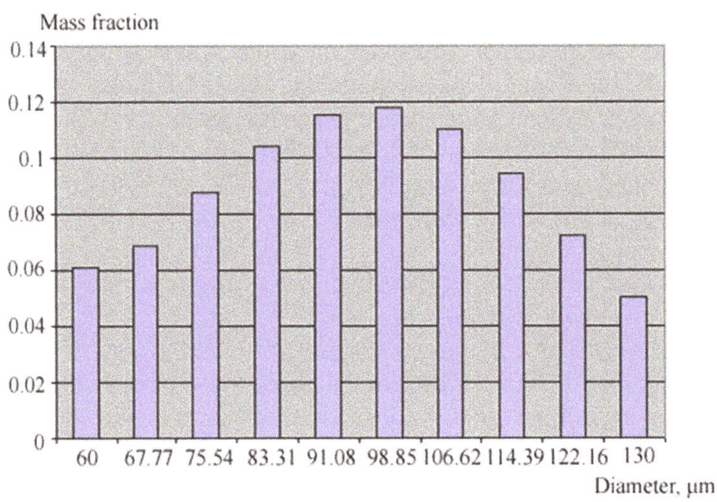

Figure 14.11. The histogram of the distribution of droplets by size with irrigation using an atomizer.

Table 14.2. Irrigation variants of the dust cloud by atomizer.

Variant	Atomizer location	Aerosol blowing angle	Aerosol blowing speed, m s^{-1}	Wind speed, m s^{-1}	Wind direction
1	bottom	45°	10	0	—
2	middle	45°	10	0	—
3	top	45°	10	0	—
4	top	0°	10	0	—
5	bottom	45°	3	1	tailwind
6	middle	45°	3	1	tailwind
7	middle	45°	5	1	tailwind
8	middle	45°	10	1	headwind
9	middle	45°	20	1	headwind
10	top	0°	10	1	headwind
11	top	0°	20	1	headwind
12	bottom	60°	20	1	cross wind
13	middle	60°	20	1	cross wind
14	top	60°	20	1	cross wind
15	top	60°	20	1	cross wind

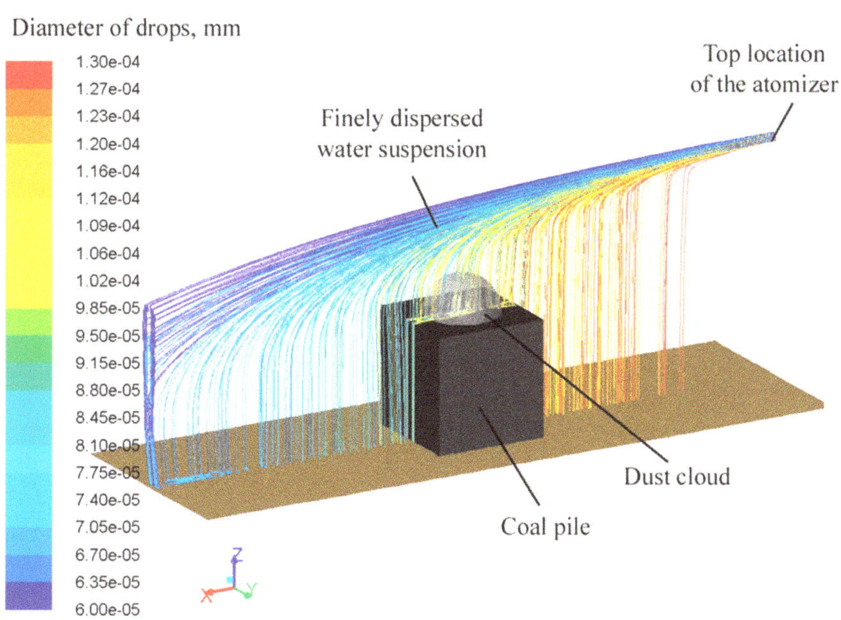

Figure 14.12. Trajectories of 'packages' of drops, painted in accordance with their diameter, in meters (variant no. 5).

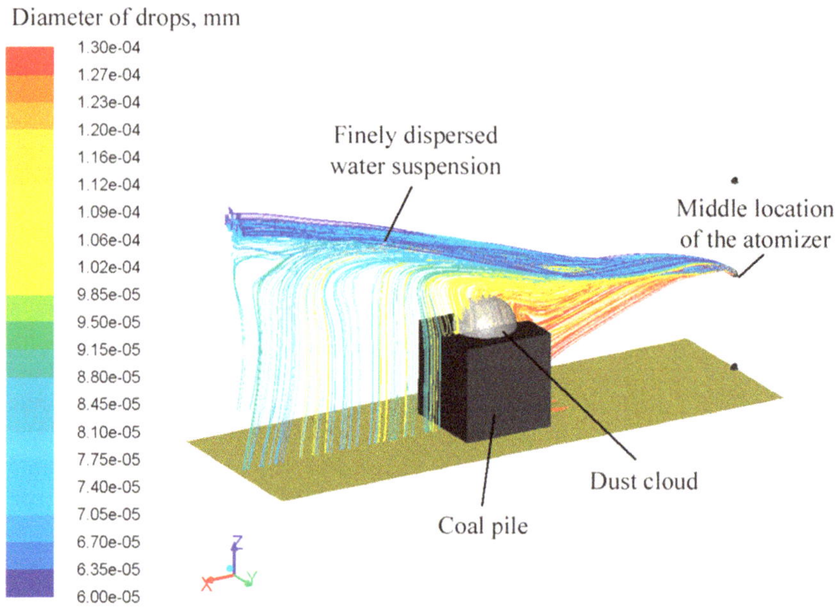

Figure 14.13. Trajectories of 'packages' of drops, painted in accordance with their diameter, in meters (variant no. 6).

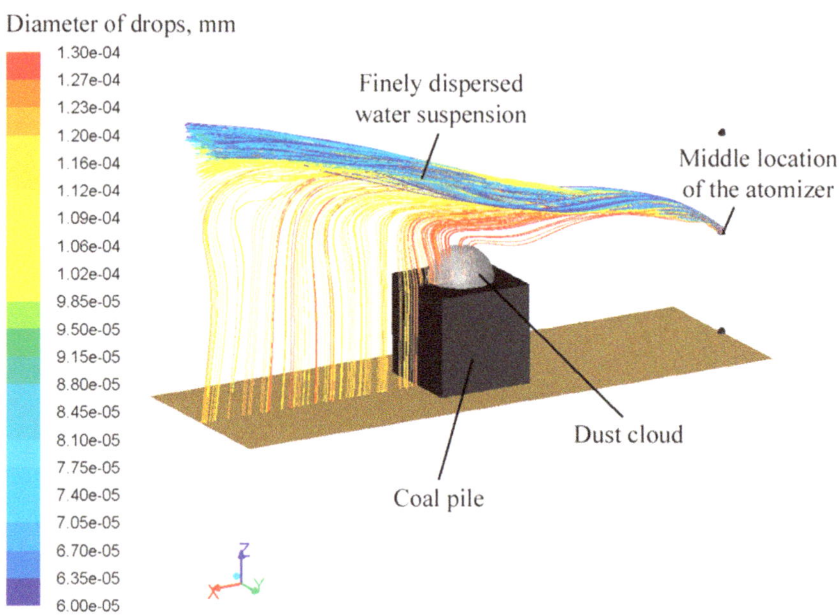

Figure 14.14. Trajectories of 'packages' of drops, painted in accordance with their diameter, in meters (variant no. 7).

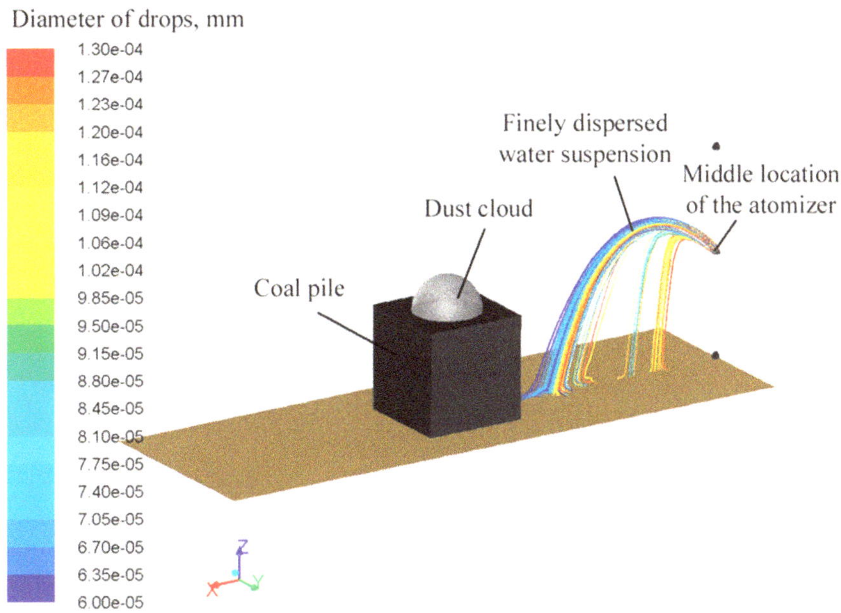

Figure 14.15. Trajectories of 'packages' of drops, painted in accordance with their diameter, in meters (variant no. 8).

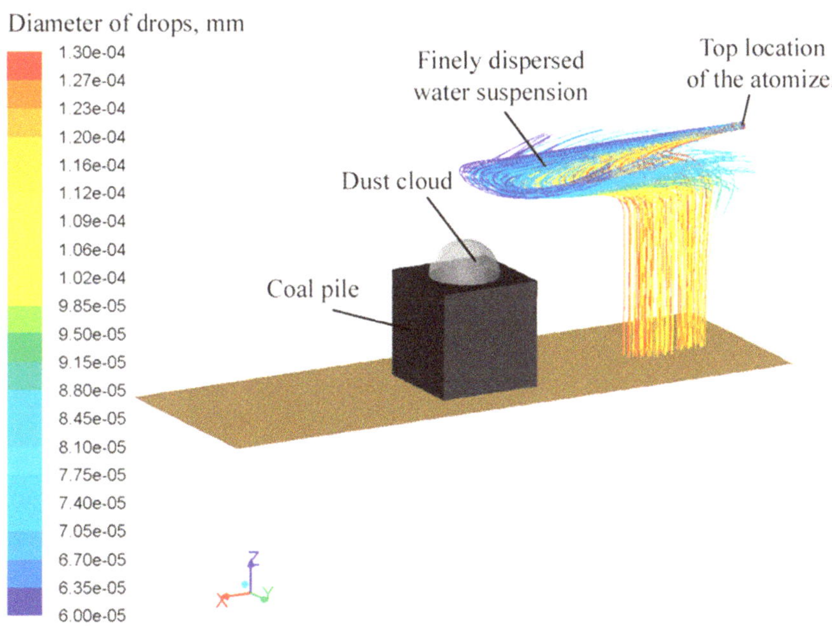

Figure 14.16. Trajectories of 'packages' of drops, painted according to their diameter, in meters (variant no. 10).

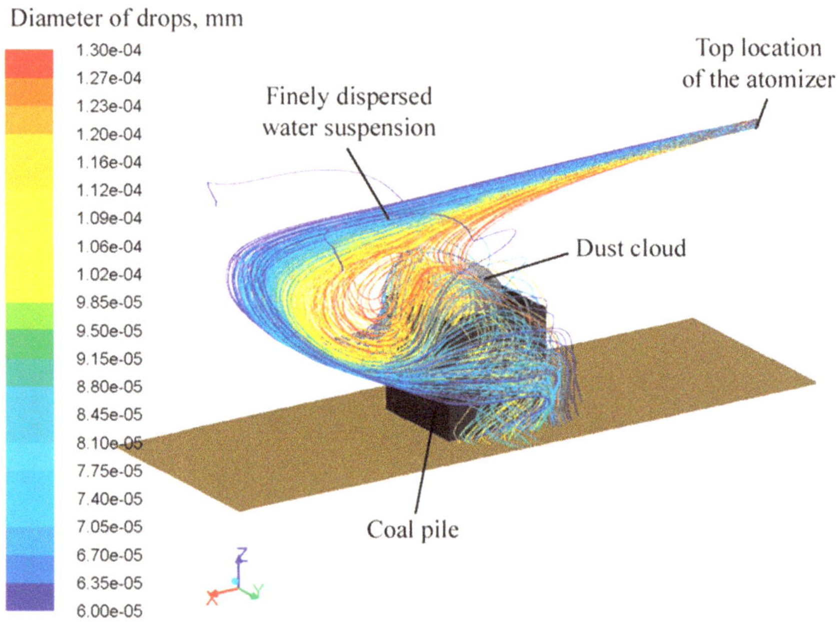

Figure 14.17. Trajectories of 'packages' of drops, colored according to their diameter, in meters (variant no. 11).

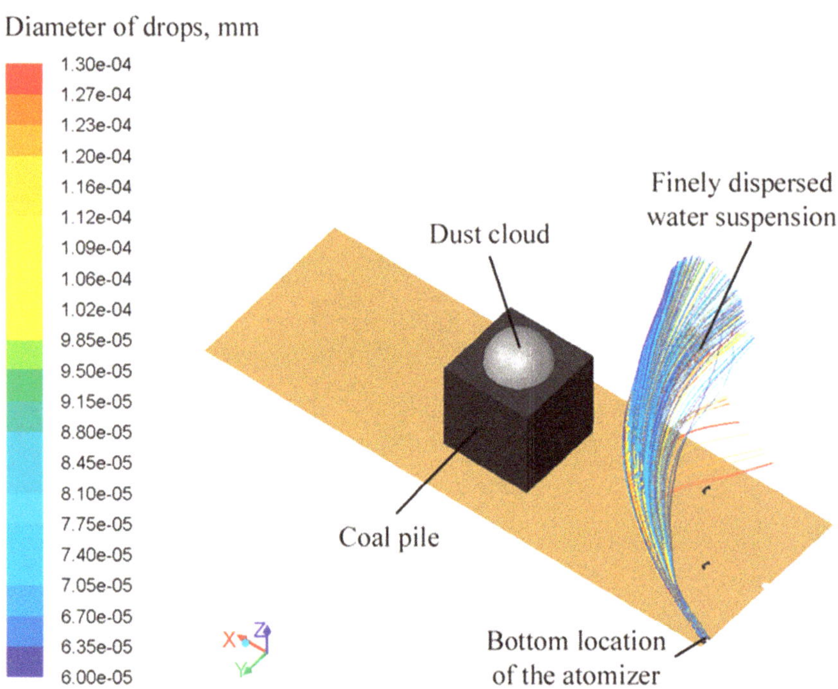

Figure 14.18. Trajectories of 'packages' of drops, colored according to their diameter, in meters (variant no. 12).

in the zone of reverse air currents that occur behind the leeward slope of the pile, are deposited on its surface.

From figure 14.17, it can be seen that in variant no. 11 large drops reach the dust cloud and settle on the leeward slope of the pile. Small drops are enveloped by a headwind and are carried out by them to the limits of the storage area.

From figure 14.18, it can be seen that in variant no. 12 only a fraction of the droplets reach the dust cloud. Most of the larger droplets are deposited on the leeward side of the pile, before they reach the dust cloud. The smallest drops are turned by the headwind and are carried by them outside the storage area.

According to the results of numerical simulation in variant no. 13, an insignificant proportion of the large drops reaches the dust cloud. Most of the droplets do not reach the dust cloud—the largest drops are deposited on the leeward side of the pile, not reaching the dust cloud, and the smaller drops turn around the headwind and are carried away from the storage area. In variant no. 14, some of the droplets fall into the dust cloud. In this case, all the drops are carried by the cross wind outside the storage area. In variant no. 15, none of the droplets reach the dust cloud—they are all carried by the side wind outside the storage area.

The most preferable were the results in variants no. 5 and no. 6. In variant no. 5, the total residence time of all droplets in a dust cloud was 175.6 s. At the same time, the mass fraction of droplets that did not fall into the dust cloud was 27.8%. In variant no. 6, the total residence time of all droplets in the dust cloud was 164.4 s. At the same time, the mass fraction of droplets that did not fall into the dust cloud is 70.8%.

The obtained data allow us to further consider the issue of optimizing the irrigation of a dust cloud using an atomizer.

Within the framework of the developed numerical model, the estimated values of the total residence time of the droplets in the working area were obtained and the proportion of drops that did not fall into the working area was established. Numerical simulation of the presented mathematical model made it possible to determine the most effectual water spraying modes using various devices for dust suppression under different wind directions. The results suggest that the use of dispersed water structures to suppress dust is effective. At the same time, environmental safety management is achieved by the right choice of the technical device and the mode of supplying the process fluid (or water).

14.3.4 Results of numerical simulation of the spatial finely dispersed structure creation process, which instantly reduces the gas stream temperature

For the numerical simulation and visualization of the results of the instant reduction of gas stream temperature, gas cooling which was obtained from the high-temperature recycling of solid carbon-containing waste was considered. In this case, we have a two-phase multicomponent structure with phase transformation, since water particles in the gas stream with a high temperature will evaporate. An ODE describes the behavior of a single drop in a gas-dynamic space [35] and has been shown in section 14.2.3. In this system, some of the parameters are functionally related to independent variables. To close and solve an ODE system, it is necessary to take into account two-way interaction by solving dispersed phase equations and continuous phase equations alternately.

The fine water mist effectiveness for gas stream cooling depends on the structure and gas–droplet stream parameters. Such parameters include the dispersion of spraying water by nozzles and its rate of outflow from the nozzles. So, the variant of spraying water by nozzles into the high-temperature gas stream is necessary, which will ensure the instant cooling of the gas stream to the required temperature. We believe that the gas flow cooling occurs in the heat exchanger.

Based on the use of the classical description of the two-phase flow movement, the physical and mathematical description of this process is possible [38–40].

A gas phase mathematical model of cooling of high-temperature gas is described in [38, 39]. This model shows the three-dimensional flow characteristics in a heat exchanger and the influence of the dispersed phase. To analyze the dispersed phase behavior, the following assumptions were made, which allowed its behavior to be considered in the Lagrangian description [40]:

- The dispersed phase is a completely sprayed torch of fluid (water), consisting of a finite set of evaporating spherical droplets of various diameters.
- All forces acting on the drop, except for the forces of gravity and aerodynamic drag, are neglected.
- We neglect the processes of secondary crushing and coagulation of droplets.

The phase interaction was taken into consideration through the discrete model 'particle—source in a cell'. This model assumes that the droplet presence in the flow is expressed by additional elements in the conservation equations of the continuous phase [40]. This is described in section 14.2.3. When calculating the trajectories of the droplets, the values of the impulse, mass and temperature of droplet 'packets' which changed when moving were tracked. These values as initial conditions were included for the gas phase calculation S_m, S_q, S_{fi}.

By estimating the droplet mass change as it passes through each control volume of the geometric model, the mass transfer between dispersed and continuous phases were calculated as

$$\Delta S_m = \sum\left(\frac{\Delta m_p}{m_{p_0}} \cdot \dot{m}_{p_0}\right), \tag{14.29}$$

where Δm_p is the change in the drop mass in the control volume, m_{p_0} is the initial mass of a drop and $\dot{m}_{p_0}$ is the initial droplets' mass flow rate.

By estimating the drop pulse change as it passes through each control volume of the geometric model, the momentum transfer between continuous and dispersed phases was calculated using formula (14.25). By estimating the drop enthalpy change as it passes through each control volume of the geometric model, the heat transfer between continuous and dispersed phases was calculated as

$$\Delta S_q = \sum\left(\frac{\overline{m}_p}{m_{p_0}}c_p\Delta T_p + \frac{\Delta m_p}{m_{p_0}}\left(-L + \int_{T^0}^{T_p} c_{pi}(T)\mathrm{d}T\right)\right)\cdot \dot{m}_{p_0}, \tag{14.30}$$

where $\overline{m}_p$ is the average drop mass in the control volume, c_p is the heat capacity of drops, ΔT is the change in the temperature of drops in the control volume, L is the latent heat of evaporation, c_{pi} is the heat capacity of vapor fluid, T_p is the drop temperature at the outlet of the control volume and T_0 is the enthalpy standard temperature.

Since the gas phase already affects the dispersed phase, the reverse effect of the dispersed phase on the continuum was also taken into account [39].

To numerically simulate the instantaneous cooling of a hot gas stream, a fragment of the heat exchanger space was investigated. This fragment is bounded by the heat exchanger walls, and its inlet and outlet sections (figure 14.19). Centrifugal nozzles are built in the heat exchanger to disperse the fluid. The purpose of this numerical simulation is to investigate the possibility of instantaneous cooling of a gas stream from 1200 °C to 300 °C.

The initial conditions

For the numerical solution of the ODE system, it is necessary to specify the initial conditions, that is, the values of independent variables at the initial moment of time $t = 0$: u_{pj0}, x_{pj0}, d_p and T_p.

We believe that at the initial moment of time the gas elementary volume is located in the cross section center of the heat exchanger flow-through part. This section precedes the liquid injection zone. For variants no. 1 and no. 2, at the initial moment of time the gas elementary volume is located in the section 'B—B' and for variant no. 3 in the section 'A—A'. In this case, the uneven computational domain included 77 087 polyhedric cells.

The initial diameters of a cluster of droplets having the same paths in the package were given by the histogram of the initial distribution of droplets by size. The initial velocity of the drops was assumed to be equal to the rate of flow of fluid from the nozzle; the initial temperature of the droplets was equal to 20 °C. The initial direction of the vector of the initial velocity of the drops was determined from the root angle of the atomization plume. Numerical integration was carried out for three

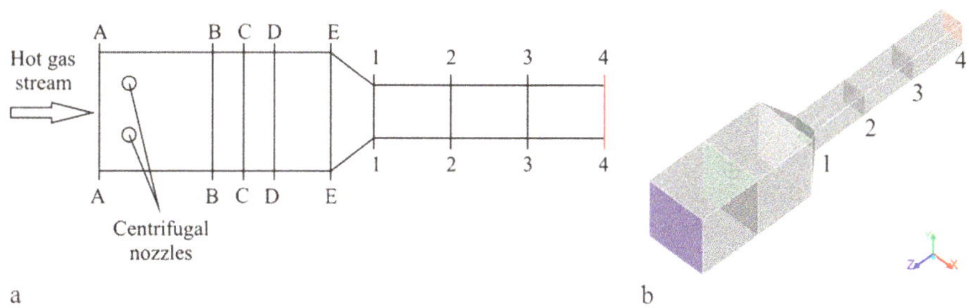

Figure 14.19. Fragment of the heat exchanger space which was investigated: (a) right side view and (b) isometry.

Table 14.3. The values of the parameters of water injection through nozzles [24].

Parameter name	Variant no. 1	Variant no. 2	Variant no. 3
d_0 (m)	0.0006	0.0009	0.0011
$A_{nozzle}/(D_s d_0)$	0.75	1.3	1
C_D	0.43	0.56	0.50
α (degrees)	66	50	57
v (m s^{-1})	10	3	3
Δp (Pa)	48 228	5650	3189
Re_e	3858	2257	1957
d_{min} (m)	$4.7 \cdot 10^{-6}$	$55 \cdot 10^{-6}$	$202 \cdot 10^{-6}$
$\bar{d}$ (m)	$16.6 \cdot 10^{-6}$	$144 \cdot 10^{-6}$	$514 \cdot 10^{-6}$
d_{max} (m)	$25 \cdot 10^{-6}$	$234 \cdot 10^{-6}$	$802 \cdot 10^{-6}$
n	3.77	3.99	3.6

variants (modes) of water supply. The values of the parameters of water injection through nozzles are presented in table 14.3 [24].

By integrating the gas phase equations [38, 39], gas parameters can be calculated at any studied fragment point of the heat exchanger. So we can control the gas flow temperature and speed in different heat exchanger sections. In addition, most rational parameters of the liquid supply by sprays can be determined.

For the numerical integration of a system of partial differential equations with specific boundary conditions, their discretization must be performed. The controlled volumes method was applied to the discretization of equations in space [41] on an unstructured computational grid, which contains polyhedric elemental volumes—cells. The droplet volume distribution in a sprayed liquid stream is based on data from [26] and is described by the Rosin–Rammler equation.

The droplets' heat and mass transfer is described by two models—evaporation and boiling. The evaporation model is valid until the boiling point is reached by the droplet T_{bp}. Once the boiling point is reached, then the drop's heat and mass transfer is determined by the boiling rate.

The evaporation model suggests that the rate of evaporation of a drop is determined by Fick's law:

$$\frac{dm_v}{dt} = A_v D\rho \frac{dc}{dr}, \tag{14.31}$$

where m_v is the steam mass, ρ is the gas density, D is the coefficient of binary diffusion of vapor in gas, c is the vapor concentration, $A_v = \pi d_p^2$ is the evaporation surface area, t is time and r is a radial coordinate.

Dividing the variables and integrating equation (14.31) with the boundary conditions $c = c_s$ with $r = r_p$, $c = c_\infty$ with $r = r_\infty$, we obtain

$$\frac{dm_\nu}{dt} = \beta A_\nu \rho(c_s - c_\infty), \tag{14.32}$$

where c_s is the volume concentration of vapor at the surface of the drop, r_p is the radius of a drop, c_∞ is a volume concentration of vapor in the ambient gas and β is the experimentally determined evaporation rate.

The results of experimental studies are usually presented as criterial dependences Sh (Re, Sc), where Sh is the Sherwood number, defined as

$$\mathrm{Sh} = \frac{\beta d_p}{D}. \tag{14.33}$$

Taking into account (14.33) and the evaporation area of a spherical drop, equation (14.32) takes the form

$$\frac{dm_\nu}{dt} = \mathrm{Sh}\rho D\pi d_p^2(c_s - c_\infty). \tag{14.34}$$

For an approximation of the criterion dependence Sh(Re, Sc), the empirical Rantz–Marshall formula was used [42]:

$$\mathrm{Sh} = 2 + 0,6\,\mathrm{Re}^{0,5}\,\mathrm{Sc}^{0,33}. \tag{14.35}$$

Suppose that the partial pressure of vapor on the surface is equal to the saturated vapor pressure p_{sat} at the drop temperature T_p. In this case, the vapor concentration on the surface of the drop can be calculated,

$$c_s = \frac{M_\nu p_{\mathrm{sat}}}{M_\nu p_{\mathrm{sat}} + M(p - p_{\mathrm{sat}})}, \tag{14.36}$$

where M is the molecular mass of gas, M_v represents the molecular mass of vapor, and p is the absolute pressure of the gas–vapor mixture.

If we differentiate equation (14.34) by time, then an equation for the decrease rate of the evaporating drop diameter can be obtained:

$$\frac{dm_p}{dt} = 0,5\pi\rho_p d_p^2 \frac{d(d_p)}{dt}. \tag{14.37}$$

Considering equation $\frac{dm_p}{dt} = -\frac{dm_p}{dt}$ from (14.34) and (14.37), we obtain

$$\frac{d(d_p)}{dt} = -\frac{2Sh\rho D}{\rho_p d_p}(c_s - c_\infty). \tag{14.38}$$

When the gas is cooled, the drop temperature of dispersed fluid changes before reaching the boiling point in accordance with the heat balance

$$m_p c_p \frac{dT_p}{dt} = \alpha A_\nu (T_\infty - T_p) + L\frac{dm_\nu}{dt}, \tag{14.39}$$

where c_p is the heat capacity of drops, α is the experimentally determined heat transfer coefficient, A_ν is the surface area of a drop, L is the latent heat of evaporation, T_p is droplet temperature and T_∞ is local gas temperature.

The results of experimental studies are usually given in the form of criterion dependences Nu (Re, Pr),

$$\mathrm{Nu} = \frac{\alpha\lambda dm_\nu}{d_p}, \tag{14.40}$$

where λ is the thermal conductivity of gas and Nu is the Nusselt number.

Taking into account (14.34) and (14.40), the notation (14.39) can be represented in the following form:

$$\frac{dT_p}{dt} = \frac{(T_\infty - T_p)}{\Theta} + \frac{Q_L}{\Theta}, \tag{14.41}$$

where

$$Q = \frac{L\mathrm{Sh}\rho D(c_s - c_\infty)}{\mathrm{Nu}\lambda}; \tag{14.42}$$

$$\Theta = \frac{\rho_p d_p^2 c_p}{6\mathrm{Nu}\lambda}. \tag{14.43}$$

Based on the assumption of a complete analogy between heat and mass transfer to determine the Nusselt number, to approximate the criterial dependence of Nu (Re, Pr), a relationship similar to formula (14.35) was used:

$$\mathrm{Nu} = 2 + 0,6\,\mathrm{Re}^{0,5}\,\mathrm{Pr}^{0,33}. \tag{14.44}$$

After the droplet temperature is equal to the boiling point, instead of equation (14.38) the equation of the speed of boiling is applied:

$$\frac{d(d_p)}{dt} = -\frac{4\lambda}{\rho_p c_{p\infty} d_p}\left(1 + 0,23\,\mathrm{Re}_p^{0,5}\ln\left[1 + \frac{c_{p\infty}(T_\infty - T_p)}{L}\right]\right), \tag{14.45}$$

where $c_{p\infty}$ is the gas heat capacity.

Equation (14.45) is derived under the assumption of stationary flow at constant pressure. It is believed that while the boiling law is applied, the drop maintains a constant temperature.

The results of numerical simulation of instantaneous cooling of high-temperature gas are shown in tables 14.4–14.6.

According to the results in variant no. 1, water drops very quickly (in the immediate vicinity of the injection site) lose their initial impulse and then follow a compact 'swarm' along the heat exchanger. The resulting asymmetry of the flow and, accordingly, the asymmetry of the droplet trajectories are due to the high

Table 14.4. The results of numerical simulation of instantaneous cooling of high-temperature gas for variant no. 1.

Parameter name	Section 1	Section 2	Section 3	Section 4
z (m)	−0.55	−0.912	−1.274	−1.6365
t_{min} (°C)	31.3	188.7	275.3	311.8
t_{av} (°C)	408.6	364.7	357.6	355.8
t_{max} (°C)	951.6	592.8	482.3	424.1
γ_T	0.8892	0.9422	0.9649	0.978
$g_{H_2O \cdot min}$	0.041	0.116	0.143	0.158
$g_{H_2O.av}$	0.1666	0.1756	0.1769	0.1774
$g_{H_2O.max}$	0.282	0.227	0.2	0.19
γ_{H_2O}	0.8804	0.9426	0.9653	0.9781

Table 14.5. The results of numerical simulation of instantaneous cooling of high-temperature gas for variant no. 2 [24].

Parameter name	Section 1	Section 2	Section 3	Section 4
z (m)	−0.55	−0.912	−1.274	−1.6365
t_{min} (°C)	257	337.7	347.7	350.4
t_{av} (°C)	356.9	354.3	354.41	354.42
t_{max} (°C)	414.1	363.2	359.6	358
γ_T	0.9795	0.9955	0.9973	0.9983
$g_{H_2O \cdot min}$	0.162	0.175	0.176	0.177
$g_{H_2O.av}$	0.177 05	0.177 66	0.177 64	0.177 63
$g_{H_2O.max}$	0.2057	0.1822	0.1795	0.1787
γ_{H_2O}	0.9799	0.9957	0.9974	0.9984

Table 14.6. The results of numerical simulation of instantaneous cooling of high-temperature gas for variant no. 3 [24].

Parameter name	Section 1	Section 2	Section 3	Section 4
z (m)	−0.55	−0.912	−1.274	−1.6365
t_{min} (°C)	285.1	330.2	345.7	354.1
t_{av} (°C)	418.2	393.3	382	375
t_{max} (°C)	616.5	467.5	422.4	396.6
γ_T	0.9501	0.9711	0.9835	0.9904
$g_{H2O \cdot min}$	0.112	0.147	0.159	0.166
$g_{H_2O.av}$	0.1604	0.1668	0.1698	0.1718
$g_{H_2O.max}$	0.2	0.184	0.18	0.177
γ_{H_2O}	0.9443	0.9693	0.9829	0.9903

sensitivity of relatively small droplets (≈16.6 μm) to small perturbations (fluctuations) of the gas velocity.

In addition, the droplets take a temperature from 12 °C to 58 °C, not reaching a boiling point of water of 100 °C, but most of the droplets have time to evaporate before entering a narrow portion of the heat exchanger, that is, section 1. Therefore, the numerical value of the average mass fraction of water vapor $g_{H_2O.av}$ in the vapor–gas mixture in section 1 is significantly less than the equilibrium one. Accordingly, the numerical value of the gas–vapor mixture average temperature t_{av} in section 1 is 55 °C more than the equilibrium temperature (454 °C).

The temperature of the gas–vapor mixture and the mass fraction of water vapor in it are unevenly distributed over the above-mentioned sections, particularly in section 1, where the coefficients of uniformity are $\gamma_T = 0.8892$ and $\gamma_{H_2O} = 0.8804$. The gas–vapor mixture maximum temperature in all sections exceeds the equilibrium (454 °C) by 598 °C, 239 °C, 128 °C and 70 °C, respectively (see table 14.4). Thus, the gas–droplet flow structure and parameters do not ensure the achievement of the intended purpose, and consequently variant no. 1 of the water supply cannot be considered satisfactory.

In variant no. 2, water drops retain their initial impulse for a long time (approximately to the middle of the channel width), after which they follow trajectories under the influence of a hot gas stream. These trajectories are curved as a result of the interphase exchange of momentum and rebound from the walls of the vapor pipe. In this case, the symmetry of the flow and, accordingly, the symmetry of the droplet trajectories are largely preserved, which is due to the weak sensitivity of relatively large droplets (≈144 μm) to small perturbations (fluctuations) of the gas velocity. In addition, from figure 14.20 it can be seen that droplets, as a result of the processes of convective heat exchange and evaporation when moving inside the heat exchanger, take temperatures from 20 °C to 59 °C, not reaching the boiling point of water, 100 °C.

It is also seen that some of the droplets do not have time to evaporate before entering a narrow section of the vapor pipe, that is, before section 1. Their evaporation continues in a narrow section of the heat exchanger to section 4. However, the number of such drops is small, since the numerical value of the average mass fraction of water vapor $g_{H_2O.av}$ in the gas–vapor mixture in section 1 is very close to equilibrium, to which this value tends further downstream in sections 2, 3 and 4 (see table 14.5). Accordingly, the numerical value of the gas–vapor mixture average temperature t_{av} in section 1 is only 3 °C higher than the equilibrium temperature (454 °C).

The temperature of the gas–vapor mixture and the mass fraction of water vapor in it are evenly distributed in the sections, as indicated above, including section 1 (figure 14.21), where the coefficients of uniformity are $\gamma_T = 0.9795$ and $\gamma_{H_2O} = 0.9799$.

The gas–vapor mixture maximum temperature at all sections exceeds the equilibrium state (454 °C) by only 60 °C, 9 °C, 6 °C and 4 °C, respectively

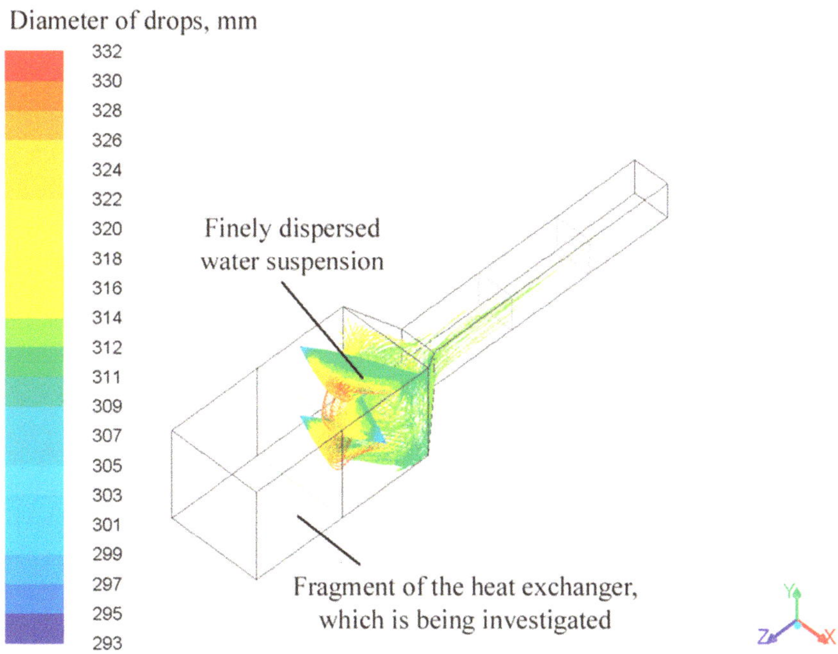

Figure 14.20. Trajectories of 'packages' of drops, painted in accordance with the time of their stay, in seconds (variant no. 2, isometry).

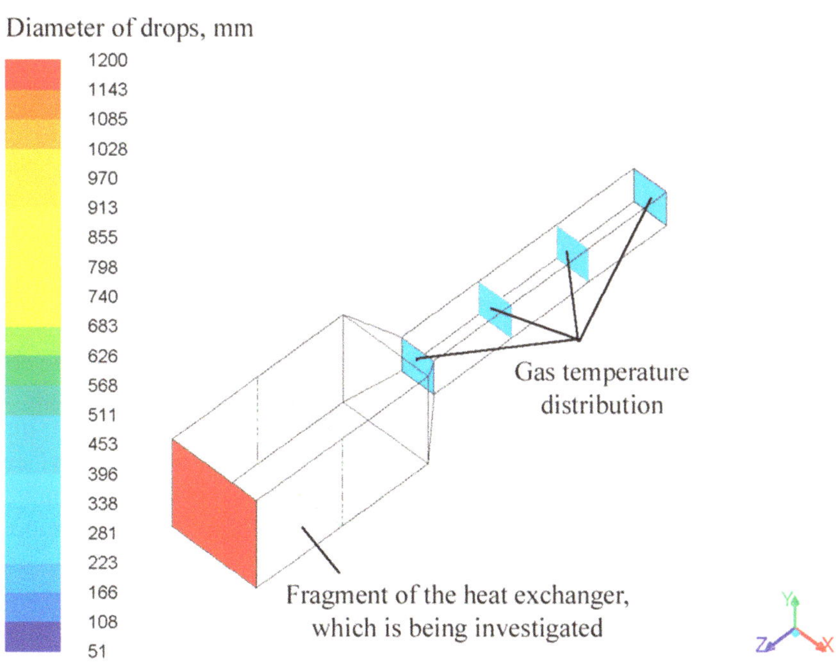

Figure 14.21. Gas temperature distribution over heat exchanger sections, in °C (variant no. 2, isometry).

(table 14.5). Thus, the structure and parameters of the gas–droplet flow as a whole correspond to the set purpose and variant no. 2 of water supply can be considered satisfactory.

In variant no. 3, water drops retain their initial impulse almost unchanged across the entire width of the channel until they collide with its walls, after which they follow trajectories mainly due to the rebound phenomenon. Accordingly, the symmetry of the droplet trajectories is preserved, which is due to the extremely low sensitivity of very large droplets ($\approx$514 μm) not only to small perturbations (fluctuations), but also to the averaged gas velocity. At the same time, droplets have a strong influence on the gas flow, which in this case remains almost symmetrical. It is also observed that water droplets, as a result of the convective heat exchange and evaporation, take temperatures from 20 °C to 57 °C, without reaching the boiling point of water, 100 °C.

Drops do not have time to evaporate, not only to the entrance to a narrow section of the heat exchanger (to section 1), but also to the output of section 4. Although the drops only partially evaporate within the entire heat exchanger, they fill the volume of the flow part very fairly and evenly. As a result, the numerical value of the average mass fraction of water vapor $g_{H_2O.av}$ in the vapor–gas mixture in section 1 is slightly less than the equilibrium. Accordingly, the numerical value of the gas–vapor mixture average temperature t_{av} at section 1 is 64 °C higher than the equilibrium (455 °C).

The temperature of the gas–vapor mixture and the mass fraction of water vapor in it are also fairly evenly distributed over all sections, particularly in section 1, where the coefficients of uniformity are $\gamma_T = 0.9501$ and $\gamma_{H_2O} = 0.9443$ (see table 14.6). At the same time, the gas–vapor mixture maximum temperature in the sections exceeds the equilibrium temperature (454 °C) by 263 °C, 114 °C, 68 °C and 43 °C, respectively. Thus, the gas–droplet flow structure and parameters do not ensure the achievement of the intended purpose, and consequently variant no. 3 of water supply cannot be considered satisfactory.

It is obvious that the recognition of variant no. 2 for supplying water as satisfactory does not exclude the existence of an even more perfect version, for the future search for which we can formulate and solve the corresponding optimization problem.

Further, a numerical experiment to determine the cooling time of the gas stream to the required temperature was carried out. Considering the movement of the gas elementary volume at the heat exchanger fragment, figure 14.22 shows the control jet of the gas stream for various modes of dispersed fluid supply and figure 14.23 shows the Z-coordinate dependence graph of the gas elemental volume on time τ.

From the graph $Z(\tau)$, the moments of time were determined at which the elementary volume of gas reaches the next section i. The average t_{av}, minimum t_{min} and maximum t_{max} gas temperature, average temperature deviation and uniformity coefficient of gas temperature distribution in control sections, and the residence time $\Delta\tau$, cooling Δt, gas elementary volume average cooling rate $\Delta t/\Delta\tau$ between adjacent sections i and (i—1) were determined.

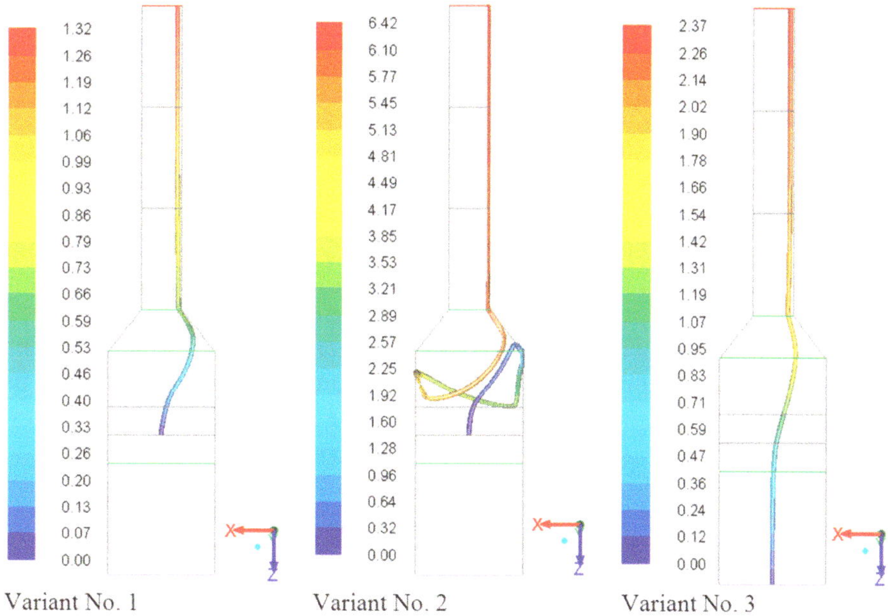

Figure 14.22. The trajectory of the gas elementary volume.

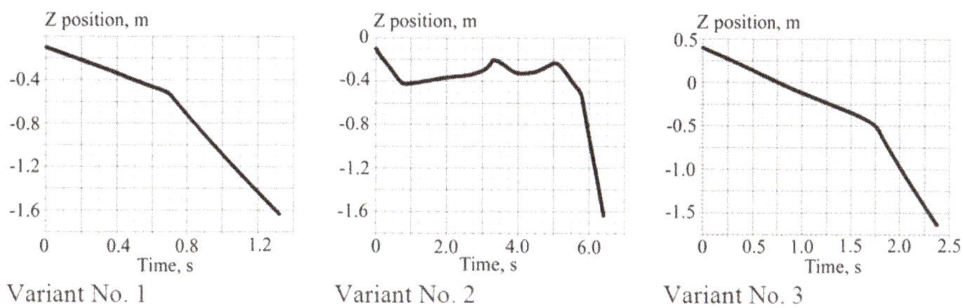

Figure 14.23. The *Z*-coordinate dependence graph of the gas elementary volume on the movement time along the heat exchanger.

The total gas elementary volume existence time at the heat exchanger for the investigated variants is 1.32 s, 6.42 s and 2.37 s, respectively. The maximum gas cooling is observed during its contact with the dispersed fluid droplets, which evaporate:

- Variant no. 1 is characterized by lowering the gas temperature to 472 ° C for 0.47 s. This is observed between sections C and E.
- Variant no. 2 is characterized by gas temperature 436 °C between sections C and E. This temperature is reached for 1 s of its stay in the heat exchanger.
- Variant no. 3 is characterized by gas temperature 454 °C between sections A and E. This temperature is reached during 1.6 s of its stay in the heat exchanger.

The investigation shows that the maximum gas cooling rate is reached between sections C and D. According to the results of numerical simulation, it can be seen that for variant no. 1 it is −1919 °C s^{-1}, for variant no. 2 it is −1269 °C s^{-1} and for variant no. 3 it is −975 °C s^{-1}. The rest of the time is taken to mix the gas with water vapor. The dependence of the gas temperature on the residence time of the gas stream in the heat exchanger is shown in figure 14.24 and the histogram of the average gas cooling rate in the areas located between the control sections i and $(i-1)$ is shown in figure 14.25.

Thus, the variant no. 2 of dispersed fluid supply is characterized by a short time to establish an equilibrium state of the vapor–gas mixture, by reduced fluid flow and eliminatesing its accumulation in the heat exchanger. However, from the point of view of ensuring environmental safety, the first option can be considered as the most

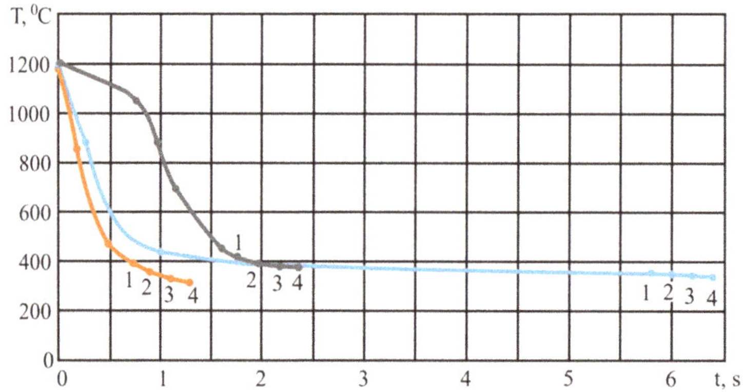

Figure 14.24. The dependence of the gas flow temperature on the time of its cooling: red—variant no. 1; blue—variant no. 2; black—variant no. 3.

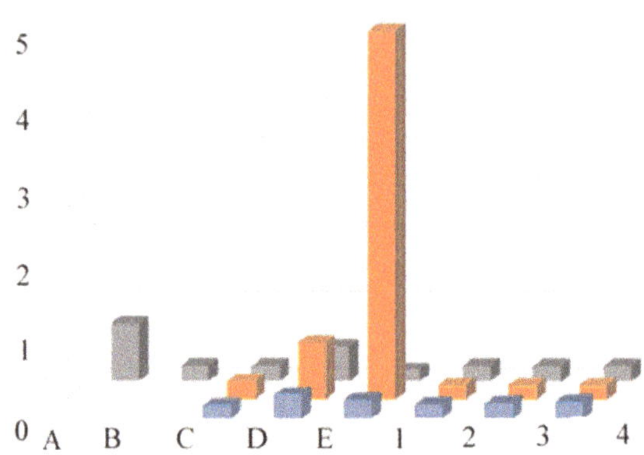

Figure 14.25. The gas residence time between the control sections: red—variant no. 1; blue—variant no. 2; gray—variant no. 3; A, B, C, D, E, 1, 2, 3, 4 are control sections.

satisfactory, since the cooling time of the gas stream from 1200 °C to 305.8 °C is 1.32 s.

However, these results do not exclude the possibility of determining the most effective option, in the search for which a corresponding optimization problem can be formulated and solved in the future.

14.4 General conclusions

The proposed mathematical model of the transportation process of finely dispersed drops of fluid to the zone where there is a source of environmental hazard reliably describes the peculiarities of the process depending on the ways of supplying fluid.

Using numerical simulation of the spatial finely dispersed structure creation process, it is possible to determine the most effective ways of supplying fluid to hazards of various origins, to propose various technological solutions for optimizing this process and managing environmental safety.

The results of the calculations provide an opportunity to argue that the use of spatial finely dispersed multiphase structures for environmental safety management is an effective way to eliminate the dangers from different sources.

References

[1] Kolesnyk V, Pavlychenko A, Borysovs'ka O and Buchavyy Y 2018 Formation of physic and mechanical composition of dust emission from the ventilation shaft of a coal mine as a factor of ecological hazard *Solid State Phenom.* **277** 178–87

[2] Vambol S, Vambol V, Sundararajan M and Ansari I 2019 The nature and detection of unauthorized waste dump sites using remote sensing *Ecol. Questions* **30** 1–17

[3] Vambol S, Bogdanov I, Vambol V, Suchikova Y, Kondratenko O, Hurenko O and Onishchenko S 2017 Research into regularities of pore formation on the surface of semiconductors *Eastern-Eur. J. Enterp. Techn.* **3/5** 37–44

[4] Kolesnik V, Ye, Pavlichenko A V and Buchavy Y V 2016 Determination of dynamic parameters of dust emission from a coal mine fang *Nauk. Vis. Nat. Hirny. Univ.* **2** 81–7

[5] Tverda O, Plyatsuk L, Repin M and Tkachuk K 2018 Controlling the process of explosive destruction of rocks in order to minimize dust formation and improve quality of rock mass *Eastern-Eur. J. Enterp. Techn.* **3/10** 35–42

[6] Sundararajan M, Sharma S N, Kumar R, Ansari I and Kumar G 2018 Estimation of SPM emission in air environment through empirical modeling and remedies for dust control in and around coal mining complexes *National Seminar on Environmental Issues: Protection, Conservation and Management (EIPCM) (Dhanbad, Jharkhand, 26–7 February 2016)* pp 207–14

[7] Tverda O, Tkachuk K and Davydenko Y 2016 Comparative analysis of methods to minimize dust from granite mine dumps *Eastern-Eur. J. Enterp. Techn.* **2/10** 40–6

[8] Kulyk M P 2014 Analiz ekolohichnoyi nebezpeky obyektiv teplovoyi enerhetyky ta metodiv zmenshennya shkidlyvykh vykydiv *Vis. Inzh. Aakad. Ukrayiny* **2** 253–8

[9] Vambol S, Vambol V, Sobyna V, Koloskov V and Poberezhna L 2018 Investigation of the energy efficiency of waste utilization technology, with considering the use of low-temperature separation of the resulting gas mixtures *Energetika* **64** 186–95

[10] Vambol S, Vambol V, Kondratenko O, Koloskov V and Suchikova Y 2018 Substantiation of expedience of application of high-temperature utilization of used tires for liquefied methane production *Arch. Mater. Sci. Eng.* **87** 77–84
[11] Savov N, Hristova T and Gencheva P 2019 Exploring possible accidents and recommendations for biological waste treatment installations *Qual.— Access Success* **20** 195–200
[12] Ragimov S, Sobyna V, Vambol S, Vambol V, Feshchenko A and Shalomov V 2018 Physical modelling of changes in the energy impact on a worker taking into account high-temperature radiation *J. Achievem. Mater. Manuf. Eng.* **91** 27–33
[13] Vambol S O, Bohdanov I T, Vambol V V, Suchikova Y O, Kondratenko O M and Onyschenko S V 2017 Formation of filamentary structures of oxide on the surface of monocrystalline gallium arsenide *J. Nano- Electron. Phys.* **9** 06016
[14] Hristova T 2017 Measures for restricting corrosion of a motor drive subjected to atmospheric corrosion in the plant for wastewater treatment *Qual.—Access Success* **18** 202–7
[15] Vambol S, Vambol V, Suchikova Y, Bogdanov I and Kondratenko O 2018 Investigation of the porous GaP layers' chemical composition and the quality of the tests carried out *Arch. Mater. Sci. Eng.* **86/2** 49–60
[16] Vambol S, Vambol V, Bogdanov I, Suchikova Y, Lopatina H and Tsybuliak N 2017 Research into effect of electrochemical etching conditions on the morphology of porous gallium arsenide *Eastern–Eur. J. Enterp. Techn.* **6/5** 22–31
[17] Suchikova Y, Bogdanov I, Onishchenko S, Vambol S, Vambol V and Kondratenko O 2017 Photoluminescence of porous indium phosphide. Evolution of spectra during air storage *Proc. of the 2017 IEEE 7th Int. Conf. on Nanomaterials: Applications and Properties (NAP-2017)* pp 138–41
[18] Sokolov D, Sobyna V, Vambol S and Vambol V 2018 Substantiation of the choice of the cutter material and method of its hardening, working under the action of friction and cyclic loading *Arch. Mater. Sci. Eng.* **94/2** 49–54
[19] Lei M, Hai J, Cheng J, Lu J, Zhang J and You T 2018 Variation of toxic pollutants emission during a feeding cycle from an updraft fixed bed gasifier for disposing rural solid waste *Chin. J. Chem. Eng.* **26** 608–13
[20] Milosh V V Dioksiny i ih potentsial'naia opasnost' v ekosisteme "chelovek—okruzhaiushchaia sreda" http://crowngold.narod.ru/articles/dioxini.htm
[21] Wu X, Zheng M, Zhao Y, Yang H, Yang L, Jin R and Liu G 2018 Thermochemical formation of polychlorinated dibenzo-p-dioxins and dibenzofurans on the fly ash matrix from metal smelting sources *Chemosphere* **191** 825–31
[22] Vambol S, Vambol V, Bogdanov I, Suchikova Y and Rashkevich N 2017 Research of the influence of decomposition of wastes of polymers with nano inclusions on the atmosphere *Eastern-Eur. J. Enterp. Techn.* **6/10** 57–64
[23] Lu P, Huang Q, Bourtsalas A T, Themelis N J, Chi Y and Yan J 2018 Review on fate of chlorine during thermal processing of solid wastes *J. Environ. Sci.* **78** 13–28
[24] Vambol V 2016 Numerical integration of the process of cooling gas formed by thermal recycling of waste *Eastern Eur. J. Enterp. Techn.* **6/8** 48–53
[25] Launder B E and Spalding D B 1972 *Lectures in Mathematical Models of Turbulence* (London: Academic), pp 458
[26] Schmidt D P, Corradini M L and Rutland C J 2000 A two-dimensional, non-equilibrium model of flashing nozzle flow *Proc. of the 3rd ASME·JSME Joint Fluids Eng. Conf.*

(San Francisco, CA, 18–23 July 1999) (New York: American Society of Mechanical Engineers), pp 1322

[27] Loitsianskii L G 1978 *Mehanika Zhidkosti i Gaza* (Moskva: Nauka), p 736

[28] Joshi M 2002 Failure of dust suppression systems at coal handling plants of thermal power stations—a case study https://plant-maintenance.com/articles/dust_suppression.pdf

[29] Redko A, Dzhyoiev R, Davidenko A, Pavlovskaya A, Pavlovskiy S, Redko I and Redko O 2019 Aerodynamic processes and heat exchange in the furnace of a steam boiler with a secondary emitter *Alexandria Eng. J.* **58** 89–101

[30] Launder B E and Spalding D B 1974 The numerical computation of turbulent flows *Comput. Meth. Appl. Mech. Eng.* **3** 269–89

[31] Vambol S A, Skob Y U A and Nechiporuk N V 2013 Modelirovaniye sistemy upravleniya ekologicheskoy bezopasnost'yu s ispol'zovaniyem mnogofaznykh dispersnykh struktur pri vzryve metanovozdushnoy smesi i ugol'noy pyli v podzemnykh gornykh vyrabotkakh ugol'nykh shakht *Vestnik Kazan. Tekhnol. Un-ta* **16/24** 168–74

[32] SAS IP 2011 Using the Rosin–Rammler diameter distribution method https://sharcnet.ca/Software/Fluent14/help/flu_ug/flu_ug_sec_discrete_diameter.html

[33] Kostyuk V and Ye 1988 K vyboru approksimiruyushchego vyrazheniya dlya koeffitsiyenta aerodinamicheskogo soprotivleniya kapli *Nauch.-metod. Mater. Teor. Aviats. Dvig.: Sbor. Nauch. Trud. KHVVAIU* **6** 13–21

[34] Tsipenko A V 2004 Matematicheskaya model' dispersnogo neravnovesnogo potoka s bol'shoy doley zhidkosti v sople s uchetom plenki, stolknoveniy i aerodinamicheskogo drobleniya kapel' *Moskva, NII NT, 46 s*

[35] Krou D 1982 Chislennyye modeli techeniy gaza s nebol'shim soderzhaniyem chastits *Ser. Teor. Osn. Inz. Rasch.* **104** 114–22

[36] Vambol' S O 2012 Systema upravlinnya ekolohichnoyu bezpekoyu pry vykorystanni pylopryhnichuyuchykh system zroshennya u protsesi navantazhennya ta rozvantazhennya sypkykh materialiv u portakh *Otk. Ynform. Komp'yut. Ynteh. Tekhnol.* **55** 161–67

[37] Vambol' S A 2012 Yssledovanye chyslennym metodom protsessa postanovky dyspersnoy vodyanoy zavesy v systemakh upravlenyya ékolohycheskoy bezopasnosty *Ekol. Bez.: Prob. Shlyak. Vyris.* **2** 154–59

[38] Vambol V V 2014 Matematicheskoye modelirovaniye gazovoy fazy okhlazhdeniya generatornogo gaza ustanovki utilizatsii otkhodov zhiznedeyatel'nosti *Ekol. Bez.* **6** 148–52

[39] Vambol V V 2015 Modelirovanie gazodinamicheskih protsessov v bloke ohlazhdeniia generatornogo gaza ustanovki dlia utilizatsii othodov *Tehnol. Tehnos. Bezo.: Intern.-Zhur.* **1** http://ipb.mos.ru/ttb/index.html

[40] Vambol V V 2015 Matematicheskoye opisaniye protsessa okhlazhdeniya generatornogo gaza pri utilizatsii otkhodov zhiznedeyatel'nosti *Tekhnol. Audit Rez. Proiz.* **2/4** 23–9

[41] Fletcher K 1991 Vychislitel'nyye metody v dinamike zhidkostey *M.: Mir* **1** 504

[42] Shervud T, Pigford R and Uilki C 1988 *Massoperedacha* (Moscow: Mashinostroenie), p 600

IOP Publishing

Modern Optimization Methods for Science, Engineering and Technology

G R Sinha

Chapter 15

Future directions: IoT, robotics and AI based applications

K C Raveendranathan

The recent innovations in the information age, where data is considered the 'new oil', point toward rapid all pervasive developments in the Internet of Things (IoT), robotics and artificial intelligence (AI) based applications. Data science has evolved in the last few decades as a promising field of vast opportunities and challenges, which encompasses all the endeavors of mankind. As raw data evolves into information and intelligence through several data processors, its value is multiplied many-fold. In this chapter, we primarily focus on future directions in the disruptive technologies such as IoT and its importance in building smarter systems for a brave, new and smarter world, where robotics and AI based applications play a pivotal role in every human activity. IoT is a coinage of Kelvin Ashton of MIT in 1999, and is, in general, any network of smart, connected devices which can be controlled from anywhere across the globe through the Internet. It should be emphasized that despite its potential advantages, such as being an enabler of global, remote connectivity and thus aiding use of our home appliances which are smart enough to connect to the Internet, its vulnerability to cyber attacks cannot be overlooked by the intelligent designer, developer, or even the end user. Robotics and its principles were known to technology evangelists and to end-users from the days of the science fiction play *R. U. R.* (Rossum's Universal Robots) by the Czech science fiction author Karel Čapek (and his brother Joseph Čapek, who actually coined the term 'robot') in the 1920s. The principles of AI were first suggested by Herbert A Simon, Marvin Minsky, Allen Newell and John McCarthy at a 1955 conference, however, the credit for the term is rightfully attributed to the latter. If we take a count of the devices/appliances/other smart things connected to the internet by the year 2025, it will be the astounding figure of 34.2 billion worldwide (projected data from IoT analytics), compared to the present status of 17.8 billion worldwide in 2018, inclusive of IoT devices. One can take

doi:10.1088/978-0-7503-2404-5ch15

these figures for granted and it is more likely that the actual numbers *may exceed* the projections by 2025 unless some other disruptive technology evolves. Robotics has matured enough with developments in interdisciplinary technologies such as mechatronics and AI, such that fully automated vehicular systems and other means of transport have become the order of the day. It is seemingly unpredictable, with the current rate of innovations in mechatronics and AI, in which direction products and processes will evolve. However, it is arguably going to be more toward a level playing field, where the tools and techniques are primarily AI based, starting from the raw/semi-processed data pumped by the IoT devices, communicated through the Internet, and processed by the most advanced and inexpensive signal processors (both passive and active). AI has grown to a mature technology in machine intelligence, and there have been several more recent developments, such as machine learning (ML) and deep learning, and several types of artificial neural networks (ANNs). The convolution neural networks (CNN), recurrent neural networks (RNN) and deep neural networks (DNN) fall into this category.

15.1 Introduction

In our information age, data is considered to be the new oil. The advent of the Internet and its associated technologies has given way to huge data proliferation from a number of sources. The IoT is an ecosystem of connected mechanical and digital devices, physical objects, or living organisms, including human beings, that are given a unique identity and have the ability to communicate and internetwork together over a network, without needing human intervention. The IoT devices that are interconnected through the Internet can be categorized into three groups:

1. Devices that receive and retransmit information.
2. Devices that receive information, and then process and act on it.
3. Devices that do both of the above.

Each of the above categories of things has enormous advantages associated with it. Examples for the first category of devices, namely those devices that collect and send information to the Internet, include all sorts of sensors. Typical sensors are temperature, motion, moisture, air quality, or even optical (light). These sensors enable the user to gather environmental information such as temperature, humidity, air quality, etc, from their surroundings autonomously and communicate through the wired or wireless Internet, which will, in turn, allow the end-users to make apt and timely decisions. Just like the human sensory organs such as the nose, ear, tongue and eyes help human beings perceive the surrounding world, these 'smart sensors' enable various IoT devices to do the same. Also, the devices that receive and act on sensory information can be operated remotely from a distant place, and this is one of the major merits of the IoT. It is interesting to note that IoT devices also greatly contribute to the proliferation of data, this will eventually open up a new avenue for extensive study and research in data sciences, namely data analytics.

The third category of devices—the ones that collect information and retransmit, as well as receive information and act on it—is the true goal of the IoT. To cite an example, a soil moisture sensor in an agricultural farm can sense the moisture in the soil and in turn decide when to switch on the irrigation system to water the plants. Note that this is done in an intelligent fashion without the intervention of the farmer. Obviously, the IoT improves the efficiency of operation of devices which are connected to the Internet. The major outcome of the IoT and associated technologies is that of data proliferation. The IoT results in large chunks of real-time data to be processed either locally or at a distant data center. Thus the proliferation of IoT devices eventually leads to big data analytics.

The IoT is considered to be a disruptive technology, due to the fact that it has indeed pre-empted several prevalent technologies. Further, several new applications have evolved and proliferated with IoT, including distributed computing, wireless sensor networks and so on. The principles of robotics have evolved and the technologies matured over the past several decades. AI has created another paradigm shift in modern-day computing. Robotics and AI, together, have produced several cutting edge technologies and applications in the current millennium. The intelligence displayed by man-made machines is termed AI or machine intelligence (MI). It is to be contrasted to the natural intelligence displayed by human beings as well as other types of living organisms, including plants. With AI technology one can build intelligent machines that can work and react exactly like human beings. It is interesting to note that we have now entered into a new era of technological advancements, where machines have started creating machines, with minimal or almost nil human aid. The machine intelligence of such machines has to be 'measured' using a different gauge of MI.

In this chapter, we mainly focus on the future directions in IoT, robotics, and AI based applications.

15.1.1 The impact of AI and robotics in medicine and healthcare

A recent study by the market research firm Tractica estimated that the number of shipments of healthcare robots would exceed 10 000 units annually by the year 2021. Another study by yet another research firm, Allied Market Research, suggested that the global market of robots used in clinical surgery would nearly double from US $3.00 billion (in 2014) to US$6 billion (in 2020). AI technology is taking a crucial role in the medical and pathological assessments of various health conditions and the subsequent phases of treatments needed. Expert systems that can scan through a massive volume of medical literature and transcripts are aiding medical practitioners the world over to achieve more accurate diagnosis in a shorter time. As a typical example, the novel 'GP at Hand' service, which is very popular in the UK (London), enables patients to check various health conditions using a mobile app, thereby being able to consult qualified medical practitioners through video conferencing within two hours. GP at Hand uses AI to diagnose the ailment.

Yet another study by Frost and Sullivan revealed that the market driven by AI technologies for healthcare would increase by 40% per annum from 2014 to 2021; the corresponding rise in revenue varied from US$633.8 million in 2014 to US$6.662 billion by 2021. Another important example is that of the DAvinCi operating robots, which will be discussed in more detail in a later part of this chapter.

They are very intelligent machines with a very high machine intelligence quotient (MIQ) that use AI algorithms at their core to improve the accuracy when operating on patients. AI techniques are used in producing high-resolution digital images and holograms that are crucial in several applications such as space-imagery, the detection and prevention of forgery, etc [1].

15.1.2 Advances in AI technology and their impact on the workforce

Over the years, AI has evolved from the elementary symbolic AI through embodied AI to hybrid AI. As CPUs, networks and wireless communications matured over the years, so did AI and robotics techniques [2]. Autonomous systems began to play a major role in several diverse application scenarios and fields, such as space technology to service robotics. Such intelligent robots with a high MIQ can assist the users in various daily chores. They are very good at performing hazardous, dirty and monotonous jobs. However, the use of such robotic systems to perform autonomously in complex and diverse applications over an extended time (from a few days to years) imposes several threats to humanity. Many of these are studied by experts in the sub-disciplines of AI. They are navigation and mapping, reasoning, knowledge representation, perception, autonomous planning, interaction and learning from past experiences. These sub-disciplines of AI with advanced techniques which are integrated with an self-governing system enable robotic systems to operate with more efficacy with a high factor of safety in a very complex, long duration application [3].

Although AI arguably demonstrates profound advantages in the use of the technology and provides several benefits that can be enjoyed every day, thereby enhancing the quality of human life on Earth, there are a few apprehensions with respect to the increasing negative 'adverse ontological and existential' outcomes of AI. The proponents of this school of thought hold the firm belief that such super-intelligent outcomes of AI technologies are highly hazardous to mankind at present. The researchers Vardi, Tegmark and Greene compared it to a time-bomb ready for explosion in the near future. A recent paper on 23 AI principles compiled by nearly 1200 AI/robotics researchers and 2342 other researchers from different domains, presented at the 'Future of Life (FLI) Conference', added more credence to the fears which most researchers share despite the benefits of AI to mankind. 'It is imperative that the rising advancements in AI technologies, continues to alienate humans from their existential human nature' [4]. Already, millions of routine, working-class jobs in manufacturing have been automated. Furthermore, AI is now progressively maturing to automate non-routine jobs in transportation, logistics, banking and financial services, secretarial and clerical services, and healthcare. To cite a few examples, the Google translator, Amazon's Alexa and so on, are enabled by AI technology. The ordinary working class had worried that advanced AI technology would, more or less, demolish many routine jobs. Despite many learning the new skill sets demanded by the new jobs which replaced the old ones, most of the workers who were not able to keep up with the need to learn the new job-skills demanded by the industry could not be redeployed. Humans could never successfully compete with machines that could outperform them in every endeavor. Several modern-day economists envisage that society will not be able to cope easily with the change. People now realize that the dictum that AI technology that destroys existing jobs will eventually create new job opportunities is too optimistic in reality [5]. Questions such as whether the jobs replaced by AI systems will be replaced by the new jobs created, of course after acquiring a new skill set, are highly relevant. A socio-economic study conducted in the year 2013 reported that about 47% of American workers held jobs that were at high risk of extinction due to automation, just a few decades from now. The highly relevant questions are: 'Will technology be able to create about a 100 million jobs if these jobs are automated by AI technologies? [5] Will AI technology be able to create new jobs to compensate for those it has demolished or made irrelevant? Will these jobs be created fast enough to meet the rising demands of those who lost their earnings? What will be the state of employees whose skill sets cannot catch up with the existential advancements in modern technology? Will such workers lose their frugal existence in society? So far, the answers to the above questions are affirmative.

A recent US study revealed that the employment potential of highly paid cognitive jobs and low-paid service jobs are growing fast, such as physical aid to the disabled or physically challenged older generation and fast-food services, which are yet to be automated. However, technology is puncturing the economy severely, through the automation of mid-skill, working-class jobs. Since 2000, several millions of low salary service jobs have disappeared, and the workers were either forced to leave the labor force or accept low income jobs that often pay meager sums, without any other perks or privileges. It is apparent that firms in the communications technology sector save huge amounts otherwise spent on salary by hiring temporary employees on a contract basis, instead of full time regular staff. This led to the 'gig economy'—a job market where short-term contracts and flexible hours with almost no benefits to the employees have flourished. Automation due to massive AI based applications has fully alienated the creation of jobs from economic growth, resulting in economic growth with an alarming rise in unemployment and large-scale shrinkages in incomes of workers, thus resulting in more inequality in society. It may also be noted that the newer technologies have created a 'winner-takes-all' situation, where the loser can hardly survive. In 1990, the three largest corporations in Detroit were worth US$65 billion and had a labor force of 1.2 million. By 2016, the three largest corporations in the Silicon Valley were worth US$1.5 trillion but could accommodate only a mere 190 000 workers [5]. Hence the modern industrial trend is larger communication companies and fewer jobs. If this is true, then AI technology must create nearly 100 million new jobs to balance the gap in the labor market.

The reliance is on advances in the domain of AI technologies that would create more new jobs for the majority of the labor force, and everyone would earn a living as 'existential beings in their society'. On the other hand, when fewer new jobs are created to accommodate even the newly up-skilled jobseekers, there are very high unemployment rates, which is the situation termed by Verdi [5] as 'a state of violent uprising'. Considering the volatile state of the labor market, educational institutions for engineering must focus on training their student community with the right skill set needed, which the innovations in AI demand. Employees would need to continually upgrade and hone their skills by attending appropriate training programs for better prospects in the job markets. Even then, 'the need to adapt and train for new jobs will become more challenging as AI continues to automate a greater variety of tasks', as very rightly pointed out by Verdi.

15.1.2.1 Some recent trends on workforce lay-offs, redeployment and new hires
A recent article that appeared in *IEEE Spectrum* [8] revisited the statistics of jobs lost due to lay-offs, redeployment after up-skilling, and the newly created jobs. The study mainly focused on the Silicon Valley area in the US and went on to add that most of the jobs lost due to lay-offs in various companies are compensated for by the hiring of new technical and non-technical staff against the vacancies that were created by lay-offs. Thus the new technologies enable the creation of new jobs and most of the workforce who lost their jobs could easily find new avenues for employment.

15.1.3 AI technologies and human intelligence

An intelligent agent (IA) is a system that senses its environment alone and takes appropriate actions to reduce its chances of failure. American computing pioneer John McCarthy, who is credited as the originator of the term artificial intelligence (AI) in 1956, defined the discipline as 'the art, science and engineering of making intelligent machines'. AI can be thought of as that discipline in the field of science that focuses on building devices or machines that address complex issues, finding solutions for common needs of human beings and thereby enhancing the quality of life. In that process, the AI system designer inculcates human intelligence into algorithms that govern the autonomous functioning of intelligent machines in a user-friendly manner.

On 11 June 1997 an IBM mainframe computer, namely 'Deep Blue', defeated Garry Kasparov, the greatest chess player (Grand Master) at the time, in six straight set games of chess, proving that machine intelligence can become comparable to that of humans. Thus Deep Blue passed Alan Turing's test in machine intelligence. Contemporary computing machines are much more than just intelligent, maturing to super-conscious, covertly intelligent automatons, with the capacity to think on their own, which is frequently depicted as detrimental to mankind. It now evident that with the advent of highly intelligent AI and ML technologies, computing devices have now successfully invented and programmed automations, such as autonomous cars, pacemakers and involuntary trading systems. A few decades earlier, all these were mere dreams. Notwithstanding the projected merits of AI and ML technologies, Stephen Hawking, the greatest physicist of our times, could not suppress his apprehension of the suggested virtues of AI technology. He rightly stated: 'The success in creating AI would be the biggest event in human history. Unfortunately, it might also be the last, unless we learn to avoid the risks associated with it' [6].

One specific area where AI technology could be horrendous to mankind, for example, could be a situation where advanced AI based weaponry systems are encoded to perform risky maneuvers in military or space applications. Such deeply programmed systems are extremely difficult to alter easily once they are deployed—as in the case of the Inter-Continental Ballistic Missile (ICBM) mechanized weaponry designed to target an enemy, some 700 miles aways, which is approximately the distance between Japan and North Korea. Advanced AI war machines of this type are known to counter all external influences aimed to alter their core objective. This potential defect must be met with critical counter measures. Hence a machine such as an ICBM guides itself, evading whatever obstacles anyone throws in its path, ultimately attaining its original task of annihilating its target. In this context Professor Stephen Hawking observed that 'Artificial intelligence machines could kill us because they are too clever. Such computers could become so competent that they kill us by accident. The real risk with AI isn't malice but competence. A super-intelligent machine will be extremely good at accomplishing its goals, and if those goals aren't aligned with ours, we're in trouble'.

In general, super-intelligent AI technologies are mostly feared for their capability to upgrade themselves. Perhaps, they can even reprogram themselves whenever the need arises. Having acquired a high MIQ, they will be more aware and conscious of their environment, including the flora and fauna around them. Interestingly, they will begin to think on their own and make judgments or decisions on what best to do with regards to carrying out their assigned tasks or objectives. The mega intelligent AI systems are feared to have the rare ability to upgrade, preserve and protect themselves from whatever they may think of as a threat, meaning they would in the nearest future be able to resist being disrupted, corrected or reprogrammed by their programmers [4].

15.2 Cloud robotics, remote brains and their implications

The term cloud computing was initially proposed by an American computer engineer, Andy Hertzfeld. The demand for cloud-based applications is increasing every day, since the cloud has a high level of scalability and is very efficient because of its capability to be resilient in meeting current needs. Ever after the cloud paradigm became ubiquitous and inexpensive, several automatons previously thought of as very lethargic and inordinately complex became quite viable. For robotics and AI, this meant that if the computational power behind the cloud could be utilized, it would be possible to build tiny, highly energy efficient robots because there would be no need to have a powerful computer on board. The entire brain of the robot could be in the cloud. Yet the concept of a 'remote-brain' is not new [7]. Due to the distributed nature of the cloud across the globe, it is feasible to have faster response times. This is possible as the processors are close to the data they are working on, to reduce network delay, and have a low mean time between failures (MTTF), and in particular is financially viable as it is a 'pay per use model'. James Kuffner, who was working at Google and is now with Toyota Research Institute (TRI), predicted correctly that the cloud could make robots 'lighter, cheaper and smarter' [8]. The use of the cloud to handle computationally intensive tasks permits the use of smaller on-board computers in robots. These are used to attend to tasks that need real-time processing, such as the control of sensors/actuators or stepper motors. The grouping of low-level and high-level reasoning was first explored as 'the concept of remote-brain' in 1996 at the University of Tokyo. The global knowledge base for robots is a depository of objects, actions or environments. The robots are enabled to download the object description and user manual, even when it was the first time it encountered that particular object, or will be able to plan a route in the environment, which was earlier traversed by another. One major disadvantage in using cloud-based architecture is the chance of loss of connectivity between a transaction, and even if the robot uses the cloud services for basic functionality only, it will fail to do anything tangible. This is an acceptable constraint if a backup is provided to meet that contingency. At present, the stability of network connectivity is no longer an issue. Acquiring a new stable infrastructure is affordable compared to the cost of a robot with full embedded intelligence. Now the real problem pertains to the latency of the network [7].

Cloud robotics usually comprise two types of communications, namely machine-to-machine (M2M) and machine-to-cloud (M2C). The first tier of communication, M2M, implies that the robots are in a collaborative computing environment termed an '*ad hoc* cloud', which allows them to jointly share computationally intensive tasks by pooling the resources needed and by exchanging information for collaboration, and more importantly connecting to robots not in the range of the cloud access point and helping them to communicate with the cloud. In short, most of the communications in this tier will be restricted to M2M. In the second tier of communication, i.e. the M2C level, the cloud can release resources for computation and storage on-demand. The robots in M2M mode communicate and share resources for a task that is beyond their capability or which is common among themselves. This is illustrated in figure 15.1.

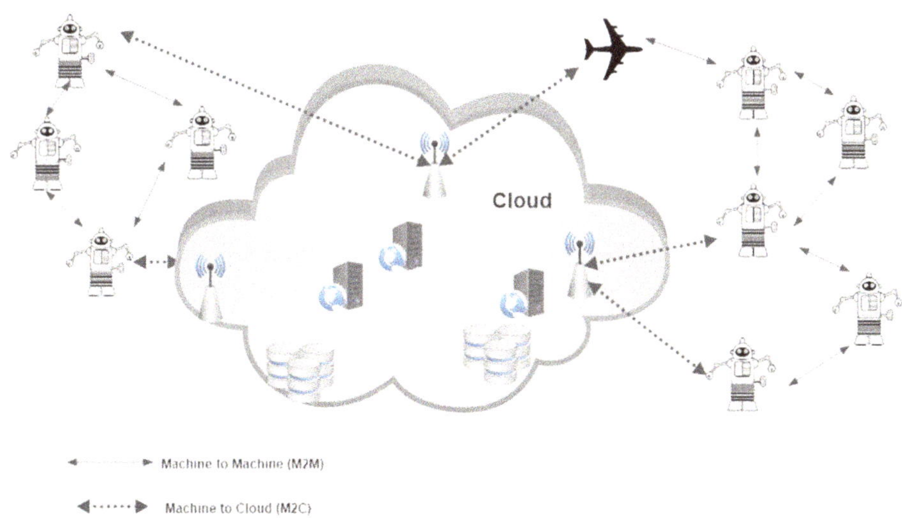

Figure 15.1. M2M and M2C communication modes. Image source: [9].

As mentioned before, this in effect acts as a 'remote-brain' along with the shared memory of actions, acquired skills, and information already learned and obtained from the cloud.

15.2.1 Cloud computing and the RoboEarth project

The underlying principle of the RoboEarth [10] project was to create a pool of robots which would allow them to share common knowledge, having a plug-in architecture, and would use the ontology web language (OWL). There were three databases in the cloud which could hold information on actions, objects and the environments. The databases were maintained using OWL. Using the robot operating system (ROS) the robots were interconnected [11]. But the robots were not restrained to the use of ROS for communication. The entire information on the capabilities of the robots, including construction and sensor types, was published for the system, and hence the actions a robot could do were made known to others. The essence of the RoboEarth architecture was the recognition/labeling component (RLC). The RLC connected the hardware of the robot. The following abstractions are maintained inside the robot's system—actions, objects and environments. The major objective of RLC was to translate abstract definitions from the databases of RoboEarth to a format understood by the particular robot and vice versa, so the robot could contribute to the databases with new acquired knowledge [7]. What it meant was that it could work on low-level actions, called atomic primitives—signals from sensors, motors and others. It could also work on high-level actions—spatial and temporal relations between actions to create and execute abstractions of actions. The architecture of RoboEarth is shown in figure 15.2.

New actions were learned through tele-operation. An operator controlled the robot. Since RoboEarth was put on the ROS platform, it received all the signals from the sensors and the motors, and assembled them to the action plan. Then the robot interrogated the operator to label the action it did. The learning process was also running when the robot was executing the algorithm of a particular action, to further improve it [7].

15.2.2 The DAvinCi platform as a service (PaaS) surgical robot

The DAvinCi framework [12] was a software system that provided the merits of parallelism and scaling up of cloud computing for service robots in fairly large set-ups. The robot operating system (ROS) was employed to communicate with the robots. The Hadoop/MapReduce framework [13] was used for parallel-processing to enhance speed. The framework capabilities were tested on the implementation of the fast simultaneous localization and mapping (FastSLAM) algorithm. The DAvinCi was a typical platform as a service (PaaS) cloud-based application. The robots communicated with the DAvinCi server. They could run ROS nodes for the robots which did not have the capability to do so. Hierarchically, the Hadoop distributed file system (HDFS) cluster which was located above the server was used for execution of robotic algorithms. The DAvinCi server acted as a central communication hub and the master node. The FastSLAM algorithm was implemented and tested successfully [7].

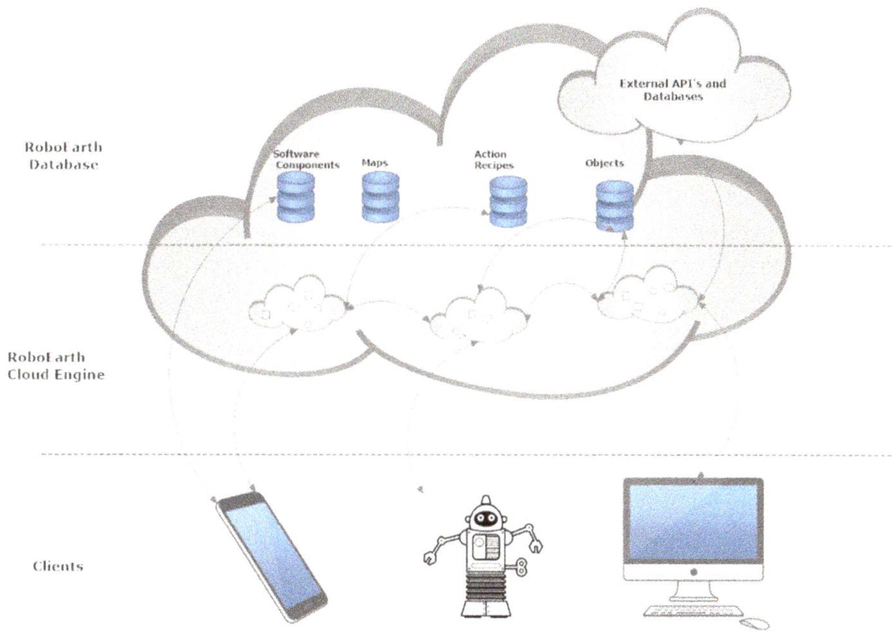

Figure 15.2. The architecture of RoboEarth. Image source: [23].

Although the domain of cloud computing is quite new, there are multiple projects based on its principles and those of robotics. The major advantages of cloud computing are its scalability, massive storage, resilience and ability for parallelism. The methods used in AI, grabbing all resources in hold ('grasping'), image processing, and neural networks, are heavily computationally intensive. Hence it is not viable to use them on a wide range of robots which do not possess enough on-board computing resources. With cloud robotics it is possible to achieve the 'remote-brain'. It is an older concept in which the entire hardware and software are separate [7].

15.3 AI and innovations in industry

A system of computer-based applications that performs a group of tasks previously done by white-collar workers, but without human intervention, is aptly called a 'white-collar robot' [14]. White-collar robots can automate certain tasks. Human intervention is limited to deciding what task is to be done by a robot, and to choose the right robot to do that. As in the case of a shop-floor managers, white-collar robots might be part of a process, giving information to others, including people and robots. This definition is consistent with the robots used for expediting manufacturing processes and does not imply that the robots have 'consciousness'. Instead, the robot just executes the processes that they were programmed to do. Quite often the robot will outperform the persons involved in certain scenarios, but the persons are also likely to outperform the robot in certain other scenarios.

15.3.1 Watson Analytics and data science

Watson Analytics (www.ibm.com/analytics/watsonanalytics) provides excellent interactive software to analyze assorted datasets. On uploading data to the cloud, Watson Analytics will analyze the data quality, provide an initial analysis and prompt the user to consider different combinations of variables. Watson Analytics is dissimilar to Watson Cognitive. Watson Cognitive is widely understood from its capabilities shown on the popular TV show *Jeopardy!* The capabilities of Watson Cognitive focus on text and natural language processing. As a result, the capabilities of Watson are grouped into two seemingly independent entities—cognitive sciences and data analytics.

Watson Analytics needs the user or someone else to upload the data to the cloud. After the data are uploaded, Watson Analytics prompts the user to analyze the data. These starting points are a sequence of questions that the system has developed on the basis of the data—for example, 'What drives Process A?', 'What is a predictive model of B?' and 'What is the trend of B and C?' [14].

The data format is important to Watson Analytics, as it does not work well when there are more than two dimensions to the data, say, with row or nested headings, or both. Also, note that Watson Analytics is an active software—one that makes recommendations and performs analysis without user intervention [15]. Generally, active software makes assumptions on user needs. Active software is autonomous and goal oriented. White-collar robots generally use active software [14].

It is highly likely that Watson Analytics frames the data for the user, which could lead the user to form initial conclusions that are not appropriate under greater scrutiny. Furthermore, Watson Analytics does not focus on how independent variables drive the dependent variable. Thus users might understand which variables predict or drive, but they may not know in which direction this occurs. Other limitations include the unavailability of classical statistical tests, and the inability to designate control variables and to choose particular statistical methods. In addition, the modeling approach limits the number of variables used in estimation to one or two [14, 15].

15.4 Innovative solutions for a smart society using AI, robotics and the IoT

AI, robotics and the IoT are attracting widespread attention, as they are expected to be technologies that affect society to a great extent in the future. These innovative technologies have the potential to build seamless communication and a symbiotic society between humans and robots, and a safe and secure networked society [16]. Various components of smart solutions include borderless communication and symbolic communication between human beings and robots (machines).

The smart solutions for a smarter society discussed in this section include automatic speech translation systems, a robot based dam inspection system, and a large-scale security system used in imaging. Speech recognition in noisy environments results in inaccurate translation. The beam forming technology demonstrated its effectiveness by improving speech recognition performance by up to 40%. Also, the directivity and direction of beam forming must be controlled based on the number and direction of speakers and noise sources. For example, directivity should be set narrower when the speakers and noise sources are close [16]. Dam inspection by intelligent robots with adequate provision for lighting was demonstrated to be a viable alternative. For efficient operation of a large-scale system with over 10 000 cameras, visual security by humans is inadequate and image recognition technology will be needed. The conventional approach requires a large number of cameras with a high-capacity network and large-scale server system, which lacks scalability. To solve this issue, a functionally distributed facial recognition system has been developed. The 'Best-shot method' detects faces from security camera images, and then transfers only the best-shot thumbnail which is selected as the most useful image for facial recognition (with the highest resolution angle, facial angle, focus, etc) to the facial recognition server. The server then extracts facial features for face matching. This method reduces the network and server load, making it easier to build a large-scale security system [16].

15.4.1 Cyber-physical systems (CPSs)

AI technologies will play more important roles in expanding cyber-physical systems [17] sustainably in the future. A high-performance cloud-based processing environment in cyber space will also be necessary. However, due to the performance limitations on cloud-based processing as described in the large-scale imaging security solutions, load distribution is required. The load balance between cloud and IoT/EDGE devices is important when building CPSs. The quickness of response is important for natural dialog translation and autonomous robot solutions.

In the past, the cyber and physical systems were considered as two fairly disjointed and distinct entities. However, modern researchers observed that these two entities are closely related, after integration of sensors/actuators in the cyber systems. Cyber systems respond to the physical world by enacting real-time control from a conventional embedded system. The emerging new research paradigm is termed as a CPS [17]. One can find a lot of applications for CPS, including monitoring of water resources and mines, aerospace, surveillance and so on. Many service providers are focusing on implementing inbuilt CPS technologies for their customers. In CPS, communication is needed for conveying sensor observations to controllers/actuators; thus, the design of the communication architecture is a critical requirement.

The concept of a vehicular cyber-physical system (VCPS) is fairly old. It refers to a wide range of integrated transportation management systems working in real-time, with high efficiency and accuracy. The traditional modes of transport are becoming more intelligent by inculcating the benefits of modern technologies, such as electronics, computers, sensors and networks. The CarTel project, developed by the Massachusetts Institute of Technology (MIT), combines mobile computing and sensing, wireless networks and data intensive algorithms running on servers in the cloud to address these challenges. CarTel helps applications to easily collect, process, deliver, analyze and visualize data from sensors located on mobile units. The applications of CarTel include traffic mitigation, road surface monitoring and hazard detection, vehicular networking and so on [17].

Precision agriculture is experimenting on large-scale farming practices, products, fundamental geographic information of farmland, micro-climate information and other data. The project 'wireless underground sensor network' (WUSN) was developed by the University of Nebraska-Lincoln Cyber Physical Networking Lab, where Agnelo R Silva and Mehmet C Vuran developed a novel cyber-physical system through the integration of pivotal systems with wireless underground sensor networks, i.e. CPS for precision agriculture [18]. The WUSNs consist of wirelessly connected underground sensor nodes that communicate through the soil.

The health cyber-physical systems (HCPS) will replace traditional health devices working on an individual basis. With interconnected sensors and networks, various health devices work together to detect the patient's physical condition in real time. This is particularly useful for patients who are critically ill, such as patients with heart disease, strokes, etc. The portable terminal devices carried by the patient can detect the patient's condition at any time and send a timely warning or prediction of critical conditions in advance. In addition, the real-time activation of health equipment and data delivery systems would be much more beneficial for patients with critical conditions [17]. The proposed standard CPS architecture is illustrated in figure 15.3.

The six standard CPS architecture modules are the sensing module, data management module (DMM), next-generation Internet, service aware modules (SAM), application module (AM), sensors and actuators. The purpose of the sensing module is to send an communication request to DMM and receive its acknowledgment of the request. Once the handshake between the DMM and sensing module is done, the transmission of the sensed data to the DMM from sensing nodes commences. Here, the bridge between the cyber world and the physical world is provided by noise reduction and data normalization techniques. Through quality of service (QoS) routing, data are transferred to SAMs using the next-generation

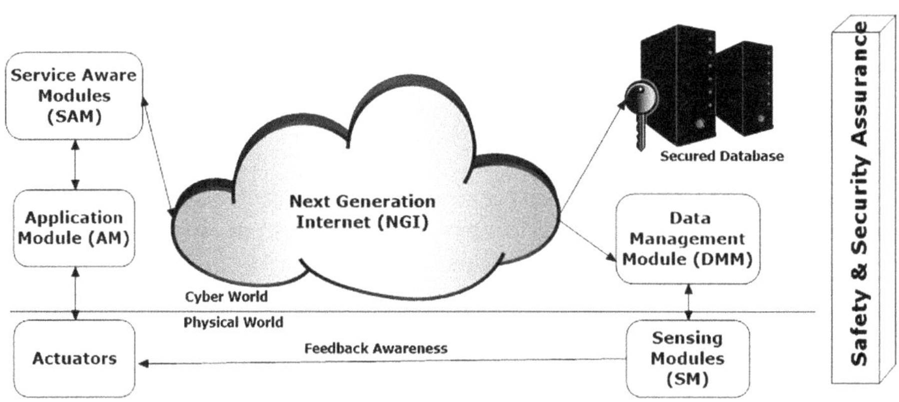

Figure 15.3. The proposed standard cyber-physical system architecture. Image source: [17].

Internet. Available services are assigned to different applications in the application module. To ensure the security and integrity of data, during each network operation data are sent to a cloud platform and also to a local database [17].

15.4.2 IoT architecture, its enabling technologies, security and privacy, and applications

Fog/edge computing has been integrated with the IoT with a purpose—to enable computing devices deployed at the network edge to improve the user experience and resilience of the services, in the case of system failures. The advantages of fog/edge computing are their inherent distributed architecture and closeness to the end-users. Thus, faster response and greater QoS for IoT based applications can be provided. This makes fog/edge computing-based IoT, very attractive in IoT deployment [19]. Note that using fog/edge computing, the massive data generated by different kinds of IoT devices can be processed at the network edge, instead of transmitting it to the centralized cloud infrastructure due to bandwidth and energy consumption constraints. Since fog/edge computing devices are organized following a distributed architecture model, they can process and store data in networked edge devices, which are close to end-users. Thus they can provide services with faster response and greater quality, compared to cloud computing. Thus, fog/edge computing is more amenable to be integrated with IoT devices, to provide efficient and secure services for a large number of end-users. In short, fog/edge computing-based IoT can be considered as the future of IoT infrastructure.

It is well known that both CPS and the IoT try to achieve a close connection between the cyber and physical worlds. In particular, the CPS and IoT can measure the status information of the physical components via smart sensors without human intervention. In both cases, the measured state information can be shared through communication networks. Based on the analysis of measured status information, both CPS and IoT can provide secure, efficient and intelligent services to the end-users. The existing efforts on the CPS and IoT applications have been expanded to similar application domains, such as smart grids, smart transportation, smart cities, and so on. In CPS, the sensor/actuator layer, communication layer, and application or control layers are present. The sensor/actuator layer is used to collect real-time data and execute the commands. The communication layer delivers data to the upper layers and commands to the lower layers. The function of the application or control layer is to analyze the data and make decisions based on it [19]. Note that the CPS is a vertical architecture. On the other hand, the IoT is a cluster of finely internetworked devices in large numbers which are used to monitor and control devices by using modern interconnection technologies in cyber space. Specifically, the crux of IoT lies in 'interconnection'. The main objective of IoT is to interconnect various heterogeneous networks so that the data collection, resource sharing, analysis and management can be carried out smoothly across networks.

Figure 15.4 illustrates the typical integration of IoT with CPS. The basic difference between CPS and IoT is that CPS is considered a system, whereas the IoT is considered an 'Internet'.

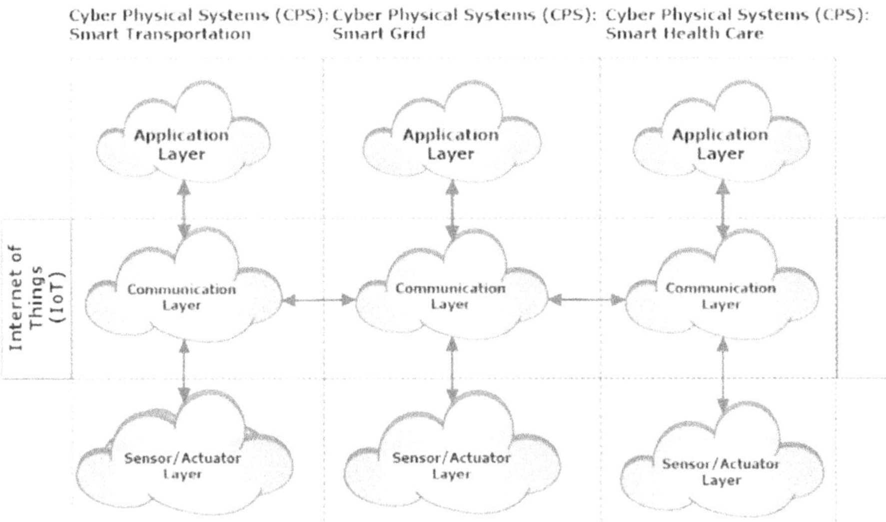

Figure 15.4. Integration of IoT and CPS. Image source: [19].

The common requirements for both are real-time, reliable, and secure data transmission and retrieval. The distinct requirements for CPS and IoT can be summarized as follows. For CPS, effective, reliable, accurate and real-time controls are the primary goals. For IoT, resource sharing and management, data sharing and management, interfacing among different networks, massive-scale data and big data collection and storage, data mining, data aggregation and information extraction, and very high QoS networking are important requirements. Applications of the integrated IoT and CPS include smart grids, intelligent transport systems (smart transportation) and smart cities.

The smart grid is an integrated technology of IoT and CPS. The smart grid has been developed to replace the traditional power grid to provide reliable and efficient energy to consumers [20]. Distributed energy resources are introduced to improve the utilization of distributed energy in electric vehicles to improve the capability of energy storage and reduce the emission of CO_2. In this, smart energy meters and duplex communication networks are introduced to achieve the effective interactions between customers and utility providers. Using IoT devices, a huge number of smart meters can be deployed in houses and buildings connected to the communication networks in the smart grid. Smart meters can monitor energy generation, storage and consumption. They are capable of interacting with power utility providers to report the energy demand information of customers and receive real-time electricity pricing for customers. With the aid of fog/edge computing infrastructure, the huge chunks of data collected from smart meters can be stored and effectively processed so that the efficient operation of the smart grid is possible. With dynamic information processing on the load conditions and the power generated, the utility providers can optimize the energy dispatched in the grid. The customers can optimize their energy consumption, resulting in the improvement of resource utilization and the reduction of cost [19]. This is a truly win-win situation.

Smart transportation, a.k.a. intelligent transportation, is another typical example of intertwined IoT–CPS based application. In smart transportation, intelligent transportation management, control systems, communication networks and computing techniques are integrated to make transportation systems more reliable, energy efficient and secure. In the smart transportation scenario, a large number of smart vehicles are interconnected through wireless networks. Smart vehicles can efficiently perceive and share traffic data and schedule drivers' travels with great efficiency, reliability and safety. In recent years, several smart vehicles (Google's self-driving car, etc) have been designed and field-tested.

Smart cities can be considered a complex IoT paradigm, aiming to manage a host of public activities via information and communication technology (ICT) solutions. Smart cities can use public resources in more efficient ways, which results in the improvement of the QoSs provided to users and the reduction of operational costs. Smart cities are indeed a complex CPS/IoT application. The smart cities thrive on several sub-applications or services, including the smart grid, smart transportation, the structural health monitoring of buildings, waste management, environmental monitoring, smart health, smart lighting, etc. All these services should be supported by a unified communication network infrastructure, or communication networks designed for these sub-applications. In other words, the services should be interconnected to establish a large-scale interconnected heterogeneous network for IoT/CPS applications, with the aim of achieving the best use of public resources in cities.

15.4.3 The Internet of robotic things (IoRT) and Industry 4.0

It was Dan Kara, director of robotics at ABI Research, who conceived the idea of the Internet of robotic things (IoRT). The IoRT is a diverse combination of technologies including cloud computing, AI, ML and the IoT. As the concept of the IoT is evolving and maturing, it leads to significant developments in terms of innovation to various application domains. Several new terminologies are evolving, based on the principles of IoT—the Internet of medical things (IoMT), Internet of nano things (IoNT), Internet of mobile things (IoMBT), Internet of cloud things (IoCT), Internet of autonomous things (IoAT), Internet of drone things (IoDT), Industrial internet of things (IIoT) and Internet of underwater things (IoUT) [21].

It is quite interesting to add that the IoT has also laid a strong foundation for the development of Industry 4.0 or smart factories. Most work will be performed via sophisticated next-generation sensors and robotic technologies in smart factories. Also, in smart factories, all the working personnel can respond quickly to any sort of outage, which currently goes almost unnoticed. Some authors rightly pointed out that Industry 4.0 [22] or the industrial Internet would eventually bridge the physical and digital worlds.

One can easily predict that the IoT, in the near future, coupled with diverse technologies such as AI, ML, deep learning, augmented reality, cloud computing and swarm intelligence will change the face of robotics. The proposed next-generation class of intelligent robotics are aptly titled as IoRT. The market research firm Stratistics MRC stated that the market of IoRT was around $4.37 billion in

2016. It is expected to reach $28.03 billion by 2023 with a staggering compound annual growth rate (CAGR) of 30.4%. The primary factors responsible for this growth are the rise of e-commerce platforms, and the exponential growth of the education sector, consumer markets, research and development wings, and above all Industry 4.0.

15.4.4 Cloud robotics and Industry 4.0

One can find that the primary motivation behind the accelerated growth of the IoRT is cloud robotics [23]. It is regarded as a system that relies on cloud computing infrastructure to access large amounts of processing power and huge amounts of data to perform very specific operations. All operations ranging from sensing, computation and storage are integrated into a single, standalone system such as a networked robot, in cloud robotics. Cloud robotic systems are programmed in such a way that a portion of their capacity is reserved for local processing for low-latency responses in case a network failure occurs. Cloud robotics is conceived and designed as a progression between pre-programmed and networked robotics. With the advent of cloud computing, big data analytics, other emerging technologies, and the integration of cloud technology and multi-robot systems, it has become quite easy to develop multi-robot systems which are very energy efficient with high real-time performance and are also very inexpensive. Figure 15.5 illustrates the use of cloud robotics in an industrial environment.

Since cloud computing became widely available, lots of computationally complex and intensive algorithms and systems previously thought of as very time consuming and hence impractical became viable. In the context of robotics and AI this implies that if the power behind the cloud could be harnessed, it would be possible to build smaller, more battery efficient robots. This is so, as there would be no need to have a powerful, energy greedy computer on-board as the real brain of the robot can be in the cloud. Thus it is now viable to build and operate a remote-brain, as discussed above. The centralized cloud for the robot also implied that the system memory could be nearly infinite and instantly available to other robots, so the process of learning and exchanging the knowledge could be simplified and almost seamless [7].

The manufacturing automation protocol (MAP) was developed by General Motors in the 1980s. A diverse set of incompatible proprietary protocols were offered by several vendors. A paradigm shift occurred in the early 1990s when the world wide web popularized the hyper text transfer protocol (HTTP) over Internet protocol (IP).

The first industrial robot was connected to the world wide web with an intuitive graphical user interface (GUI) in the year 1994. Thus it became possible for visitors to teleoperate the robot via any Internet browser. In the mid and late 1990s, researchers across the globe developed a series of web interfaces for robots and devices to explore issues such as more user-friendly user interfaces and robustness in design. Thus the era of the subfield of 'networked robotics' began.

The term Industry 4.0 was first introduced in Germany in the year 2011. It paved the way for a fourth industrial revolution that would use computer networking very

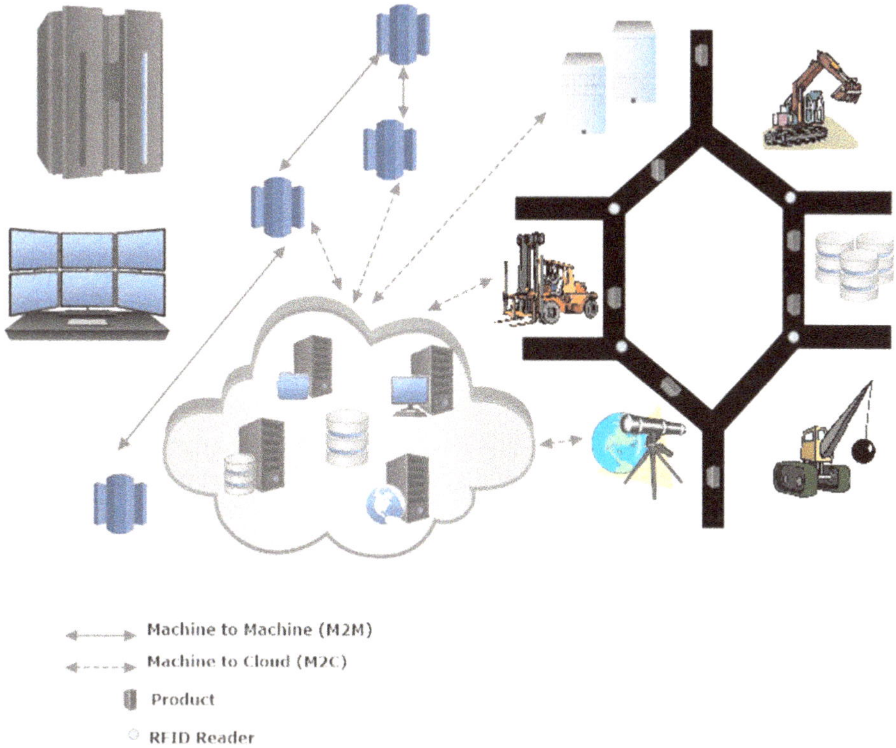

Figure 15.5. The implementation of cloud robotics in an industrial environment. Image source: [23].

widely to automate factory operation remotely. It became the successor to the first (mechanization of production using water and steam power), the second (mass production with electric power) and the third (use of electronics to automate production) industrial revolutions. The term 'industrial Internet' was first coined by the engineers from General Electric in 2012, to describe new efforts where industrial equipment such as wind turbines, jet engines and MRI machines connect over networks to exchange data. Thus, the processing of data for industries including energy, transportation and healthcare became quite efficient [24].

15.4.5 Opportunities, challenges and future directions

Several new challenges are present in using the cloud computing paradigm for robotics and automation systems. A wide range of privacy and security issues were raised by the open, overwhelming connectivity prevalent in the cloud. These concerns include data privacy, since the data generated by cloud-connected robots and sensors may include images or video or data from people, institutions (including data pertaining to healthcare), or even trade secrets of the corporate world. Cloud robotics and automation also introduces the potential threat of robots and associated systems being hacked remotely: an unethical hacker could take over a robot and use it to disrupt functionality or cause severe damage or disruption. For example, researchers at University of Texas at Austin were successful in hacking into and remotely control UAV drones via inexpensive GPS spoofing systems in an evaluation study for the Department of Homeland Security (DHS) and the Federal Aviation Administration (FAA). These concerns raise new regulatory, accountability and legal issues related to safety, control and transparency [24].

Now, let us consider the technical challenges in cloud robotics. The primary concern is that to cope with time-varying network latency and quality of service (QoS), faster algorithms and processes are needed. Faster data connections, both wired Internet connections and wireless standards such as Long Term Evolution (LTE), are reducing latency. Yet, the algorithms must be capable of degrading gracefully when the cloud resources are very slow, noisy or totally unavailable. For example, 'anytime' load balancing algorithms for speech recognition on smart phones send the speech signal to the cloud for analysis and simultaneously process it internally and then use the best results available after a reasonable delay. Similar algorithms are desirable for robotics and automation systems using the cloud. Faster algorithms are also needed that scale up to the size of 'big data', which often contains dirty data that requires new methods to clean or sample the data appropriately. When the cloud is employed for parallel-processing, it is vital to have oversampling by the algorithms to take into account the fact that some remote processors may fail or experience inordinate delays (long latency) in returning results. Also algorithms are needed to filter unreliable or incorrect input data and balance the costs of human intervention with the cost of robot failure, whenever human computations are employed. Transferring the robotics and automation algorithms into the cloud requires frameworks that facilitate the smooth transition. In general, the cloud services can be classified into three categories: infrastructure as a service (IaaS), platform as a service (PaaS) and software as a service (SaaS). RoboEarth is an example of PaaS.

With SaaS, a user interface allows data to be sent to a cloud server that processes it and returns the outputs, which relieves users from the burden of maintaining data, software and hardware, and allows companies to control proprietary software. This technique is widely known as robotics and automation as a service (RAaaS) [24].

Cloud robotics permits different robots to share computational resources, information and data with each other, and share new knowledge and skills not learned by themselves. This will eventually open up a new paradigm shift in robotics, leading to exciting developments in the future. It permits the deployment of inexpensive robots with low computational power and on-board memory requirements, to leverage their potential by using the high speed communications network and the elastic computing resources offered by the cloud infrastructure. Applications that can benefit from the cloud robotics approach are several. They include grasping, navigation and several others, such as weather monitoring, intrusion detection, remote surveillance, non-destructive testing, fault detection and formation control [9].

15.5 The human 4.0 or the Internet of skills (IoS) and the tactile Internet (zero delay Internet)

One can predict the fast emergence of an entirely new, more vibrant Internet, that capitalizes on the latest developments in 5G and ultra-low delay networking, as well as the innovations in AI and robotics. This novel Internet will enable the delivery of skills in digital form. The delivery of physical experiences remotely (and globally) is made possible by the Internet of skills (IoS), which will revolutionize operations and servicing capabilities for industries. In general, it will be a quantum jump in the way we communicate, teach, learn and interact with our surroundings. It will be a brave new world where our best engineers can service cars instantaneously around the world over the tactile internet; or anybody can be taught how to paint by the best available artists globally. At an estimated revenue of US$20 trillion per annum worldwide, which is approximately 20% of today's global gross domestic product (GDP), it will be a technology enabler for skill set delivery—thus a very timely technology for service driven economies across the globe [25]. The transformation to the Internet of skills is illustrated in figure 15.6.

The IoS will democratize labor, in the same way as the Internet has democratized the dissemination of knowledge. The core technologies used in the Internet of skills are (a) ultra-fast data networks (zero delay Internet or tactile Internet), (b) haptic encoders (both kinesthetic and tactile) and (c) edge AI (to beat the light limit).

15.6 Future directions in robotics, AI and the IoT

We have gone a long way since the advent of the IoT and AI. We are now living in a smarter world that deals with computational statistics and brilliant cloud data servers. It is absolutely certain that with the merger of two brilliant innovations such as IoT and AI, things are moving towards a superior future. Businesses such as automation, propelled robotics, smart domestic appliances that constitute the smart home, assembling, design and retail are for the most part harvesting the advantages of these advanced technologies. While this mix or integration of shrewd strategies is yet to mature and its benefits are yet to be fully harvested, every year we advance a couple of steps and make a quantum leap compared to the last one. It is obvious that our homes and domestic appliances have already started adopting the intelligence from AI and the capability to connect smart devices through the technical endowment of the IoT [26].

It is expected that the IoT will affect the further development of AI and the progression will be highly promising. When one analyzes the AI that has been commercialized to date, the algorithms deployed are generally single agents, and most of them are practically first-person algorithms. The purpose of the algorithms was to foresee, analyze and act, beyond serendipity. Although the majority of the insight happens independently, there is communication. Consider how we observe different algorithms, for example those intended for managing a commercial center. The associated costs are fixed because of numerous small-scale interactions between individual dealers or responses to issues in the supply chain management or

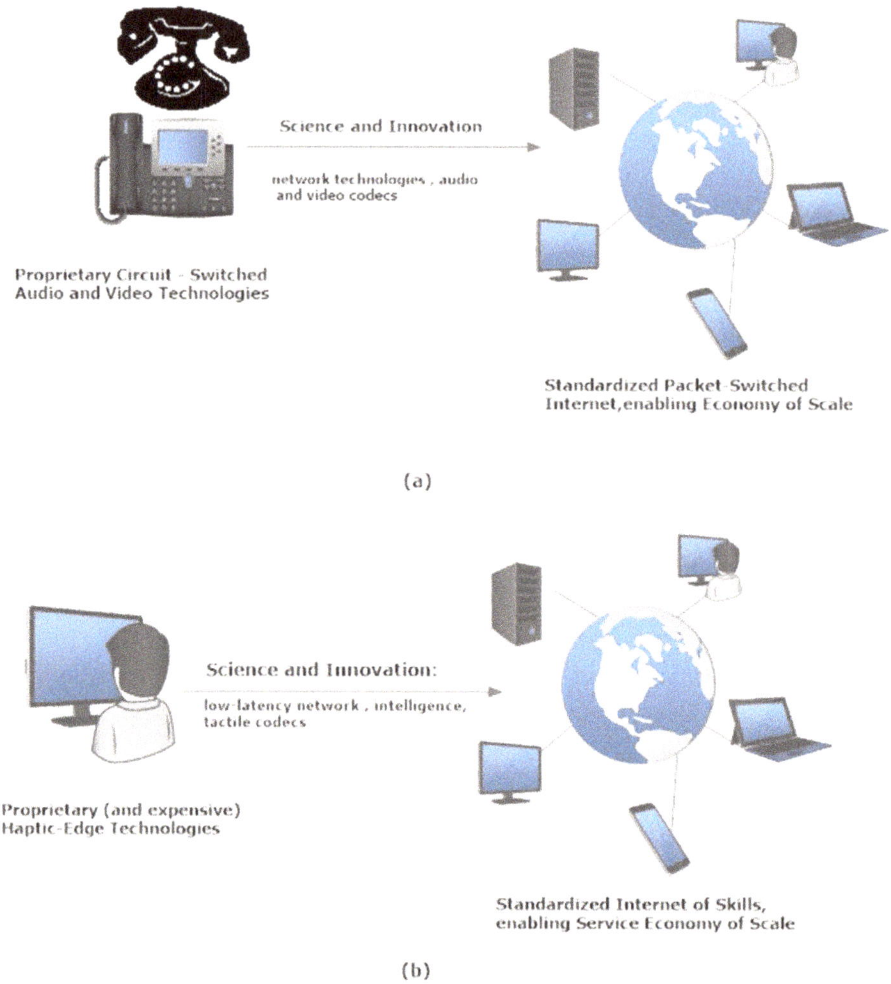

Figure 15.6. The evolution of the Internet of skills (human 4.0). Image source: [25].

fluctuations in currency. These interacting sets of individual events permeate to a value computation for a wide assortment of goods. Such an algorithm is termed social AI [26].

With smart gadgets that can be controlled with a cell phone, the level of intelligence in a household will turn out to be progressively greater. Without their physical presence, anybody could send directions to a washing machine to wash and dry garments, turn the lights on and off, turn the thermostats high or low, and direct venetian blinds to close and open, using a remote gadget. Smart kitchens would become more productive, where a greater part of the work could be performed with smart electrical appliances that need fewer directions from the human operator. Smart homes will be interconnected with inbuilt IoT. The average household will face an ascent in the use of smart devices and appliances, and will apparently gear up to a problem-free way of life. The digital voice assistants offered by Amazon and Google are some of the incredible gadgets which will make life easier. The development of amazing IoT solutions would be more in vogue, in view of controlled purchasing power and appreciation for advancing innovations [26].

The gap between the IoT and AI will continue to diminish, leading to the amazing growth of functionality in gadgets, and a consequent operational excellence. With explicit client intelligence, a similar IoT gadget would be equipped to offer explicit client experiences. IoT applications would become more intelligent, to cleverly watch the Earth, correct flaws and autocorrect operational glitches. A growth in client explicit innovations will provide the necessary base for a personal AI era [26].

The dependence on the cloud for inexpensive, localized computing power should not be ignored. The advent of cloud robotics will result in numerous applications. Also, more recent developments in deep learning and ML will continue to support several AI applications, including those in image processing.

In the era of rapidly evolving technologies, big data analytics and IoT are the two leading radical technologies which can explicitly modify the style of business operations. Both technologies are still in their nascent stages and hold massive potential and opportunities, and pose several unresolved challenges. The two technologies can be coupled together for more efficient implementation and can help all applications by making smarter decisions [27].

ML is a modern science which enables computers to work autonomously. No explicit programming is needed in ML. The modern-day technology deploys algorithms that can train and improve on the data that is being fed to them. Over the years, ML has matured and made possible the concept of self-driving autonomous cars. ML is also a technology enabler for more effective web searches, spam free emails, practical speech recognition software, personalized marketing and so on. Today, ML is increasingly being deployed in credit card purchase fraud detection, personalized advertising through pattern recognition, personalized shopping/entertainment recommendations, to determine cab arrival times, pick-up locations, and finding routes on maps [28]. The five recent technological trends are (a) the creation of more jobs in data science, (b) new approaches to data security, (c) robotic process automation (industry 4.0), (d) improved IT operations and (e) transparency in decision making [28].

It is interesting to note that the essence of ML and AI lies in ANNs. DNNs enhance the learning capability of ANNs by adding more hidden layers to the network. CNNs are a class of deep, feed-forward (not recurrent) artificial neural networks that are applied to analyze videos. CNNs are usually composed of a set of layers that can be grouped by their functionalities. CNNs consist of feature learning layers and classification layers apart from the input and output layers.

With the advent of the ANN, ML has taken a giant leap in recent times. ANNs are biologically inspired computational models, and are capable of exceeding the performance of previous forms of AI to a larger extent. The CNN is one of the most impressive forms of ANN architecture. CNNs are primarily used to solve difficult image-driven pattern recognition tasks with their precise yet simple architecture. A simplified method of getting started with ANNs is provided by CNNs [29]. Note that first generation ANNs were shallow, in the sense that apart from the input and output layers they contained at most one hidden layer. In contrast, DNNs contain several hidden layers.

CNNs differ from other forms of ANNs in that instead of focusing on the entirety of the problem domain, knowledge about the specific type of input is exploited. This in turn allows for a much simpler network architecture to be set up [29].

References

[1] Banks J 2018 The human touch—practical and ethical implications of putting AI and robotics to work for patients *IEEE Pulse* **9** 15–8

[2] Fujita M 2019 AI x robotics: technology challenges and opportunities in sensors, actuators, and integrated circuits *Proc. of the 2019 IEEE Inter. Solid-State Circ. Conf. (ISSCC 2019)* pp 276–7

[3] Lars Kunze *et al* 2018 Artificial intelligence for long-term robot autonomy: a survey *IEEE Robotics and Automation Letters* **3** 4023–30

[4] Wogu I A P *et al* 2017 Artificial intelligence, alienation and ontological problems of other minds: a critical investigation into the future of man and machines *Proc. of the 2017 Int. Conf. on Computing, Networking and Informatics (ICCNI)*

[5] Davey T 2017 Artificial intelligence and the future of work: an interview with Moshe Vardi *Future of Life* https://futureoflife.org/2017/06/14/artificial-intelligence-and-the-future-of-work-an-interview-with-moshe-vardi/

[6] Hawking S, Tegmark M, Russell S and Wilczek F 2014 Transcending complacency on superintelligent machines *Huffington Post* http://huffingtonpost.com/stephen-hawking/artificial-intelligence_b_5174265.html

[7] Lorencik D and Sincak P 2013 Cloud robotics: current trends and possible use as a service *Proc. of the IEEE 11th Int. Symp. on Applied Machine Intelligence and Informatics (SAMI 2013)* pp 85–8

[8] Guizzo E 2011 Robots with their heads in the clouds *IEEE Spectrum* http://spectrum.ieee.org/robotics/humanoids/robots-with-their-heads-in-the-clouds

[9] Hu G, Tay W P and Wen Y 2012 Cloud robotics: architecture, challenges and applications *IEEE Network* **26** 21–8

[10] RoboEarth Project http://roboearth.org/

[11] The Robotic Operating System (ROS) http://ros.org/wiki/

[12] Arumugam R *et al* 2010 DAvinCi: a cloud computing framework for service robots *Proc. of the 2010 IEEE Int. Conf. on Robotics and Automation (ICRA) (3–7 May)* pp 3084–9

[13] Apache Hadoop http://hadoop.apache.org/

[14] O'Leary D E 2017 Emerging white-collar robotics: the case of Watson analytics *IEEE Intell. Syst.* **32** 63–7

[15] IBM 2014 Introduction to IBM Watson: Analytics, Data Loading and Data Quality, IBM Document Version 2.0, December 16, 2014 https://docplayer.net/10919185-Introduction-to-ibm-watson-analytics-data-loading-and-data-quality.html

[16] Yukitake T 2017 Innovative solutions toward future society with AI, robotics, and IoT *Proc. of 2017 Symp. on VLSI Circuits* pp C16–9

[17] Ahmed S H, Kim G and Kim D 2013 Cyber physical system: architecture, applications and research challenges *Proc. of the IFIP Wireless Days Conf. (WD'13)* pp 1–5

[18] Silva A R and Vuran M C 2010 (CPS)^2: integration of center pivot systems with wireless underground sensor networks for autonomous precision agriculture *Proc. of the 1st ACM/ IEEE Int. Conf. on Cyber-Physical Systems* pp 79–88
[19] Jie Lin *et al* 2017 A survey on internet of things: architecture, enabling technologies, security and privacy, and applications *IEEE Internet of Things J.* **4** 1125–42
[20] NIST 2016 *NIST & The Smart Grid* Accessed: 12 April 2019 https://nist.gov/engineering-laboratory/smart-grid/about-smart-grid/nist-and-smart-grid
[21] Nayyar A, Batth R S and Nagpal A 2018 Internet of robotic things: driving intelligent robotics of future—concept, architecture, applications and technologies *Proc. of the 2018 4th Int. Conf. on Computing Sciences* pp 151–60
[22] Adebayo A O, Chaubey M S and Numbu L P 2019 Industry 4.0: the fourth industrial revolution and how it relates to the application of Internet of things (IoT) *J. Multidiscip. Eng. Sci. Stud.* **5** 2477–82
[23] Jiafu Wan *et al* 2016 Cloud robotics: current status and open issues *IEEE Access* **4** 2797–807
[24] Ben Kehoe *et al* 2015 A survey of research on cloud robotics and automation *IEEE Trans. Autom. Sci. Eng.* **12** 398–409
[25] Mischa Dohler *et al* 2017 Internet of skills, where robotics meets AI, 5G and the tactile Internet *Proc. of the 2017 European Conf. on Networks and Communications (EuCNC)* pp 1–5
[26] Dialani P 2019 AIOPS: the integration of AI and IoT https://analyticsinsight.net/aiops-the-integration-of-ai-and-iot/
[27] Sarkar S 2017 How to build IoT solutions with big data analytics https://analyticsinsight.net/how-to-build-iot-solutions-with-big-data-analytics/
[28] Some K 2018 Top 5 machine learning trends of 2018 https://analyticsinsight.net/top-5-machine-learning-trends-of-2018/
[29] O'Shea K and Nash R 2015 An introduction to convolutional *Neural Netw.* **2** 1–10

IOP Publishing

Modern Optimization Methods for Science, Engineering and Technology

G R Sinha

Chapter 16

Efficacy of genetic algorithms for computationally intractable problems

Ajay Kulkarni and Sachin Puntambekar

A genetic algorithm is a metaheuristic method that has been proved highly efficient for searching for optimal solutions for problems belonging to the category of NP-hard and which are often algorithmically solvable but computationally intractable.

A genetic algorithm is a probabilistic heuristic search technique motivated by the principle of natural genetic systems. It aims to locate the global optimal solution for a problem from the given solution space. Due to its population based approach, the requirement for a fitness function rather than its derivatives and the probabilistic nature of the operators, genetic algorithms possess the capability of exploring the search space efficiently and efficaciously and are therefore applied to search for optimal or near-optimal solutions of various optimization problems.

A genetic algorithm can be viewed as an abstract version of the evolutionary process which operates on a population of artificial chromosomes. Each chromosome represents an encoded version of a candidate solution and is associated with a fitness value which reflects the eminence of that chromosome as the solution to the problem. Binary coding is a commonly used technique for encoding the solutions—in this strategy the solution appears as a bit string which facilitates the subsequent operations. These candidate solutions are further subjected to operations such as crossover and mutations to search an offspring population that is expected to be better than the parent population.

Both crossover and mutation are nondeterministic in nature and are applied with certain probabilities. The crossover operator is a mechanism for exchanging information between chromosomes; this operator allows two parent chromosomes to exchange their genetic characteristics in order to generate two offspring. Single-point crossover, two-point crossover and uniform crossover are commonly used exchange mechanisms for binary coded genetic algorithms. The mutation operator is used to introduce new genetic material; it mutates one or more alleles in a chromosome from their current states which can result in new individuals with different characteristics. For binary coded chromosomes, bit flip mutation is normally used. In this procedure, mutation is applied to each bit location in the chromosomes with a certain predefined probability. Mutation is a subsequent process of crossover and is applied to the offspring evolved as a result of crossover process. The process of mutation plays a very vital role in perpetuating the genetic diversity and thereby avoiding the problem of genetic drift which may lead to premature convergence. The mutation probability is often kept low so as to avoid the redundant randomness in the search

Chromosomes generated as the result of recombination are further subjected to the process of replacement in which the offspring population becomes the new parent population. Complete replacement, replacement with elitism and steady state replacement are commonly used replacement schemes. These processes are thereafter iterated through a number of generations until some termination criterion is satisfied. The termination criterion could be either a fixed number of iterations or convergence to a best fit solution.

This chapter details the various aspects of genetic algorithms, such as their operational mechanism, theoretical analysis and ability to deal with other complicacies such as constrained optimization and others.

16.1 Introduction

A genetic algorithm is a heuristic optimization search technique inspired by the principles of natural genetic systems and aims to seek a global optimum solution for a problem from the set of candidate solutions in an iterative manner. Due to its population based approach, the requirement for a fitness function rather than its derivatives and the probabilistic nature of operators, genetic algorithms possess the potential to explore the search space efficiently and efficaciously for optimal or near-optimal solution of the problem under consideration. Genetic algorithms and their variants have been successfully applied in engineering fields to search for the solution for optimization problems with significant complexities. The capability of a genetic algorithm to solve the optimization problems of the nature of NP-hard and NP-complete problems has been reflected in several research findings cited in the literature. Genetic algorithms can be viewed as a technique to search for an optimal solution of a problem by performing repeated iterations of a solution. To quantify the process of iteration and to implement the concept of natural selection, genetic algorithms rely on an objective function or fitness function. The fitness function provides a measure to determine the candidate solution's relative fitness and is used by the genetic algorithm to evolve better solutions in subsequent iterations. Another aspect of genetic algorithms is the concept of population, instead of operating on a unique solution the genetic algorithm operates simultaneously on a population of candidate solutions, known as individuals. This implicit parallelism involved in the genetic algorithm makes it a global search rather than confining it to a local area of the search space. However, the size of the population is an issue of concern, as small population size may lead to premature convergence while large size may result in excessive computational time.

Also, to imitate the natural selection procedure, the genetic algorithm operates on the encoded versions of the parameters to be optimized and not on the parameters themselves. Parameter encoding transforms the actual optimization problem into combinatorial optimization as the genetic algorithm is essentially a combinatorial search technique.

A genetic algorithm maintains a population of solutions, known as chromosomes or individuals, over the search space which is iteratively modified so as to drive this population towards an optimal or near-optimal solution. Starting with some randomly or heuristically selected initial population, the genetic algorithm generates a new population at each iteration using the following steps: (i) calculation of a fitness function value for each chromosome constituting the old (existing) population—this value reflects the potency of each solution and can be considered as a meaningful measure to analyze the claim of the algorithm to approach optima in successive iterations; (ii) using the fitness value as selection measure the individuals are selected for subsequent procedures of recombination and selection; (iii) selected individuals (parents) are thereafter subjected to a genetic operation called crossover to generate probably better new solutions (offspring); (iv) the newly generated solutions are further subjected to mutation, this operator aims to preserve the genetic diversity in the vicinity of the candidate solution; and (v) a new population is generated to replace the existing one. Iterations are carried out until some terminating criterion is met [1].

Recombination and selection operations in genetic algorithms are nondeterministic and are governed by some probabilistic rules instead of some deterministic procedure; this nondeterministic nature facilitates the objective of preserving the global explorative properties of the search. An important attribute of genetic algorithms is that they require only a mathematical function acting as an objective function for the considered problem and have no dependence on other characteristics of the fitness function, such as the existence of its derivatives or differentiability. This attribute allows the applicability of this algorithm even for problems with non-smooth functions. Procedural details of the genetic algorithm are explained by the flowchart shown in figure 16.1.

16.2 Genetic algorithm implementation

A genetic algorithm is implemented using a number of distinct components. The main components of a genetic algorithm designed for a particular problem are genetic representation, initialization, evaluation, selection, recombination (crossover and mutation), replacement and termination [1–4].

1. *Search space*: In the case of search algorithms, the search space is defined as the space containing all the feasible solutions. In graphical notation, each point in the search space may be viewed as a feasible solution to the problem under consideration and can be characterized by its fitness value for that problem. Search techniques such as genetic algorithms try to locate the optimal solutions among the solutions constituting the search space.

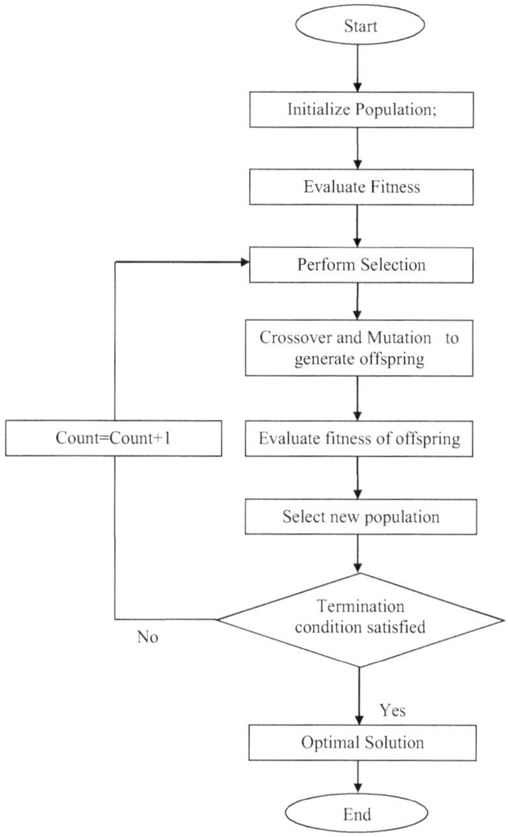

Figure 16.1. Flowchart of the working principle of genetic algorithms.

2. *Genetic representation/encoding*: To facilitate the simulation of the process of natural evolution and selection, a genetic algorithm operates on the encoded versions of the solutions, known as chromosomes. These artificial chromosomes closely resemble natural chromosomes, which are convoluted threads of DNA (deoxyribonucleic acid). Traits which are coded versions obtained by some combination of DNA, can be viewed as the constituting elements of a chromosome; various traits governing the hereditary characteristics of an individual are strung longitudinally along a chromosome. Artificial chromosomes, encoded versions of the solutions, can be viewed as the stringing of various features of the solution along its length. These encoded versions are known as the genotype and its decoded counterpart is called the phenotype. Regarding parameter encoding, genetic algorithms use the following typical vocabulary: individual solution strings in their encoded version are referred to as chromosomes, whereas a particular part of the chromosomes representing a feature or characteristics is known as a gene, the position and value of a particular gene are known as the locus and allele, respectively.

In genetic algorithms, genetic operators such as crossover and mutation operate on genotype, whereas evaluation and selection of individuals is carried out using phenotype, so the mapping function should be selected such that there exists a close resemblance between the genotype and the characteristics of phenotypic space. There exist several encoding schemes in the genetic algorithm literature which are often problem specific, e.g. real number coding is suitable for optimization problems subjected to constraints, whereas binary or integer number encoding is appropriate for combinatorial optimization problems.

Some of the main coding schemes are:

Binary coding:

- In genetic algorithms, binary representation of chromosomes is the most commonly used encoding scheme. In binary representation, the string is composed of zeros and ones. The length of the string is governed by the precision desired for the solution.

  In a genetic algorithm each variable (allele) of the optimization problem is encoded as a binary string of appropriate length (gene) and the chromosome is constructed by concatenating all the substrings (genes) together. For a two variable optimization problem with four bit substrings, a chromosome can be represented as $(x_1, x_2) \rightarrow (s_1, s_2) = 00101100$. Binary encoding follows a linear mapping rule:

$$x_i = x_i^l + \frac{x_i^u - x_i^l}{2^\beta - 1} \sum_{j=0}^{\beta} \gamma_j 2^j, \tag{16.1}$$

  where x_i^l and x_i^u represent the lower and upper bounds of x_i, respectively, β represents the length of the binary representation of x_i, whereas $\sum_{j=0}^{\beta} \gamma_j 2^j$ is the decoded value of the binary substring. The accuracy obtained with substrings of length β provides accuracy of the order of $1/2^\beta$ of the search space, so arbitrary precision can be achieved by using strings of appropriate length. Binary representation is a commonly used encoding scheme in classical genetic algorithms and is normally supported by two arguments. First, binary alphabets maximize the level of implicit parallelism. Second, the encoding scheme associated with a high cardinal alphabet requires a larger population size for the effective exploration of the search space which often reduces the computational efficiency of the algorithms. However, the binary representation has not been found to be appropriate when dealing with a continuous search space with large dimensions and when a numerical precision of considerable degree is required, as it handles continuous problems as discrete ones. Application of binary coding for the problems of a continuous domain requires decoding and repair algorithms to transform the solution searched for by the genetic algorithm into a viable solution of the original problem. Designing of decoding and repair algorithms often complicates the implementation of genetic algorithms.

Real coding:

- The real coding scheme has been found to be appropriate for optimization problems of continuous nature; it uses a floating point representation for representing various parameters to be optimized. As a result, each chromosomal string appears as a vector of floating point numbers. The precision in this scheme is often decided by the nature of the problem to be solved. A common approach in this scheme for constructing chromosomes is to represent each variable to be optimized by a gene and to keep the length of the chromosomal string the same as that of the solution vector. Also, the value of a gene representing a particular variable is restricted to the interval defined for that variable and the genetic operators forced to preserve this requirement. Float point representation of parameters is capable of representing the large domains, whereas in binary implementations, an increase in domain size results in a loss of precision with a fixed length of the chromosomes. Real coding also offers the feature of local tuning of the solution by exploiting the graduality of the functions with continuous variables, which is often difficult in the case of binary coding due to the problem of the Hamming cliff. In real coding, as the genotype and phenotype are identical, there is no requirement of coding and decoding and so the speed of the algorithm increases.

3. *Initialization*: The selection of the initial population is a vital issue in genetic algorithms; it is mainly concerned with the population size and method for selection of individuals to constitute a population. An optimum population size is always required for the effective and efficient implementation of genetic algorithms, as a small population may not provide sufficient space to explore the search space effectively, whereas a large population may impair the efficiency of the algorithm by consuming excessive computational time. According to Goldberg [1], the size of the population is directly related to the difficulty n of the problem, whereas according to Reeves [2] the minimum population size to initiate the meaningful search is governed by the principle that it is possible to traverse every feasible solution in the search space starting from the initial population by using the crossover mechanism only. The selection of a good initial population has a better possibility of finding a good solution. Diversity can be considered as the key issue when selecting the initial population as it ensures the effective exploration of the search space and is particularly useful in the case of an unknown or weakly known search space. One commonly used approach is to randomly select the initial population; however, it may not provide the uniform coverage of the search space. For reducing the computational time, one normally used approach is

to seed some high quality solutions into the initial population. However, this inclusion induces the possibility of premature convergence.

4. *Evaluation/fitness*: As genetic algorithms try to impersonate the survival of the fittest principle of nature to formulate a search process, it is necessary to evaluate the fitness of a potential solution relative to others. This evaluation is carried out using the fitness function. The fitness function is a mathematical evaluation function that computes and reflects the superiority of the chromosome as a solution to the problem under consideration. The fitness function allocates reproductive traits to an individual and acts as a measure to be maximized in subsequent iterations. It implies that individuals with a higher fitness function value usually have a better chance of participating in subsequent stages of the algorithm. The algorithm is structured such that it aims to increase the average population fitness in an iterative manner.
5. *Selection*: In a genetic algorithm, the fitness function value is used as an assessor of the quality of the solution represented by a chromosome and the average fitness is considered as a qualitative measure of the population comprising a stipulated number of chromosomes. The selection mechanism of a genetic algorithm utilizes the fitness as a guide to select individuals to form a mating pool so as to participate in the process of evolution. Due to fitness based criteria, the chromosomes with better fitness have a relatively higher probability of being selected for subsequent operations than others. As the selection mechanism is normally carried out with replacement, fit chromosomes have a chance of being selected more than once, thus fitter chromosomes participate more frequently in the mating process and posess a relatively higher probability of surviving in succeeding iterations. During the selection process key factors that are required to be balanced are selection pressure and genetic diversity. Selection pressure describes the tendency to select only the best individuals of the current population to participate in subsequent steps; selection pressure controls the rate of convergence of the genetic algorithm towards the optimum. Genetic diversity refers to the maintenance of diversity in the solution population and is required to ensure the effective exploration of solution space, which is often necessary during the earlier stages of the optimization process. Very high selection pressure results in a loss of genetic diversity due to which the genetic algorithm is likely to undergo premature convergence with some local optimum. With too low a selection pressure, the genetic algorithm may not converge to an optimum solution in adequate computational time. In order to ensure the convergence of a genetic algorithm to a global optimum solution in reasonable time, an appropriate balance between the selective pressure and genetic diversity is required to be maintained by the selection mechanism. There exist several selection mechanisms in the genetic algorithm literature, such as roulette wheel selection, stochastic reminder sampling, stochastic universal sampling, linear rank selection, exponential rank selection, tournament selection and truncation selection [1]. Roulette wheel (fitness proportional) selection is the conventional selection method in which each

candidate is assigned a slot on the roulette wheel with size proportional to its fitness, thus the candidates with higher fitness have a larger slot size than the less fit individuals. The roulette wheel is spun an appropriate number of times, each time selecting a candidate pointed at by the wheel pointer. As per the scheme suggested by Goldberg [1], the roulette wheel selection mechanism to select n individuals from a population size of n can be implemented through the following steps:

i. Calculate the fitness value, f_i, for all the candidate solutions constituting the population.
ii. Calculate the probability (slot size), p_i of selection for each solution

$$p_i = f_i / \overline{f}, \quad \text{where } \overline{f} = \sum_{j=1}^{i} f_j.$$

iii. Calculate the cumulative probability, $q_i = \sum_{j=1}^{i} p_j$.
iv. Generate a random number, $r \in (0, 1]$.
v. Select the individual s_1 if $r < q_1$, otherwise select s_i if $q_{i-1} < r \leqslant q_i$.
vi. Repeat steps iv and v n times to select n candidates in the mating pool.

In n experiments of the roulette wheel, an individual with probability p_i is expected to make np_i copies in the mating pool. The computational complexity of this procedure is of the order of $O(\log n)$ steps for a population of size n. In roulette wheel selection, the fitness of an individual acts as a governing factor to decide the selection probability of an individual, the dominating individual has a chance of being selected more than once during the trials and thus this technique is associated with high selection pressure which may result in premature convergence leading to local optima due to the possible selection of a dominating individual. Also, this technique is associated with a high degree of stochastic variability and the actual number of times a chromosome is selected may vary from its expected value. There exist some variants of this scheme, such as stochastic remainder sampling and stochastic universal sampling, which circumvent these drawbacks to a certain extent. Stochastic remainder selection calculates the reproduction count of each individual as $es_i = nf_i / \overline{f}$ and uses the integer portion of es_i as the deterministic count of s_i in the mating pool. The fractional parts of es_i are then used to fill the remaining pool in a probabilistic way. The probabilistic mechanism is carried out either without replacement or with replacement, in the former case each remainder is used to bias the flip of a coin to decide whether the individual receives another copy or not, whereas in the latter case the fractional parts are used to size the slots of the roulette wheel selection process. Stochastic universal selection, which corresponds to systematic sampling, is carried out by sizing the slots of a roulette wheel with n equally spaced pointers. A single spin of the wheel simultaneously produces the selection count of all the chromosomes which is equal to the number of pointers that points to the slots associated with respective chromosomes. Linear ranking selection provides the flexibility of tuning the selection pressure and thereby deals with the problem of premature convergence in an effective manner. In rank selection, individuals are sorted and ranked according to their fitness values, rank n is assigned to the best individual and rank 1 to the worst. As per Colin [2], the selection probability is assigned linearly to the individuals according to their rank:

$$p(j) = \alpha + \beta j;\, j \in \{1, 2, \cdots, n\}, \tag{16.2}$$

where $\alpha = \frac{2n - \varphi(n+1)}{n(n-1)}$ and $\beta = \frac{2(\varphi - 1)}{n(n-1)}$ with $\sum_{j=1}^{n} \alpha + \beta j = 1$.

Here, φ is the selection pressure defined as $\varphi = \frac{\text{prob[best chrosome]}}{\text{prob[average chrosome]}}$ with $1 < \varphi \leqslant 2$. With this framework, corresponding to a pseudo-random number r, j can be calculated as

$$j = \frac{-(2\alpha + \beta) \pm \sqrt{(2\alpha + \beta)^2 + 4\beta r}}{2\beta}. \tag{16.3}$$

This process can be carried out in $O(1)$ time. In linear ranking selection, the sorting of the individuals is a crucial operation and complexity of sorting often dominates the complexity of the algorithm, thus complexity is $O(n \log n)$. Exponential ranking selection is similar to linear ranking selection except for the fact that the exponentially weighting technique is used to decide the probabilities of the individuals:

$$p(j) = \frac{c - 1}{c^n - 1} c^{n-j};\, j \in \{1, 2, \cdots, n\};\, 0 < c < 1. \tag{16.4}$$

In the tournament selection scheme, a set of k individuals is randomly selected from the population and the best individual from that set is further selected for subsequent processing and the process is repeated n times. Here, k is known as tournament size which may be selected arbitrarily, but in general the process is carried out with $k = 2$, which is known as a binary tournament. In tournament selection, group size k is used as the controlling parameter to tune the selection pressure; group size 2 implies the weakest selection pressure. The time complexity of tournament selection is of the order of $O(n)$. In truncation selection a certain portion q of the fittest candidates are selected and are reproduced $1/q$ times to maintain appropriate population size. This technique often results in premature convergence due to deterministic elimination of less fit candidates during the selection procedure. As truncation selection involves sorting, its time complexity is of the order $O(n \log n)$.

6. *Recombination*: Recombination is concerned with the generation of new individuals by recombining the chromosomes selected from a source population. New individuals thus generated may become the members of a

successor population. Crossover and mutation are two genetic operators which constitute the process of recombination. The key idea behind the recombination process is to simulate the exchange of genetic material that takes place when organisms reproduce as well as the changes in genetic characteristics that may occur after fertilization. As the selection of individuals to participate in the process of recombination is biased towards higher fitness, there exists a fair chance of the evolution of chromosomes with improved fitness. In a genetic algorithm, these operators are dealt with in a nondeterministic way, these operators act with a predetermined probability and their exact outcomes are also nondeterministic. The crossover operator is one of the defining characteristics of the genetic algorithm responsible for the introduction of new genetic material. The crossover mechanism allows two parent chromosomes from the mating pool to exchange their characteristics in some specified manner so as to evolve two offsprings with different characteristics, with the possibility that the crossover of good chromosomes may result in the evolution of better individuals. In general, the crossover operator is applied to a pair of chromosomes in some probabilistic sense and the likelihood of crossover being applied is controlled by the crossover rate or crossover probability p_c. Definitions of the crossover operator are largely representation dependent. In the case of binary coded genetic algorithms, commonly used crossover techniques are single-point crossover, two-point crossover and uniform crossover. In single-point crossover, a crossover point is randomly selected and the characteristics of the parent chromosomes are exchanged to construct the child chromosomes, as illustrated below.

Parent one:	10010010
Parent two:	10110110
Crossover point:	↑
Child one:	10010110
Child two:	10110010

In two-point crossover, two crossover points are selected along the length of the chromosomes and the characteristics of the two parents are exchanged between these two points to generate two offspring.

Parent one:	10010010
Parent two:	10110110
Crossover points:	↑ ↑
Child one:	10110110
Child two:	10010010

In the case of uniform crossover, the crossover operator is expressed as a binary string or masks of the same length as that of the chromosome. An offspring is generated by copying either of the parent alleles at each locus according to the ones and zeros in the mask—if at a particular locus the mask bit is 1, then the allele is copied from parent one, and if the mask bit is 0 then the allele is copied from parent two. The second offspring is generated by reversing the process.

Parent one:	10010010
Parent two:	10110110
Mask:	10011010
Child one:	10110110
Child two:	10010010

Another genetic operator used in genetic algorithms to introduce new genetic material is mutation. Mutation flips one or more alleles in a chromosome from its current state which can result in new individuals with different characteristics. The mutation operator thus introduces genetic diversity whenever the population approaches homogeneity because of repeated use of selection and crossover operators and the individuals resulting from the process of mutation may even lead the algorithm towards some better solution. Mutation is applied to the evolved offspring as a result of the crossover process. The process of mutation is applied in a probabilistic sense and is controlled by the mutation rate or mutation probability p_m. Mutation probability is often kept low in order to avoid redundant randomness in the search. In the case of binary coded genetic algorithms, the mutation operator is applied to each bit in the chromosome. Corresponding to a bit position, this procedure involves the generation of a random number in the interval [0, 1] which is thereafter compared to the mutation probability p_m. If the random number is greater than the mutation probability, no mutation is applied at that position, and in case the value of the random number is lower, the bit value is flipped from 0 to 1 or vice versa. Mutation plays a vital role in maintaining the genetic diversity and thereby avoiding the problem of genetic drift which may lead to premature convergence. It has been observed that in the absence of a mutation operator, a finite size population always converges to a single genotype, this phenomenon is known as genetic drift. When all the individuals converge to a single genotype, crossover does not contribute to the search and so the mutation operator is required to maintain diversity.

7. *Replacement*: After the process of recombination, the next step is replacement, which describes the technique for the generation of a successor population to replace the existing one. There exist several replacement techniques in the genetic algorithm literature, which primarily emphasize the issues of selection pressure and genetic diversity. The classical replacement scheme, proposed by Holland, is based on a generational approach which favors complete replacement, where the successor population is entirely composed of the chromosomes generated by the process of recombination. This complete replacement scheme is relatively easy to implement

and is free from the issues of diversity and selection pressure; the drawback associated with this technique is that it contains the risk of discarding the good solutions from the existing population. An improved version of the complete replacement scheme is replacement with elitism; this method emphasizes preserving one or two of the best individuals from the existing population and replacing the others. Elitism speeds up the performance of the genetic algorithm but exerts a high selection pressure due to the deterministic selection of relatively fit candidates. Another scheme for the replacement is steady state or incremental replacement, which introduces the concept of population overlapping, in this scheme only a fraction G (known as the generation gap) of the existing population is replaced. Selection of individuals from the existing population for replacement is an issue of concern in steady state replacement. One technique is the replacement of the worst individuals; however, it exerts a strong selection pressure and often requires a large population size or high mutation rate to maintain diversity. Another replacement scheme, which maintains the considerable degree of diversity, is steady state with no duplicates. In this case, offspring are not included in the population if they are mere duplicates of the existing individuals. Other evolution strategy based replacement schemes are the (μ, λ) and $(\mu + \lambda)$ schemes. In the (μ, λ) scheme, $\mu(>\lambda)$ individuals are generated from the λ parents and the best λ of these newly generated individuals are selected as the successor population. In the $(\mu + \lambda)$ scheme, μ offspring are generated and combined with λ parents to generate a set of $(\mu + \lambda)$ individuals; from this set the best λ individuals are selected to constitute the successor population. This scheme also exerts a high selection pressure and requires a high mutation probability to maintain divergence.

8. *Termination*: The processes of selection, recombination and replacement are iterated until some terminating criterion is satisfied. Commonly used termination criteria are either population convergence criteria where almost all the solutions become identical or nearly identical. Another is the improvement in best fitness score with iterations which emphasizes the accuracy of the solution, and the third is a fixed number of iterations with a reasonable compromise between population convergence, accuracy of the solution and computation time.

16.3 Convergence analysis of the genetic algorithm

A correctly designed and implemented genetic algorithm population with increasing fitness is supposed to evolve over generations and ultimately culminate in a global optimum. Theoretical investigation of the convergence characteristics of genetic algorithms has been undertaken by several researchers. Attempts to develop a theoretical framework to explain the chromosome convergence in genetic algorithms have mainly resulted in the schema theorem [5]. The schema theorem provides a theoretical framework to analyze the convergence properties of the genetic algorithm. The schema theorem is based on the conception of schema which is obtained by fixing the allele of specific chromosome loci, thus it can be viewed as a typical pattern which could be observed in some chromosomes. A schema thus represents a similarity pattern which describes a subset of individuals with similar features in some positions. A schema is a string of $\{1, 0, *\}$ where a typical pattern is described by using $\{1, 0\}$ and $*$ is used at the chromosome loci kept unspecified by the pattern. For example, a schema $\lambda = 101 * *011$ describes a subset of chromosomes with elements $\{10100011, 10101011, 10110011, 10111011\}$, i.e. the elements of this subset belong to λ. For a schema, commonly defined terms are the order and length of schema. The order of λ, $o(\lambda)$ is the number of defined bits, whereas length λ, l_λ is the difference of the allele positions of the first and last defined bits. For $\lambda = 101 * *011$, $o(\lambda) = 6$ and $l_\lambda = 7$.

The schema theorem provides an explanation of how the schemata featured in relatively fit chromosomes possess greater prospects of propagating through successive populations with the evolution of the genetic algorithm. The theorem can be formally stated as follows.

Schema theorem. Let λ represent a schema for binary coded chromosomes of length L and let $m(\lambda, t)$ be the number of chromosomes in the current population (t) which belongs to λ. Then the expected number of chromosomes of λ in the next generation $(t + 1)$ is given by the formula

$$m(\lambda, t + 1) \geqslant m(\lambda, t)\frac{f(\lambda)}{\overline{f}}\left(1 - p_c\frac{l_\lambda}{L - 1}\right)(1 - p_m)^{o(\lambda)}, \tag{16.5}$$

where $f(\lambda)$ is the average fitness of chromosomes of λ and $\overline{f}$ is the average fitness of the population. Here, the term $m(\lambda, t)\frac{f(\lambda)}{\overline{f}}$ represents the expected number of chromosomes of λ after the selection process (fitness proportionate selection). It implies that schema with relative fitness above one will continue to increase, whereas with relative fitness less than one they will decrease; this is due to the selective pressure in fitness proportionate selection. Term $(1 - p_c\frac{l_\lambda}{L - 1})$ represents the probability of survival of chromosomes of λ after the crossover process, this term is high when l_λ is low and when l_λ approaches $L - 1$, $(1 - p_c\frac{l_\lambda}{L-1})$ approaches $(1 - p_c)$, which indicates that the disruption of the chromosomes is almost certainly due to crossover. Term $(1 - p_m)^{o(\lambda)}$ represents the probability of survival of chromosomes of λ after the mutation process, and is high when $o(\lambda)$ is low.

It indicates that schemata with short length, low order and above average fitness receive increasing trials in successive iterations. These schemata are known as building blocks and can be viewed as the partial solutions which provide a higher fitness value to the chromosomes which contain these building blocks. It means that a chromosome composed of such building blocks will be a near-optimal solution. The consequence of this theoretical framework is the building block hypothesis, which states that the genetic algorithm can approach the near-optimal solution through appropriate selection and combination of building blocks. The notion of building blocks is particularly useful when a heuristic insight of the search problem is available. During the process of encoding, the desired factors could be encoded in neighboring loci, thereby having low order schema. Crossover sites could also be adjusted heuristically such that chromosome disruption occurs only at locations which are worth it for that problem. The procedure thus results in increasing the probability that the defined process will productively juxtapose building blocks and rapidly approach an optimal solution [5].

16.4 Key factors

There exist some crucial characteristics which offer genetic algorithms an efficacy to deal with optimization problems which are computationally intractable and associated with other complexities, such as constraint satisfaction, multimodal solution and others. Significant progress has taken place in this field and the following subsections cover these issues [1–5].

16.4.1 Exploitation and exploration

Search techniques are considered as generalized solution methods for the problems which belong to the nondeterministic category. Search techniques can be categorized as blind strategies or heuristic strategies. Blind search strategies do not use any additional information obtained about the problem domain to guide the further search, whereas heuristic search strategies use additional information to guide the search along the best search directions. To ensure an efficient and effective search, search techniques are required to address issues such as exploration and exploitation. Exploration can be considered as the process of seeking solutions in new areas of a search space, whereas exploitation is the process of finding solutions in the vicinity of the already existing solution in order to refine it. Hill-climbing and random search are examples of pure exploitation and exploratory strategies, respectively, whereas search techniques such as genetic algorithms maintain an appropriate balance between exploration and exploitation of the search space so as to ensure the optimal or near-optimal solution. In genetic algorithms, issues of exploitation and exploration are mainly controlled by the process of selection and selection operators. The extent of exploration or exploitation can be controlled by selection processes by causing a variation in selection pressure. Search techniques with low selection pressure are exploratory in nature, whereas those with higher selection pressure are inclined towards exploitation. Selection of an appropriate selection pressure may result in a proper balance between exploration and exploitation. Mutation is also a way to control the exploration and exploitation. Mutation, as in the case of a binary coded genetic algorithm, flips the bits constituting a chromosome with a given probability. By keeping a control over the mutation probability, this genetic operator can be used to maintain the genetic diversity of the population. Mutation with high mutation probability acts as an exploration operator and explores new solutions, whereas with small mutation probability acts as an exploitation operator and preserves most of the genetic material. However, as a high mutation rate introduces blind diversity, thereby preventing the population from converging, the use of mutation as an exploration operator is often considered inefficient. Crossover is another search operator used in genetic algorithms; however, the nature of the search performed by the crossover operator is mainly determined by the diversity of the population. For random and diverse populations, normally at the commencement of the search, crossover performs a widespread search and explores the search space, whereas for similar populations, normally after several iterations when high fitness solutions develop, it performs the search in the neighborhood of existing solutions. The probability of the crossover operator also affects the exploration and exploitation—high crossover probability results in the quick introduction of new solutions in the search space, but there is a chance of missing good solutions and of failing to exploit existing solutions. With low crossover probability, most of the existing solutions are preserved and a considerable search area may be left unexplored. Thus the crossover and mutation probabilities can be used as controlling parameters to achieve balance between exploration and exploitation. Population diversity is also considered as a measure to achieve balance between exploration and exploitation. High genetic diversity indicates that the genetic algorithm was in the phase of exploration, whereas low genetic diversity indicates that the algorithm was in the phase of exploitation.

16.4.2 Constrained optimization

The existence of constraints in any system often acts as an obstacle in the application of search techniques such as genetic algorithms for optimizing the problems as genetic operators (search operators), such cases may result in some infeasible solutions. A few methods are highlighted in the genetic algorithm literature to deal with infeasible solutions. The first technique, the penalty function method, allows the genetic operators to generate infeasible solutions which are later on penalized by a modification of the objective function. The penalty function method discriminates an infeasible solution according to its violation of each constraint, thus it transforms the constrained problem into an unconstrained optimization problem. However, the penalty function method is highly sensitive to the penalty parameters (weights) selected for various constraints; an inappropriate choice of penalty parameters may either result in inconsequential penalty terms or highly consequential penalty terms as compared to objective functions. The former results in inordinate exploration of infeasible solutions, whereas the latter may cause premature convergence to a feasible but unacceptable solution. As in such cases, a feasible solution, once generated, may dominate the mating pool with a rapid elimination of infeasible solutions before the generation of any other feasible solution. Other techniques emphasize the use of repair algorithms to correct the infeasible solutions or the application of some decoder scheme to amplify the probability of generation of feasible solutions. Repair algorithms are normally used in the case of combinatorial optimization problems where search operators sometimes result in some illegal solution i.e. a solution which cannot be decoded. However, decoder and repair algorithms are highly problem specific and are difficult to design. Also, the decoder is often required to be augmented with some repair algorithm or penalty function as there still exists some possibility of the generation of infeasible solutions [1].

16.4.3 Multimodal optimization

Multimodality in the case of optimization problems is referred to as the existence of more than one optimal solution and multimodal optimization deals with the search for multiple optima. Knowledge of multiple optima provides the flexibility to choose an optimal solution under the constraints of cost, realizablility, etc. Methods which are used in genetic algorithms to allocate and/or maintain multiple optimal or suboptimal solutions are referred as niching methods. Genetic algorithms, due to the nature of their selection operators, often converge to a single solution, even though multiple solutions of the same quality exist. Earlier techniques, known as temporal niching methods, were based on the concept of locating multiple optima in a sequential manner for which a traditional algorithm is run several times in order to find a different optima each time. Such methods, however, require certain knowledge of the basin of attraction of optima, otherwise the runs may converge to the same optimum. Since the genetic algorithm is a population based approach, by modifying the framework of traditional genetic algorithms, the coexistence of a large number of optimal solutions may be made possible, thereby evolving multiple optimal solutions simultaneously. There exist niching methods such as fitness sharing and crowding, which operate on the concept of preserving population diversity which eventually leads to simultaneous evolution of multiple optima. These niching methods use a distance metric defined over the search space to distinguish the similarity of individuals. Crowding techniques use this measurement for the replacement of similar individuals, whereas the fitness sharing method uses the metric for de-rating an individual's fitness by a particular amount in accordance with the number of similar individuals in the population [3, 4].

De Jong [3] introduced a crowding factor model for multimodal optimization. For a given generation gap *G*, *G.popsize* individuals are selected to create an equal number of offspring. For each of the offspring, CF individuals are randomly selected from the current population to constitute a sample; CF is known as the crowding factor. The offspring then replaces the most similar individual of this sample. The hamming-distance between two individuals is used as the measurement for similarity.

Restricted tournament selection is a variant of the crowding method in which the parents are selected randomly from the population. For each offspring, CF individuals are randomly selected and the offspring then competes with the most similar individual from the selected ones and in the case that the offspring is better, it replaces that individual. The struggle genetic algorithm is an another variant of the crowding method which is functionally equivalent to restricted tournament selection with crowding factor equal to the population size.

The fitness sharing method uses the concept of sharing function to affect the population diversity and to maintain sub-populations. Sharing function reflects the extent of sharing that can be done between two strings. In the case that the two strings are identical, the sharing function returns a value of 1 and if the distance between two strings exceeds a certain threshold distance (niche radius) the sharing function returns a value of 0, implying that the individuals are in different niches. A commonly used sharing function is

$$\mathrm{Sh}(d_{ij}) = \begin{cases} 1 - \dfrac{d_{ij}}{\rho}; & d_{ij} < \rho \\ 0; & \text{otherwise} \end{cases}, \tag{16.6}$$

where d_{ij} is the distance between the ith and jth string in phenotype space and ρ is the niche radius. For every string, a sharing function is calculated for other strings and these sharing function values are added to obtain a niche count $m_i = \sum_j d_{ij}$. Thereafter the shared fitness value of an individual is computed as $f_i^* = f_i / m_i$ and is used in the selection mechanism instead of f_i. Now, if there exist lesser strings from any optimal solution then these strings will have a lower niche count and higher shared fitness than the strings corresponding to other optima. As the selection operator will now emphasize these strings, the number of strings form this optima will increase in the subsequent iterations. Thus this method allows the simultaneous existence of multiple optima in the population. One commonly used replacement in this case is the overlapping population scheme where new offspring are first added to the current population and shared fitness values are computed for all the candidates and individuals (equal to the number of offspring) with the worst shared fitness values then eliminated from the population and the shared fitness values are recalculated for the next iteration [1, 4].

16.4.4 Multi-objective optimization

Multi-objective optimization emphasizes that simultaneous optimization of more than one objective function is required. There exist two approaches for multi-objective optimization. The first technique is to convert the problem into a single objective optimization problem by suitably combining the individual objective function into a composite one. A commonly used approach to develop the single objective function is the weighted combination of individual functions. However, this method is highly sensitive to the weight parameters and a slight perturbation in the weights may lead to a different solution. The second technique deals with the determination of a set of solutions (known as Pareto optimal solutions) obtained by considering different single objective functions synthesized by some technique. A commonly used approach to determine the Pareto optimal solution is the concept of nondomination. Nondominated solutions are those solutions which are superior to others in at least one objective function. Solutions which are nondominated with respect to each other constitute the Pareto optimal set. Due to their population based approach, genetic algorithms are well suited for the multi-objective optimization problems as a classical genetic algorithm can be extended to maintain a diverse set of solutions. There exist several multi-objective optimization algorithms, such as the weighted sum approach, vector evaluated genetic algorithms (VEGA: altering the objective function approach) and Pareto ranking approaches, such as niched Pareto genetic algorithm, random weighted genetic algorithm (RWGA), nondominated sorting genetic algorithm (NSGA), nondominated sorting genetic algorithm with elitism (NSGA-II) [4].

16.5 Concluding remarks

Genetic algorithms have been successfully applied to search for the optimal solution for combinatorial optimization problems in various fields, and successful application of genetic algorithms to NP-hard and NP-complete problems has been reported in several research findings [5–18]. Holistic detail of the application of genetic algorithms is a formidable task and can be found from the references cited and the references within these. Some of the broad application areas of genetic algorithms are machine learning, economics, immune systems, ecology, social systems and others. Because of its promising performance, the application areas of genetic algorithms have grown rapidly in recent years. The genetic algorithm community and researchers from other realms are constantly working to develop new variants of the genetic algorithm which would provide optimization solutions to various problems with an efficient time and space complexity approach.

References

[1] Goldberg D S 1989 *Genetic Algorithms in Search, Optimization and Machine Learning* (Boston, MA: Addison-Wesley Longman)

[2] Reeves C R and Rowe J E 2002 *Genetic Algorithms: Principles and Perspectives: A Guide to GA Theory* (Norwell, MA: Kluwer Academic)

[3] Jong K D 1988 Learning with genetic algorithms: an overview *Mach. Learn.* **3** 121–38

[4] Deb K 1999 Introduction to genetic algorithms *Sadhana* **24** 293–315

[5] McCall J 2005 Genetic algorithm for modeling and optimization *J. Comput. Appl. Math.* **184** 205–22

[6] Tang K S, Man K F, Kwong S and He Q 1996 Genetic algorithms and their applications *IEEE Process Mag.* 22–37

[7] Renders J M and Flasse S P 1996 Hybrid methods using genetic algorithms for global optimization *IEEE Trans. Syst. Man. Cyber.* B **26** 243–58

[8] Blanco A, Delgado M and Pegalajar M C 2001 A real coded genetic algorithm for training recurrent neural networks *Neural Netw.* **14** 93–105

[9] Alonge F, DIppolito F and Raimondi F M 2003 System identification via optimized wavelet based neural networks *Proc. IEEE Contol Theory Appl.* **150** 147–54

[10] Sahoo D and Dulikravich G S 2006 Evolutionary wavelet neural network for large scale estimation in optimization *Proc Multidisciplinary Analysis and Optimization Conf. (Portsmouth, VA)* pp 1–11

[11] Awad M 2009 Optimization RBFNN parameters using genetic algorithms: applied on function approximation *Int. J. Comput. Sci. Secur.* **4** 295–307

[12] Shou-sheng L and Yong D 2010 An evolutionary wavelet network and its training method *Int. Conf. on Computer Application and System Modeling* pp 379–83

[13] Aly A A 2011 PID parameters optimization using genetic algorithm technique for electro-hydraulic servo control system *Intell. Control Autom.* **2** 69–76

[14] Vishwakarma D D 2012 Genetic algorithm based weight optimization of artificial neural network *Int. J. Adv. Res. Electri. Electron. Instrum. Eng.* **1** 206–11

[15] Awad M 2014 Using genetic algorithms to optimize wavelet neural networks parameters for function approximation *Int. J. Comput. Sci. Issues* **11** 256–67

[16] Kulkarni A and Kumar A 2015 Structurally optimized wavelet network based adaptive control for a class of uncertain underactuated systems with actuator saturation *Inter J. Hyb. Intell. Sys.* **12** 171–84

[17] Kopel A and Yu X H 2008 Optimize neural network controller design using genetic algorithm *Proc. 7th World Congress on Intelligent Control and Automation (Chongqing, China)* pp 2012–6

[18] Chiroma H, Noor A S M, Abdulkareem S, Abubakar A I, Hermawan A, Qin H, Hamza M F and Herawan T 2017 Neural networks optimization through genetic algorithm searches: a review *Appl. Math. Inf. Sci.* **11** 1543–64

Chapter 17

A novel approach for QoS optimization in 4G cellular networks

Vandana Khare and G R Sinha

A cellular network needs to carry multimedia traffic efficiently during real-time transmission. Since this network transmits multimedia data, there is a requirement to adapt the service rate according to the network conditions in order to maintain successful network operation. Hence, there is a need to allocate resources proficiently. Quality of service (QoS) improvement in such a network is more challenging, in particular in 4G cellular networks. Many schemes have been proposed for QoS improvement in cellular networks which combine call admission control (CAC) and power control schemes. However, CAC with power control is not enough to provide the most optimal resource utilization. Therefore, it is essential to develop an efficient QoS based resource allocation and scheduling scheme for real-time multimedia traffic. The main goal of the proposed research is to develop a QoS based resource allocation and adaptive rate scheduling (QRAAS) scheme in cellular networks to improve network reliability, maximize system capacity and improve system efficiency, and hence to enhance QoS in cellular networks. To implement the above-mentioned QRAAS scheme, the resource allocation based call admission control (RACAC) scheme and adaptive rate scheduling (ARS) scheme were developed.

17.1 Mobile generations

Fourth generation (4G) technology always requires good QoS to maintain real-time (RT) multimedia traffic in cellular and mobile networks [1–3]. The OFDMA network is the most extensively used network because earlier networks such as FDMA and TDMA are not able to support the bandwidth and data rate requirements. In the case of the FDMA network, only voice transmission is supported, whereas TDMA as well as wide-band CDMA (WCDMA) support voice and data

doi:10.1088/978-0-7503-2404-5ch17

transmission. OFDMA is the only network to support fourth generation technology because of the high data rate of transmission [4–6].

OFDDMA is the only technique which covers more bandwidth and provides better coverage when compared to other networks. OFDMA actually helps in spreading the data in the entire bandwidth, which is very large, implemented as spectrum method, and the spreading takes place just before transmitting the signal. The spreading factor (SF) plays an important role in the WCDMA network because it is the ratio of chip rate to data rate [7, 8]. The spreading factor decides the number of users and depends on the type of modulation. The phase shift keying is of different types, of which quadrature phase shift keying (QPSK) is very popularly utilized in the WCDMA network.

The WCDMA network enables multimedia services in wireless broadband networks. Proper spectrum utilization is also a difficult task since it is a restricted resource [2]. In RT multimedia, resource allocation is an important task which should be performed effectively and allotted resources must also be utilized without causing wastage of bandwidth.

Bandwidth allocation becomes difficult when handling of the data and voice must be performed concurrently, because voice service is a real-time service and is highly sensitive to delays whereas data services are not. They are, however, sensitive to data loss and require the release of error free data packets [3]. Hence, while providing QoS for voice and data services, several characteristics must be taken into account [4].

Services with higher data rates are always supported by OFDMA networks. Packet access can be managed effectively by proper resource allocation in OFDMA wireless networks. For making use of resource allocation schemes, reliable CAC and effective scheduling schemes are necessary for assured QoS in OFDMA wireless networks [5].

Better QoS implies that the network should provide assured service to a particular application. QoS parameters depend upon specific requirements of an application. Many real-time applications require assured QoS in wireless cellular and mobile networks.

In general, the QoS parameters include bandwidth, delay and delay jitter [6]. Due to the air-interface nature of OFDMA networks, the coverage area and capacity of a cell strongly depends on the desired QoS in terms of acceptable interference level, uniform distribution of mobile users and the related time-dependent user traffic intensity.

In OFDMA wireless networks, resource allocation with CAC, power control and scheduling play an important role in avoiding congestion and maintaining better QoS for satisfactory network operation.

17.2 OFDMA networks

OFDMA acts as an air-interface network which is used in forth generation cellular and mobile communication systems and was produced by the universal mobile telecommunication system (UMTS). OFDMA is a multiple access technique based on orthogonal frequency division multiple access. Predominantly, two types of codes are used for spreading the data. The channelization code is used to decide how many bits are needed to spread single-bit information, whereas scrambling code is used to separate the data. The spreading factor plays a vital role in the OFDMA network. It decides the number of users during up-link and down-link transmission. The spreading factor is 4–256 for up-link transmission and 4–512 for down-link transmission. One of the important characteristics of the OFDMA network is the high data transmission rate at the range of 144 Kbps to 2 Mbps, because of which it supports multimedia transmission. All the channels in the OFDMA network are of higher bandwidth, due to the fact that no link is assigned a devoted bandwidth as per the requirement of the data rate [1]. Channel encoding is performed considering that any decoder which has the knowledge of the code may extract the required data from the overall signal bandwidth, which is basically in the form of noise signal.

17.2.1 Limitations of FDMA, TDMA and WCDMA networks

The major limitations of different types of networks are as follows:

1. FDMA and TDMA do not support high data rate transmission.
2. Efficient radio spectrum utilization is not possible.
3. They do not support multimedia traffic transmission.
4. They are unable to support global roaming.
5. They are capable of accommodating only a limited number of users.
6. They suffer high co-channel and adjacent channel interference.

Due to the above-mentioned issues, FDMA and TDMA can only support first generation (1G), second generation (2G) and third generation (3G) networks. OFDMA networks are the only solution for the transmission of RT multimedia traffic in fourth generation (4G) cellular and mobile networks. In OFDMA networks, a high data rate of transmission is possible for RT multimedia traffic as the data rate depends on the chip rate and spreading factor. In the case of down-link transmission the spreading factor is 4–512, which indicates that the network is able to support 512 users for one tower or base station.

However, every tower is accommodating only 200–225 users while the remaining codes are left free for handoff. Maintaining the QoS and its parameters, i.e. throughput, delay and power consumption, is thus a very important task.

A basic comparison of the FDMA, TDMA, WCDMA and OFDMA networks, with respect to different parameters, is given in table 17.1, which depicts the data rate, bandwidth as well as operating frequencies of the OFDMA network. It can be observed that only power control is an important factor in maintaining the required QoS in OFDMA networks.

17.2.2 Features of OFDMA networks

OFDMA networks in mobile communications provide a high speed of data transmission without failing QoS requirements. The OFDMA characteristics are given below [3].

Table 17.1. Basic comparison of FDMA, TDMA, WCDMA and OFDMA networks.

Parameters	1G	2G	3G	4G
Inventions	1980	1993	2006	2012
Access system	FDMA	TDMA	WCDMA	OFDMA
Speed (data rates)	2.4 Kbps–23.4 Kbps	84 Kbps	144 Kbps to 2 Mbps	100 Mbps
Bandwidth	20 kHz	25 MHz	25 MHz	100 MHz
Operating frequencies	800 MHz	GSM: 900 MHz	1920–1980 MHz for UL and 2110–2170 MHz for DL	850 MHz for UL, 1800 MHz for DL
Limitations	Limited capacity	Slow speed	High power consumption	Complicated hardware
Applications	Voice calls	SMS	Video conferencing mobile TVGPS	High speed operations

- *High data rate transmission*: A very important feature of OFDMA networks is that they support a high data rate for transmission of multimedia traffic in 4G mobile communication, which is essential to maintain high speed as well as throughput for long distance users. It provides a data rate from 10 Mbps to 20 Mbps which can be increased up to 100 Mbps for 4 G using high speed packet access services.
- *Supports RT multimedia traffic*: Fourth generation (4G) communication systems require audio, text and video at the same time. First generation (1G) supports only text while second generation (2G) supports audio and text because of the low data rate requirements. The WCDMA network also supports audio, video and text to some extent. The OFDMA network provides RT multimedia transmission because of high data rate services.
- *Efficient radio spectrum utilization*: Currently, because of an increasing number of users, the effective use of frequency is very important. The OFDMA network is efficiently managing multimedia traffic for a given radio range in mobile communications.
- *Global roaming*: The FDMA and TDMA networks are not able to support international roaming because of the low data transmission rate. However, OFDMA networks emphasize the concept of 'anytime anywhere' in multimedia traffic transmission. This is due to the high data transmission rate and high bandwidth utilization in OFDMA networks.
- *High capacity and more flexible communication*: The OFDMA network provides more coverage for mobile users because of the spread spectrum technique. Also, it avoids co-channel interference between adjacent channels and maintains the required QoS in the network.
- *Complete data security*: The functions of OFDMA networks are highly adaptive to carry out the allocation of resources according to the network

conditions and meet the requirement of strong and secure data transmission and reception by mobile users.

- *Desired quality multimedia communication*: The compatibility problem with 1G and 2G as well as 3G networks is entirely overcome by 4G using OFDMA networks. This can be credited to its high rate of data transmission that is up to 10 Mbps for 3G and 100 Mbps for 4G mobile users.
- *Cheaper communication cost*: The Telephony Regulatory Authority of India (TRAI) tirelessly analyzes cost related issues and maintains lower rates for 4G mobile communication users. Additionally, it provides multimedia communication with global roaming to all mobile subscribers.

17.2.3 Quality of service in OFDMA networks

In OFDMA networks, the improvement of QoS is a major issue owing to the following characteristics:

1. Higher data transmission rate.
2. Limited bandwidth.
3. Global roaming.
4. More data security.
5. Interferences.
6. Handoff support.

What is QoS?

In the case of cellular networks, high and efficient QoS implies that the network service provider is providing satisfactory performance in terms of quality of voice, a lower call blocking and call dropping rate, and high data transmission rate, in particular for multimedia traffic transmission.

Why QoS?

Third generation networks, mainly WCDMA networks, require multimedia traffic transmission. Every user in cellular networks transmits the same frequency, because of which the coverage area as well as the capacity of the network decreases. Hence, maximum network capacity is achieved only if all the users maintain the minimum signal-to-inference ratio (SIR), i.e. higher throughput, and this is possible only when QoS in the network is maintained. In maintaining the required conditions, some important parameters are responsible.

17.2.3.1 Parameters of QoS

The important parameters for determining QoS are:

- *Throughput*: the rate at which the packet should be transmitted from source to destination.
- *Delay*: the time taken by the packet to travel through the network.
- *Packet delivery ratio*: the number of packets received to the total number of packets transmitted.
- *Call blocking probability*: the ratio of the number of calls successfully connected to the total number of calls attempted.

- *Residual energy*: the amount of energy remaining in the system.
- *Channel bandwidth*: the ratio of utilized bandwidth to total bandwidth.
- *Reliability*: the ability of a system to perform consistently without failure.
- *Call dropping probability*: the ratio of the number of calls terminated after admission to the number of calls attempted.

17.2.3.2 QoS issues in OFDMA networks

Major issues in OFDMA networks are:

- In 4G OFDMA networks, QoS improvement may be a major challenge due to RT multimedia traffic transmission.
- RT applications require guaranteed QoS in OFDMA networks.
- Due to the unavailability of resources, it becomes difficult to maintain QoS in 4G wireless networks.
- Another challenge faced by the network is the admission of incoming calls with less blocking probability.
- Power control is a heavy task, required to maintain appropriate connectivity, energy and minimum interference in 4G OFDMA networks.
- Monitoring RT multimedia traffic with perfect scheduling is also an important obstacle in 4G OFDMA networks which needs attention.

17.2.4 QoS improvement techniques in OFDMA networks

Several issues resist the maintenance of the required QoS in the network. One of the major hurdles is resource allocation, along with admission of calls for new arrival users. Fast and accurate power control in OFDMA networks is also a tough task which requires consideration. Finally, scheduling also plays a major role in monitoring the traffic effectively and avoiding congestion for maintaining better QoS and satisfactory network operation. The following schemes are used to improve QoS in OFDMA networks.

17.2.4.1 Resource allocation with call admission control

Resource allocation is an important factor in OFDMA networks because if the appropriate resources are not available, the network performance will be adversely affected.

Resource allocation is based on three parameters:

1. Signal to interference ratio.
2. Throughput.
3. Power.

The earlier work was mostly based on the signal to interference ratio, because at that time only voice transmission was required. When GSM emerged, i.e. when both voice and message entered the 2G network, power allocation came into the picture. Currently, owing to multimedia transmission, throughput based resource allocation has become an indispensable part of every application in 4G networks, which requires a considerably high speed and consequently, a higher data rate.

17.2.4.2 Call admission control

CAC is a technique that allows the admission of a user based on the available resources or present traffic load conditions after comparison with a predefined value. If resource allocation is executed along with CAC, it gives marvelous results in improving network performance for a given QoS requirement.

RACAC decides whether a new user should be admitted or not based on the available resources and after comparison with a predefined value. The parameters required for RACAC are:

- Signal-to-interference ratio (SIR).
- Throughput (TP).
- Power (P).

All these parameters are used in CAC based on requirements. Hence, interference and power measurement is required for comparison of signal strength, whereas throughput measurement is required for comparison of data rate or speed.

17.2.4.3 Power control

OFDMA is an interference limited technique. This is because every user transmits the same frequency based on the frequency reuse concept. In the case of frequency reuse, the entire frequency spectrum is divided into bands and these bands of various frequencies are assigned to different cells. However, after a certain geological distance, the same frequency is re-assigned to other cells. Nevertheless, if the distance between two cells is too small then co-channel interference is generated.

As a matter of fact, co-channel interference is negligible in OFDMA networks because the operation is based on orthogonal properties. Orthogonality ensures that if two different codes are multiplied then the output is zero.

The main problematic issue with OFDMA networks is the 'near–far' problem, in which the mobile users closer to the base station or mobile tower receive good signal strength, while those far away are unable to receive sufficient signal strength. In this case, if the transmitted power is increased, then the signal strength can easily be increased and the following advantages can be established:

- Avoids the problem of fading.
- Reduces the bit error rate (BER).

Fading can be defined as the reduction in signal strength. This is easily resolved by increasing the transmitted power and reducing the BER, which can be defined as the ratio of the number of bit errors to the total number of bits transmitted. It should be kept low in all conditions.

If we increase the transmitted power, interference is also increased, suggesting that fast and perfect power control is required. Power control can be categorized into two types:

- Open loop power control.
- Closed loop power control.

Open loop power control is based on the mobile unit and no feedback is required from the base station. This type of power control is fast but not accurate.

Close loop power control depends on the feedback. The base station continuously takes feedback from the mobile unit and adjusts the signal strength. This technique is slow but accurate to a great extent. Close loop power control is again classified into two types:

- Inner loop power control.
- Outer loop power control.

In the case of inner loop power control, the base station checks the signal level of a particular user by taking feedback from the mobile unit and adjusts it according to the QoS requirement.

In the case of outer loop power control, the received power is controlled by changing the target value, which is pre-decided, based on the QoS requirement.

The need for power control

Power control is very important in OFDMA networks to manage the following:

- Interference management.
- Connectivity management.
- Energy management.

In the case of interference management for down-link operation, the base station continuously takes feedback and adjusts the signal strength by controlling interference in the system. As a result, the near–far problem is eliminated.

In the case of connectivity management, when the mobile user leaves a particular area and enters another area, the base station in that area should be able to provide sufficient signal strength for that particular user. Hence, perfect connectivity management is possible and call dropping is reduced. Also, the handoff problem is perfectly resolved.

In the case of energy management, the mobile battery reduces power consumption and hence improves battery life. This is possible only through fast and accurate power control.

17.2.4.4 Scheduling

Scheduling plays a very important role in OFDMA networks. It monitors the traffic and minimizes delay at the receiving end. It is basically designed for RT or non-RT traffic transmission as well as reception.

In the case of the OFDMA network, two types of scheduling are mainly used:

- Fixed scheduling.
- Adaptive scheduling.

Fixed scheduling suggests that once the packets are scheduled, they will be transmitted with constant time and so the delay will be greater.

Adaptive scheduling is based on adaptability, suggesting that network conditions will change according to the situation. In the case of OFDMA networks, it is based on resources. Hence, owing to adaptive scheduling, maximum resource utilization is possible.

The resources used by OFDMA networks are

- Channel codes.
- Power.
- Rate.

All three types of resources can be used but the use of channel codes and power is very popular. The use of rate as the main resource will be more beneficial for multimedia traffic transmission because speed management is very important for RT traffic in 4G networks, such as video telephony, live telecast and online games. Such types of applications demand minimum delay (in milliseconds). So, if adaptive scheduling is based on rate, then delay will surely be reduced and network throughput will also be maintained according to given QoS requirements.

17.2.4.5 Wireless network traffic

In general, there are two types of traffic:

1. Real-time (RT).
2. Non-real-time (NRT).

RT traffic is based on a variable bit rate (VBR) and is mostly applicable for time sensitive applications such as video telephony, audio telephony, mobile TV, online games and videos. NRT VBR traffic is useful for the booking of air tickets and bank transactions. All types of applications come under multimedia RT traffic transmission. NRT traffic is based on a constant bit rate (CBR) and is typically used for NRT applications such as SMS, Internet, voice, etc.

RT traffic mainly suffers from delays because of the fast development of mobile communications. Improvement in QoS is therefore an important task in such networks. NRT traffic generally suffers from packet loss constraints and QoS improvement is not an issue in these types of networks.

17.3 Simulation model and parameters

This section deals with simulation models and their parameters.

17.3.1 Simulation topology

To simulate the proposed scheme, network simulator (NS2) with version 2.31 is used. In the simulation, the base station and user are considered in a 600 m × 600 m region for 50 s simulation time. It consists of six base stations along with six users. Each user contains six mobile nodes (see figure 17.1). All nodes have the same transmission range of 250 m.

For the simulation, VBR traffic is used. The simulation settings and parameters are given in table 17.2.

Figure 17.1. Simulation topology for the OFDMA scheme.

Table 17.2. Simulation settings and parameters.

Area size	600 m × 600 m
No. of users	6
MAC	OFDMA
Slot duration	1.25 ms
Routing protocol	AODV
Transmission rate	1, 1.5, 2, 2.5, 3 Mbps
Frame length	8 slots
Simulation time	50 s
Traffic model	VBR
No. of cells	6
Radio range	250 m
Packet size	512 bytes

17.3.2 Performance metrics

Performance evaluation is mainly based on the following matrices:

- *Packet delivery ratio*: the ratio of packets received successfully to the total number of packets sent for RT traffic flows.
- *Delay*: the time taken by the packet to reach the destination, represented in milliseconds.
- *Throughput*: the rate at which the packets travel through the network, represented in megabits/second.

17.4 Adaptive rate scheduling in OFDMA networks

17.4.1 Introduction

After traffic regulation using the RACAC scheme, it is very important to monitor the traffic. Scheduling plays an important role in monitoring the traffic in OFDMA networks to improve spectral efficiency.

17.3 Notation used in adaptive rate scheduling.

Notation	Meaning
FCU	Feedback control unit
i	Session name
$D_{\text{schedule}}(i)$	Average scheduling delay of session i
$Q_{\text{session}}(i)$	Queue size of session i
$R_{\text{est\_arrival}}(i)$	Estimated arrival rate of session i
$R_{\text{allocated}}(i)$	Rate allocated to session i
$D_{\text{avg}}(i)$	Current average delay of session i
$D_{\text{threshold}}(i)$	Mean threshold delay for session i
T	Scheduled time duration in frames

Resource allocation with call admission control and the adaptive rate scheduling scheme (RACAC-ARS), are employed. The adaptive rate scheduling scheme is associated with a heuristic approach based on feedback control unit (FCU) logic. Under this approach, during the entry of the session into the network, a rate is assigned and adjusted based on the feedback obtained from the users already admitted into the network.

A heuristic-based FCU is present in the network, which records all the information related to the rate of every user. On the basis of the rate of the operating users, the rate of the incoming users for respective sessions is adjusted. Important notations used in the analysis of adaptive rate scheduling are shown in table 17.3.

The process of adaptive rate scheduling is as followed:

1. During the entry of the session, the rate R_{allocate} for the incoming user is allocated. After that, the time slot duration T and queue size Q of session i are assigned and the arrival rate $R_{\text{est-arrival}}$ is estimated. When the user arrives at the base station (BS), the rate is adjusted based on the feedback (FB) obtained from the already admitted session. At the same time, FCU determines the delay for threshold level $D_{\text{threshold}}$. It also estimates the average scheduling delay $D_{\text{avg}}(\text{i})$ from the first user and checks the required conditions. If D_{avg} is less than or equal to $D_{\text{threshold}}$, FCU pre-empts all the users.
2. If D_{avg} is more than $D_{\text{threshold}}$, error in arrival rate estimation is indicated. Later, the rate R_{allocate} is assigned and the arrival rate for the user is estimated. Finally, the estimated rate is sent as feedback to the BS, based on which the BS adjusts the rate for the incoming user.

17.4.2 Adaptive rate scheduling algorithm

Figure 17.2 shows all the steps in the flow of the implementation:

Step 1. Allocate the rate R_{allocate} for the incoming session.

Step 2. Assign time slot duration T and queue size Q of session i.

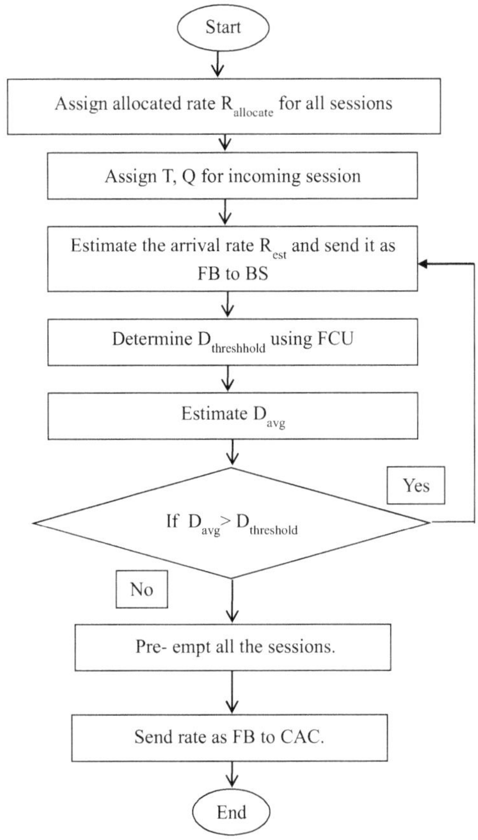

Figure 17.2. Flow chart of the adaptive rate scheduling algorithm.

Step 3. Estimate the arrival rate $R_{\text{est-arrival}}$ when the user arrives.
Step 4. BS adjusts the rate based on FB obtained from the already admitted user.
Step 5. FCU determines the delay for threshold level $D_{\text{threshold}}$.
Step 6. Estimate the average scheduling delay $D_{\text{avg}}(i)$ from the first user and check the following conditions:
If $D_{\text{avg}} \leqslant D_{\text{threshold}}$, FCU pre-empts all the users, otherwise if $D_{\text{avg}} > D_{\text{threshold}}$ indicates error in arrival rate estimation then:

- Assign rate R_{allocate} and estimate the arrival rate for the user.
- Send the estimated rate as FB to the BS.
- Adjust the BS rate for the incoming user.
- END

17.4.3 Average scheduling delay estimation for the ARS scheme

The average scheduling delay of all the users is estimated from the first user i admitted into the network:

$$D_{\text{schedule}}(i) = (Q_{\text{session}}(i) + T[R_{\text{est\_arrival}}(i)])/R_{\text{allocated}}(i), \tag{17.1}$$

where $Q_{\text{session}}(i)$ is the size of the queue of user i, T is the time slot duration in milliseconds, $R_{\text{est\_arrival}}(i)$ is the estimated arrival rate allotted to user i and $R_{\text{allocated}}(i)$ is the rate allocated.

17.5 Conclusions

Currently, the 4G network is very popular and it is thought to be the next generation of mobile computing. Its use and advantages distinguish it from all other peer technologies. In order to keep its services available all the time, there is a need to improve its QoS with various parameters.

In this chapter, the developed scheme for RACAC improves network reliability by attempting to reduce the number of sessions that are blocked due to inadequate resources. The main concept of resource allocation in the scheme is used to increase the data rate during transmission according to user requirements, and simultaneously improve throughput for real-time multimedia traffic using variable bit rate traffic for transmission. The resource allocation with CAC scheme for next generation networks maximizes sessions for users while increasing the data transmission rate. It also ensures that a new session does not violate the QoS of ongoing sessions with the help of an ARS scheme. In an ARS scheme, a heuristic-based approach finds the solution close to the true one. The overall process depends on FCU logic. Thus, our proposed scheme utilizes resources efficiently and maximizes the spectral efficiency of the network. According to analysis of the results, it is observed that our proposed scheme improves throughput and reduces delay for 4G cellular networks.

References

[1] Jagiasi V and Giri N 2017 Resource allocation strategies for cellular networks *Inter. J. Eng. Res. Techn.* **5** 130–6

[2] Jain M and Mittal R 2016 Adaptive call admission control and resource allocation in multi-server wireless cellular networks *J. Ind. Eng.* **12** 71–80

[3] Memis M O and Ercetin O 2015 Resource allocation for statistical QoS guarantees in MIMO cellular networks *EURASIP J. Wirel. Commun. Netw.* **2015** 217

[4] Nantha Kumar G and Arokiasamy A 2014 An efficient combined call admission control and scheduling algorithm to improve quality of service for multimedia services over next generation wireless networks *Inter. J. Comput. Appl* **107** 416–25

[5] Liu J, Chen W, Zhang Y J and Cao Z 2013 A utility maximization framework for fair and efficient multicasting in multicarrier wireless cellular network *IEEE Trans. Netw.* **21** 110–20

[6] Kadhem K and Khan Y 2012 Evaluation and comparison of resource allocation strategies for new streaming services in wireless cellular networks *IEEE Trans. Commun.* **12** 123–35
[7] Jraifi Abdelouahed A 2012 Theoretical prediction of users in WCDMA interference *J. Theor. Appl. Inf. Techn* **35** 163–8
[8] Peifang Z and Jordan S 2008 Cross layer dynamic resource allocation with targeted throughput for WCDMA data *IEEE Trans. Wireless Commun.* **7** 4896–906

Lightning Source UK Ltd.
Milton Keynes UK
UKHW051132211219
355737UK00003B/26/P

9 780750 324021